Gas Hydrate in Water Treatment

Gas Hydrate in Water Treatment

Technological, Economic, and Industrial Aspects

Bhajan Lal and Sirisha Nallakukkala
Universiti Teknologi PETRONAS

WILEY

This edition first published 2022

Registered office
John Wiley & Sons, Inc., 111 River Street, Hoboken, NJ 07030, USA

Editorial Office
111 River Street, Hoboken, NJ 07030, USA

For details of our global editorial offices, customer services, and more information about Wiley products visit us at www.wiley.com.

Wiley also publishes its books in a variety of electronic formats and by print-on-demand. Some content that appears in standard print versions of this book may not be available in other formats.

Library of Congress Cataloging-in-Publication Data
Names: Lal, Bhajan (Senior lecturer), author. | Nallakukkala, Sirisha, author.
Title: Gas hydrate in water treatment : technological, economic, and industrial aspects / Bhajan Lal and Sirisha Nallakukkala.
Description: Hoboken, NJ : John Wiley & Sons, 2022. | Includes bibliographical references and index.
Identifiers: LCCN 2021061620 (print) | LCCN 2021061621 (ebook) | ISBN 9781119866114 (hardback) | ISBN 9781119866138 (pdf) | ISBN 9781119866077 (epub) | ISBN 9781119866145 (ebook)
Subjects: LCSH: Natural gas--Hydrates. | Water--Purification.
Classification: LCC TN884 .L35 2022 (print) | LCC TN884 (ebook) | DDC 665.7--dc23/eng/20220217
LC record available at https://lccn.loc.gov/2021061620
LC ebook record available at https://lccn.loc.gov/2021061621

Cover image: © Upklyak/Freepik.com
Cover design by Wiley

Set in 9.5/12.5pt STIXTwoText by Integra Software Services Pvt. Ltd, Pondicherry, India

10 9 8 7 6 5 4 3 2 1

MIX
Paper from responsible sources
FSC® C013604

Contents

Preface

Bhajan Lal and Sirisha Nallakukkala

Universal access to clean water is one of 17 Global Goals that make up the United Nation's 2030 Agenda for Sustainable Development. Water is an abundant natural resource that covers three-quarters of the earth's surface. However, only about 3% of all water sources are potable. According to the World Watch Institute, more than two-thirds of the world's population may experience water shortages by 2025. Even countries not presently facing water shortages may have to tackle the problem of freshwater scarcity soon. Water desalination and wastewater treatment offer a solution.

Various processes have been proposed and implemented for desalination and wastewater treatment over the decades. More recently, the application of gas hydrate formation phenomena to treat water has received attention as a green process, requiring only water and CO_2. Thus, there is a pressing need to have a well-written textbook that focuses on gas hydrate-based desalination fundamentals and other industrial aspects. Such a book would serve a large sector of the desalination community, including process engineers, designers, students, and researchers.

The fundamentals of the gas hydrate-based desalination process are based on physical principles that include mass and energy conservation, mass, momentum, heat transfer, kinetics, and thermodynamics. The authors firmly believe that a sound understanding of these fundamentals is necessary to analyze or evaluate the performance of any existing and known desalination processes, including gas hydrate-based desalination. Moreover, understanding the fundamentals allows for critical evaluation of novel schemes and the innovation of new schemes. This book consists of twelve chapters that discuss different aspects of conventional and novel desalination techniques and propose gas hydrates-based desalination as a potential research area to be explored further.

Chapter 1 presents a brief introduction of water scarcity concerns and potential water source alternatives. The composition of seawater and wastewater, the

Gas Hydrate in Water Treatment: Technological, Economic, and Industrial Aspects,
First Edition. Bhajan Lal and Sirisha Nallakukkala.

estimated usage of water in several industries, and the surging need to drill deeper for groundwater are also discussed. It also gives an overview regarding the application of gas hydrates in desalination processes by providing insights on the thermodynamics and kinetics of hydrate formation and the dissociation process.

Chapter 2 focuses on the various desalination methods implemented in the past and present times. Conventional and gas hydrates-based desalination are discussed, along with their function and mechanism.

Chapter 3 presents the applications of gas hydrate phenomena in the desalination sector. Patented designs and possible improvisations are briefly discussed, along with the advantages and disadvantages of different hydrate formers.

Chapter 4 discusses the chemicals used in conventional desalination processes. Chemicals have traditionally been the crux of water treatment methods. The future of water treatment is allied with innovative and physical treatment approaches, thus decreasing the surging demand for chemical additives used in traditional water treatment. The chapter also discusses process chemicals used to promote gas hydrate formation and criteria for choosing economical hydrate formers.

Chapter 5 explores the evolving thermodynamic and kinetic modeling of seawater through the gas hydrate-based desalination process. Further application of artificial intelligence in desalination efficiency is also discussed. A literature review of kinetic models and suggestions on the applied prediction models are also included.

Chapter 6 presents wastewater as a potential alternative water source. The effects of wastewater on the environment and the means of treating wastewater using gas hydrate technology are also discussed.

Chapter 7 focuses on intelligent process optimization for water treatment plants, which enables automated treatment process control and real-time decision support. It also covers the application of Artificial Neural Networks in wastewater and freshwater treatment processes.

Chapter 8 presents quantitative evaluations of wastewater quality. Techniques for sampling and methods implemented for analyzing and determining them are briefly presented. It covers the various analytical techniques used to determine the organics, inorganics, and radioactive materials. It also covers the toxicity test systems, requirements, evaluation, and their implementation.

Chapter 9 summarizes the techno-economical aspect of conventional desalination technologies and the results of various studies on specific energy consumption in gas hydrate-based desalination.

Chapter 10 discusses the application of renewable energy in desalination. A detailed discussion on renewable energy sources used in desalination is provided, and the potential of hybrid technologies is explored.

Chapter 11 presents brine/waste product management methods. The disposal of the waste product is an essential aspect to consider to avoid adverse environmental effects. The recovery of useful products from the waste brine is also discussed.

Chapter 12 discusses the post-treatment of desalinated water using conventional and modern technologies for remineralization of produced permeate. The chapter also provides an overview of alternative disinfection systems and guidance for their use in desalination applications.

It is a great pleasure to thank Wiley for providing this opportunity to share this book with the scientific community. This book could not have been completed without the support and contribution of the authors involved. In the end, we would like to invite you to share your thoughts and feedback in order to improve our future work.

Bhajan Lal, Seri Iskandar, Malaysia.

1

Introduction to Desalination

Jesa Singh, Vinayagam Sivabalan, and Bhajan Lal

1.1 Coping with Water Scarcity

Climate change and population growth have significantly altered global water resources in the twentieth century and are likely to pose more challenges in the future. One of the greatest of these challenges is freshwater security and equitability. Freshwater is used primarily for sustenance and in various industries that enable today's high living standards, such as mining, agriculture, and power generation. As evident in Figure 1.1, irrigation consumes the most water among the industrial uses [1]. In the agricultural sector, maintaining an adequate water supply is of foremost importance in ensuring optimal conditions for managing livestock and raising crops, both of which are vital to the wellness of the general population and economies. Lack of freshwater can also affect material and energy-related industries by ultimately decreasing product quality.

Groundwater is an essential source of freshwater, meeting the domestic needs of about half of the global population and more than a third of global consumptive irrigation demand. In the United States (US), groundwater resources provide over fifty billion gallons per day for agricultural needs and drinking water to about half the total population. Nearly all of the US rural population depends on groundwater resources. In areas where surface water such as lakes and rivers are scarce or inaccessible, groundwater also supplies much of the habitant's hydrologic needs.

Groundwater depletion, defined as long-term water-level declines caused by sustained groundwater pumping, is a key issue associated with groundwater use. Groundwater depletion is significant in most irrigated regions of the world where millions of people reside [2]. Many areas of the United States are experiencing groundwater depletion, which can result in water table lowering,

Gas Hydrate in Water Treatment: Technological, Economic, and Industrial Aspects,
First Edition. Bhajan Lal and Sirisha Nallakukkala.

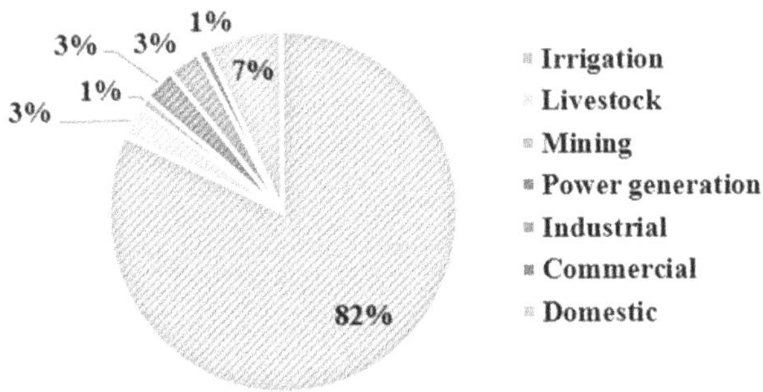

Figure 1.1 Freshwater consumption.

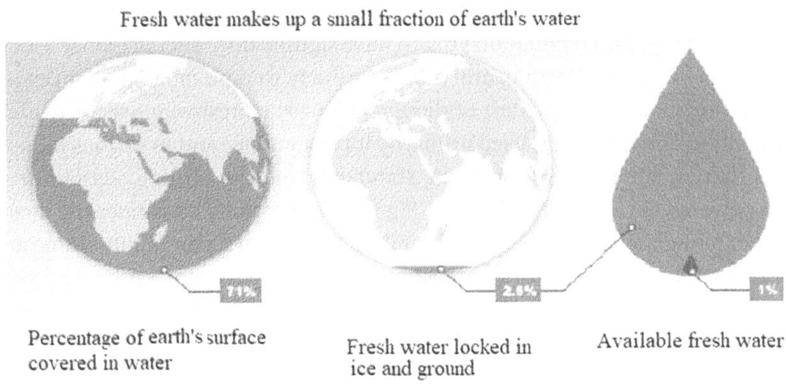

Figure 1.2 Earth's freshwater resources.

leading to the reduction of freshwater in streams and lakes [2]. While the expansion of groundwater-based irrigation has helped to meet increased food demands, it has resulted in several environmental implications. Groundwater depletion has now become one of the most prominent challenges for food and water security.

Figure 1.2 shows the natural sources of water that make up 71% of our planet [3]. However, out of this 71%, a mere 2.5% is freshwater. From this 2.5%, we only have easy access to approximately 1%, while the remaining 1.5% are trapped in glaciers, ice caps, and deep underground, making it economically unfeasible to access efficiently [4]. The accessibility of freshwater poses a clear problem, as the finite supply may not be enough to upkeep the ever increasing human population.

Water scarcity is a situation where there is not enough water to meet essential needs, like drinking or growing food. Total water consumption has been increasing at a rate of more than twice the rate of population increase in the last century. By 2025, approximately 1.8 billion people will be living in arid regions with absolute water scarcity, while two-thirds of the remaining population will be living under the looming shadow of water scarcity [5]. Figure 1.3 shows the trend in water scarcity around the globe [6].

In the event of water scarcity, the most typical symptom is the rise of price in freshwater, due to its inability to meet the required demands and expectations. Compounded with the population growth, it will eventually arrive at a point where there simply is not enough to go around. The general populace tends to not recognize the limits of water supplies, and would rather focus on acquiring additional supplies rather than managing existing supplies. This is repeated in many basins around the world. Because we overbuild and expect too much, we arrive at a situation where there is not enough water to go around. Moreover, when water is reallocated, it tends to meet the needs of the wealthier rather than the poor or marginalized. Cases of successful water management that have effectively tackled water scarcity exist and must be critically examined to prevent the devastating effects of abject resource mismanagement [7].

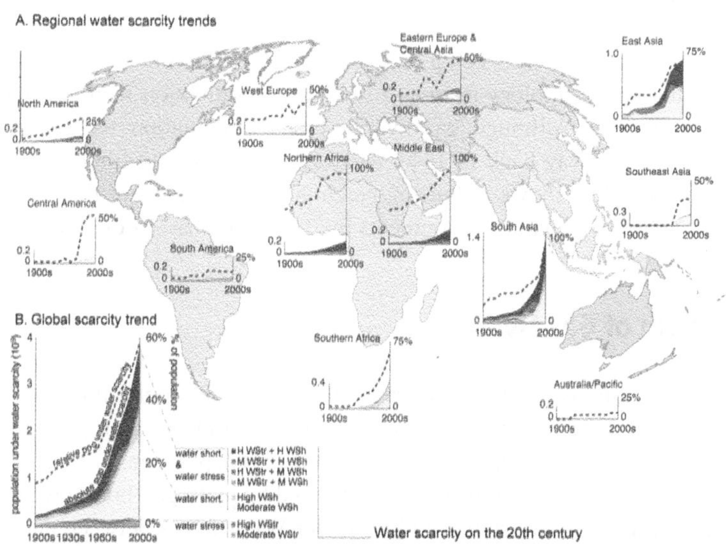

Figure 1.3 Water scarcity trajectories. Filled graphs represent the absolute population under water scarcity (in billions) while dashed lines represent the population relative to the total regional population [6].

A process called desalination can mitigate freshwater scarcity by converting seawater, which is unsuitable for either human consumption or industrial and agricultural operations, into freshwater. Various desalination methods have been developed over the last several decades to augment freshwater supplies in arid regions [4] including, multistage flash distillation (MSF), multi-effect distillation (MED), reverse osmosis (RO), solar evaporation, and freeze and thaw (cryodesalination). These processes are thoroughly discussed in Chapter 2.

Currently, most desalination plants worldwide are RO plants, while some use MSF. Together, these two processes account for 90% of the global desalination capacity. Generally, the seawater desalination process separates seawater into two streams. One stream is freshwater, while concentrated brine makes up the other. Achieving this separation requires a certain amount of energy and equipment specific to the process. Commercially, the two most important processes are MSF and RO, with MSF being the more dominant of the two in the twentieth century. In 1998, MSF plants accounted for 78% of the global desalination capacity, while RO only accounted for 10%. With time and advancements in technology, RO has grown in importance as a desalination process due to its lower cost and simplicity [4]. Conventional desalination methods may seem to be the answer everyone is looking for to solve water scarcity. However, that could not be further from the truth as conventional desalination methods incur high costs in terms of operation and maintenance [9]. Due to this, many arid regions in the world are unable to afford these technologies as an augmented freshwater source, hardly solving their water insecurity issue. Continuous research is being done in the desalination fields, with gas hydrate formation technology as one of the latest desalination technologies currently in testing. Although gas hydrate-based desalination technology began development way back in 1940, details regarding the technology are very scarce, though it shows promising potential. The contents of this book aim to highlight the growing developments in gas hydrate-based desalination.

1.2 Origin of Gas Hydrates

Gas hydrates are nonstoichiometric crystalline complexes containing water and guest molecules that form under low temperature (T) and high pressure (P) conditions. Guest molecules, such as methane (CH_4), propane (C_3H_8), and carbon dioxide (CO_2), are bundled in cages formed by hydrogen-bound water molecules. These cages are stabilized by van der Waals forces between guest molecules and water molecules. Gas hydrates are at the center of research within sustainable chemistry because of their innovative applications in a wide range of scientific and industrial contexts, such as gas exploration and production, energy storage, and CO_2 sequestration.

Under favorable conditions of temperature and pressure (P–T), i.e. low T and high P values, the hydration reaction of the guest species, or hydrate former F, is given by Eq. (1.1):

$$F + xH_2O = F'' \, xH_2O, \tag{1.1}$$

where "x" is the hydration number [10].

So far, three key kinds of gas hydrate structures have been determined, namely s1, sII, and sH structures. The shape, type, and size of the guest molecule greatly impact the type of gas hydrate structure formed. The shape and size of the hydrate cavities in the cages govern the structure difference, while the type and size of gas molecules indicate the type of hydrate formed. A cubic meter of hydrate constitutes about 180 m^3 of gas. A unit cell of sI hydrate consists of six large ($5^{12}\,6^2$) cavities and two small (5^{12}) cavities made up of 46 H_2O molecules. Guest molecules with a diameter of less than 6 Å, such as CH_4, C_2H_6, CO_2, and hydrogen sulfide (H_2S), form the sI structure. An sII hydrate unit cell contains 16 small (5^{12}) and eight large cavities ($5^{12}6^4$) formed by 136 H_2O molecules and guest molecules such as C_3H_8, isobutane. Smaller molecules like nitrogen, whose diameter is less than 4.2 Å, are exceptions and form sII structure when used as pure hydrate formers. sH structure consists of three cavities that are basic 5^{12}, $4^35^66^3$, and $5^{12}6^8$ cages. Smaller molecules, such as CH_4, xenon (Xe), or H_2S, help to stabilize the structure by occupying the small cages. Molecules that are larger in size (7 Å $< d <$ 9 Å), such as isopentane and neohexane, can form sH structures when accompanied by smaller molecules such as N_2, H_2S, and CH_4 [10].

1.3 Concept of Hydrate Formation

Hydrate formation is a physical process (crystallization) that includes nucleation, growth, and dissociation. The nucleation process starts once the essential requirements for the formation hydrate formation are accessible, i.e. high pressure, low temperature, a hydrate former, and the necessary amount of water. The stability of hydrate formation mostly relies on the pressure, temperature, type, composition, and thermodynamic behavior of the hydrate former. During the nucleation process, the water molecules group up on the sides of the hydrate former to form a whole or an incomplete embryo crystal. During this phase, certain variables like driving force, activation energy, the critical size of nuclei, and nucleation rate are critical. Based on pressure–temperature

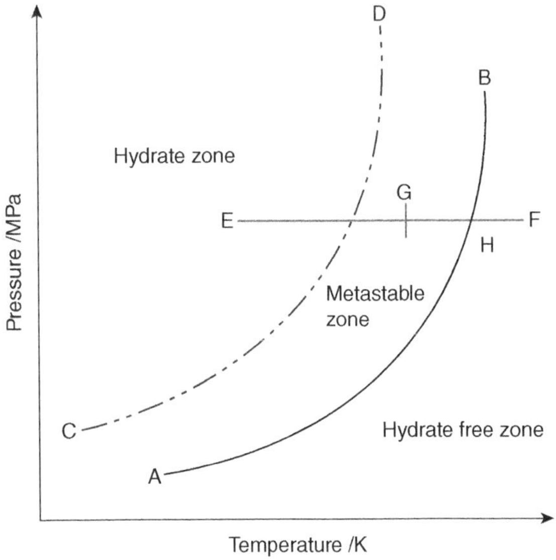

Figure 1.4 The formation of hydrate as a function of driving force (subcooling).

(P–T) plots, the formation of hydrate, nucleation, and the metastability of the hydrates can be estimated, as shown in Figure 1.4.

In Figure 1.4, the AB curve signifies the equilibrium curve and curve CD signifies supersaturation that demonstrates the limit of the metastability region. At position F, superheated state, there is no sign of the growth of the crystal. If nucleation starts at position H, then hydrates will start to form at position G. If the driving force (subcooling temperature) increases, then hydrate formation increases from position G toward E. Nucleation initiates rapidly and crystallizes left of the curve CD [10]. The cages tend to become unstable after nucleation completes and dissolve or grow into hydrates, thus forming metastable nuclei. Phase equilibrium data makes it possible to predict whether the hydrate former inhibits or promotes hydrate formation. The development of a hydrate depends on the interaction among the two fluids, temperature, pressure, mixing pattern, water history, and subcooling, as extensively discussed by [10, 11]. The knowledge of phase equilibrium conditions of various hydrate formers in wastewater samples can successfully guide the choice of experimental temperature and pressure. Several researchers have tried to identify the necessary parameters for hydrate-based water treatment by calculating the hydrate phase equilibria of refrigerants [12] and refrigerants in different salts like $MgCl_2$, NaCl, $CaCl_2$ [13, 14]. These feasibility studies used a single solute at low concentrations, but there is a need to experimentally investigate the hydrate phase equilibria with mixed solutes at high concentrations.

1.4 Application of Gas Hydrate in Desalination

Gaseous hydrate-based promoters can be used to treat saline water, wastewater, and produced water. The pure water molecules, now in the solid form, can be recovered through the melting (dissociation) and separation process. During the formation of hydrate, impurities or metal ions are rejected into the residual water resulting in the formation of pure hydrate, which on melting results in pure water and gas. The rejected water is rich in metal ion content and can be reused in subsequent desalination runs. This approach has been investigated extensively in the past using CH_4 and CO_2 hydrates, but no industrial solutions presently exist for a variety of practical and economic issues. However, recent advances in clathrate promoter research have opened the way to more appropriate, resilient, and less expensive alternatives to CH_4 or CO_2 for desalination applications.

1.5 Phase Behavior and Thermodynamic Measurement

Hydrate phase equilibrium studies are generally performed at a constant volume approach. It includes filling the reactor vessel with a sufficient amount of water. The reactor is then submerged in the coolant bath consisting of ethylene glycol and water. Excess air present inside the reactor is purged with the hydrate former (CO_2) gas. Later, the reactor is further pressurized with CO_2 guest gas and stabilized at the desired experimental pressure conditions. Once the pressure of the system is stabilized, the temperature is reduced to the experimental temperature to enable hydrate formation. A sharp rise in temperature or reduction in pressure indicates that hydrate formation is occurring. As more gas is consumed, pressure continues to drop until the completion of hydrate formation. Subsequently, the temperature of the bath encompassing the reactor gradually increases, leading to the dissociation of the hydrates. As dissociation proceeds, the pressure inside the reactor increases due to the release of gas. When dissociation is complete, the reactor pressure restabilizes. These changes in the temperature and pressure throughout the experiment are recorded by the data acquisition system. Hydrate phase equilibrium of the

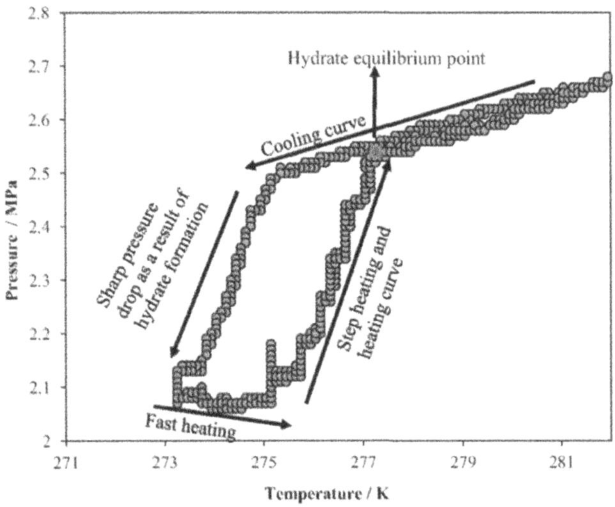

Figure 1.5 Pressure–temperature plot for CO_2 hydrate formation.

three phases (hydrate, liquid, and vapor) is achieved by sketching the intersection of (P–T) profiles of hydrate formation and dissociation mechanism, as shown in Figure 1.5 [15].

1.6 Kinetics of Hydrate Formation

A thorough understanding of the kinetics of hydrate is beneficial to (i) improve the formation of hydrates, such as storage of gas, separation of gas, and sequestration of gas [16, 17]; (ii) prevent hydrate development, particularly in the context of flow assurance [18]; and (iii) understand the rate of hydrate formation [19]. In comparison to gas hydrate dissociation [20], the investigation of hydrate formation is more confusing, as it involves nucleation followed by the growth phase.

The different steps of the hydrate nucleation process can be achieved as shown in Figure 1.6. Due to the gas solubility in liquid, the development of hydrates begins with an increase in gas intake. After the dissolution of gas into a liquid, the supersaturation phase begins when the operating temperature and pressure favor the formation of hydrates. There is a critical radius that must form during the induction time for nucleus stabilization; the hydrate nuclei must be larger than the critical radius. Any nuclei with a radius smaller than the critical radius would dissolve in the liquid phase. These critical radius

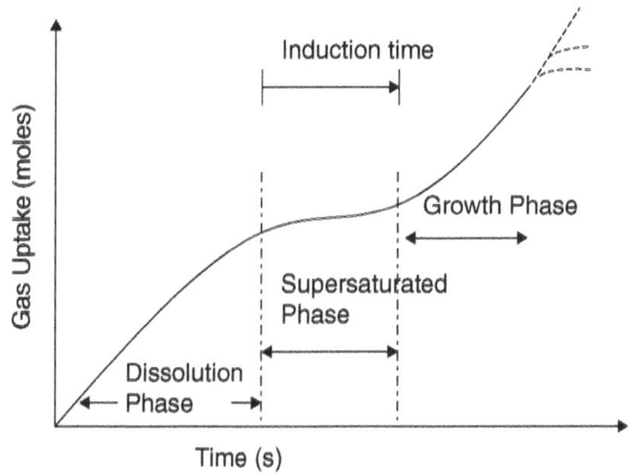

Figure 1.6 Phases in the hydrate formation process.

nuclei serve as the focal point for hydrate growth [21, 22], which is accompanied by a rise in gas intake.

Finally, depending on the mass transfer resistance, the gas intake may end at various phases, as illustrated in Figure 1.6 by the dotted lines. It is possible to break the mass transfer constraint at the gas–liquid interface using a stirrer or mixing mechanism. Hence, a further additional critical factor to be considered is the heat transfer mechanism to form hydrates faster. For faster hydrate formation phenomena to happen, the water surface area, gas dissolution rate, and contact must be enhanced. Hydrate formation is an exothermic reaction, so the heat generated during cooling should be eliminated in a batch or continuous approach to optimize the hydrate formation rate. The kinetics estimation variables are primarily utilized to comprehend and measure the formation kinetics following experiments. These variables are induction time, water to hydrate conversion, rate of hydrate formation, moles of gas consumed, and water recovery.

Generally, the evaluation of kinetic measurements is measured in a constant cooling batch approach. In the high-pressure reactor, a sufficient quantity of water is added. After that, the cell is transferred to the water bath containing water and ethylene glycol as coolant. Then the reactor is pressurized with the hydrate forming gas (CO_2) to the desired experimental pressure, and the temperature of the bath is set to the initial experimental condition. The system is left to stabilize before starting the experiment. After that, the system temperature is reduced to the desired experimental temperature by cooling. As cooling continues, the pressure reduces due to the hydrate formation. Simultaneously, due to the exothermic nature of hydrate formation, the temperature also increases.

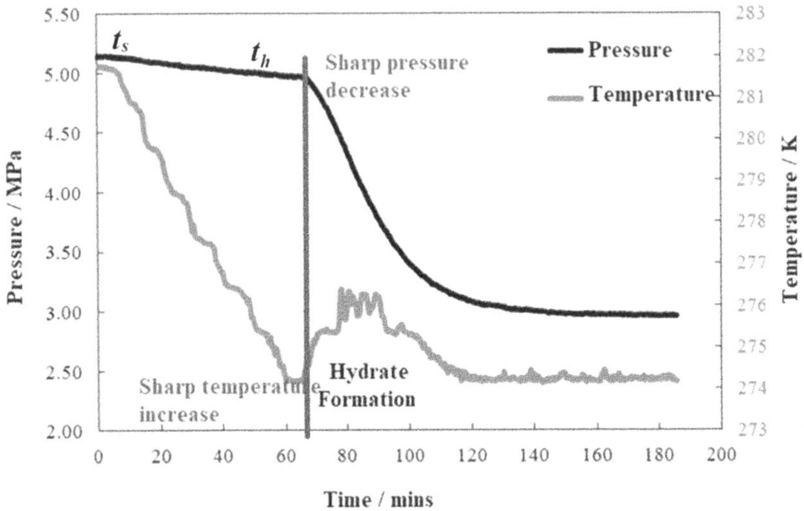

Figure 1.7 Temperature–pressure vs. time for CO_2 hydrate formation.

These changes in pressure and temperature are recorded by the data acquisition system. Once the hydrate formation is complete, the temperature and pressure become stable, as shown in Figure 1.7. Some of the kinetic parameters evaluated in terms of hydrate-based desalination are induction time, moles of gas used up, rate of hydrate formation, and water to hydrate conversion (mol%).

1.6.1 Induction Time

The time between the commencement of hydrate formation and the initial nucleation is known as the induction time. It is characterized by a rapid rise in temperature due to the exothermic nature of hydrate formation and a concomitant decrease in pressure owing to hydrate formation. It is written as presented in Eq. (1.2):

$$I_t = t_h - t_s, \tag{1.2}$$

where t_s is the system pressure and temperature to stabilize the initial experimental operating conditions, and t_h denotes the time taken for the noticeable hydrate to form [10].

1.6.2 Moles of Gas Used Up

The moles of gas used up is a critical variable that aids in determining the kinetics of the entire process. It reflects the quantity of gas spent throughout the method, which is equivalent to the quantity of CO_2 gas that could potentially be stored in hydrates. It can be estimated from the experimental pressure and temperature using the real gas equation, where Δn_H is the difference between moles of gas consumed at time t and 0, as presented in Eq. (1.3):

$$\Delta n_H = \left[\frac{PV}{zRT}\right]0 - \left[\frac{PV}{zRT}\right]t, \tag{1.3}$$

where z, R, T, V and P are Pitzers correlation, the gas constant, temperature, gas phase volume and pressure, respectively.

1.6.3 Rate of Hydrate Formation

The forward difference approach is used to calculate the starting rate of hydrate formation, as given by Eq. (1.4) [22].

$$\left(\frac{d\Delta n_{H,\downarrow}}{dt}\right)\mathrm{t} = \frac{\left(\Delta n_{H,\downarrow}\right)\mathrm{t} + \Delta \mathrm{t} - \left(\Delta n_{H,\downarrow}\right)\mathrm{t}}{\Delta t} \tag{1.4}$$

1.6.4 Water to Hydrate Conversion

The proportion of water converted into hydrate (C_{wh}) is a key factor in hydrate formation kinetics and is influenced by salt. Equation (1.5) is used to determine C_{wh}:

$$C_{wh}(\%) = \frac{\Delta n_H * h^n}{n_{H_2O}} * 100, \tag{1.5}$$

where Δn_H is the moles of gas used for the hydrate formation process as calculated in Eq.(1.3), n_{H_2O} refers to the moles of water existing in the reactor, and h^n is the hydration number described as the number of water molecules essential to clathrate a single hydrate former molecule [23].

Water recovery signifies the volumetric process efficiency of the hydrate-based desalination process and can be estimated by Eq. (1.6), as given in the literature [24].

$$\text{Percentage water recovery} = \frac{(\textit{Volume of water converted to hydrate}) * (F_h)}{\textit{Volume of feed solution}} * 100 \tag{1.6}$$

Rejection of salt signifies the efficiency of the process in eliminating contaminants. The salt rejection or removal efficiency is estimated by Eq. (1.7) [24, 25].

$$\text{Percentage removal efficiency} = \frac{C_{A0} - C_A}{C_{A0}} * 100 \tag{1.7}$$

Here, C_{A0} is the initial concentration of effluent, and C_A is that in the dissociated water. From the ion concentration present in the feed sample and the generated water following hydrate dissociation, the rejection percentage of each ion might be computed. The system's performance improves as the rejection rate rises. The yield of water is estimated as:

$$\text{Water yield} = \frac{V_1}{V_0} * 100, \tag{1.8}$$

where V_0 is the initial volume of effluent, and V_1 is the volume of dissociated water.

1.7 Hydrate Decomposition

The dissociation of the hydrate is an important phase in hydrate-based desalination (HBD). Hydrate dissociation might be accomplished by three methods: (i) by depressurization [26], (ii) through thermal heating [27], and (iii) by chemical injection [28]. Depressurization involves lowering the pressure at the interface so that the equilibrium temperature of the hydrate is lower than the ambient temperature. The heat input is caused by the temperature gradient from the surroundings, and the hydrate dissociates as a result. Modeling the hydrate dissociation phenomena by depressurization is a current study focus among researchers. In thermal treatment, the pressure is kept constant by applying heat to dissociate the hydrate crystal. This is the preferred practice in research laboratories. Chemical injection involves injecting chemicals into the hydrate to dissociate it, which is risky owing to geomechanics instability and natural calamities. The hydrate dissociation necessitates the use of heat energy in order to melt the hydrate crystals by releasing water and gas.

The stability of the hydrate's formation is estimated by studying the hydrate decomposition. At the onset of the dissociation process, residual water is drained, and the freshwater is collected and analyzed post-dissociation. To begin the dissociation process, the reactor vessel is depressurized to about 10% over the hydrate equilibrium pressure after the formation of hydrates. Next, the temperature is permitted to stabilize by slowly increasing the bath temperature. As temperature increases, it passes through the hydrate equilibrium boundary conditions, and the hydrates begin to dissociate, thereby increasing the reactor pressure. As the dissociate progresses, the pressure becomes constant, indicating the completion of dissociation of hydrate. The dissociated hydrate is collected as freshwater, and the rejected ion-rich water can be further treated through the gas hydrate process.

References

1 Aitken, C.K., McMahon, T.A., Wearing, A.J., and Finlayson, B.L. (1994). Residential water use: Predicting and reducing consumption 1. *Journal of Applied Social Psychology* 24 (2): 136–158.
2 Dangar, S., Asoka, A., and Mishra, V. (2021). Causes and implications of groundwater depletion in India: A review. *Journal of Hydrology*, 596: 126103. doi: 10.1016/j.jhydrol.2021.12610352659.
3 Water In The West. (2013). Water and energy nexus: A literature review.
4 Khawaji, A.D., Kutubkhanah, I.K., and Wie, J.M. (2008). Advances in seawater desalination technologies. *Desalination* 221 (1–3): 47–69.
5 Molden, D. (2019). Scarcity of water or scarcity of management? *International Journal of Water Resources Development* 36 (2–3): 258–268.
6 Kummu, M., Guillaume, J.H., De Moel, H., Eisner, S., Flörke, M., Porkka, M., et al. (2016). The world's road to water scarcity: Shortage and stress in the 20th century and pathways towards sustainability. *Scientific Reports* 6 (1): 1–16.
7 Zheng, J., Cheng, F., Li, Y., Lü, X., and Yang, M. (2019). Progress and trends in hydrate based desalination (HBD) technology: A review. *Chinese Journal of Chemical Engineering* 27 (9): 2037–2043.
8 Karagiannis, I.C. and Soldatos, P.G. (2008). Water desalination cost literature: Review and assessment. *Desalination* 223 (1–3): 448–456.
9 Sloan, E.D. and Koh, C.A. (2008). *Clathrate Hydrates of Natural Gases*, 3e. Boca Raton: CRC Press.
10 Lal, B. and Nashed, O. (2019). *Chemical Additives for Gas Hydrates*. Springer Nature.
11 Liang, D., Guo, K., Wang, R., and Fan, S. (2001). Hydrate equilibrium data of 1, 1, 1, 2-tetrafluoroethane (HFC-134a), 1, 1-dichloro-1-fluoroethane (HCFC-141b) and 1, 1-difluoroethane (HFC-152a). *Fluid Phase Equilibria* 187: 61–70.
12 Ngema, P.T., Naidoo, P., Mohammadi, A.H., Richon, D., and Ramjugernath, D. (2016). Thermodynamic stability conditions of clathrate hydrates for refrigerant (R134a or R410a or R507) with $MgCl_2$ aqueous solution. *Fluid Phase Equilibria* 413: 92–98.
13 Ngema, P.T., Petticrew, C., Naidoo, P., Mohammadi, A.H., and Ramjugernath, D. (2014). Experimental measurements and thermodynamic modeling of the dissociation conditions of clathrate hydrates for (refrigerant+ NaCl+ water) systems. *Journal of Chemical & Engineering Data* 59 (2): 466–475.
14 Bavoh, C.B., Partoon, B., Lal, B., Gonfa, G., Foo Khor, S., and Sharif, A.M. (2017). Inhibition effect of amino acids on carbon dioxide hydrate. *Chemical Engineering Science* 171: 331–339. doi: 10.1016/j.ces.2017.05.046.
15 Babu, P., Kumar, R., and Linga, P. (2013). Pre-combustion capture of carbon dioxide in a fixed bed reactor using the clathrate hydrate process. *Energy* 50: 364–373.

16 Dashti, H., Yew, L.Z., and Lou, X. (2015). Recent advances in gas hydrate-based CO_2 capture. *Journal of Natural Gas Science and Engineering* 23: 195–207.
17 Lederhos, J.P., Long, J.P., Sum, A., Christiansen, R.L., and Sloan, E.D., Jr. (1996). Effective kinetic inhibitors for natural gas hydrates. *Chemical Engineering Science* 51 (8): 1221–1229.
18 Babu, P., Linga, P., Kumar, R., and Englezos, P. (2015). A review of the hydrate based gas separation (HBGS) process for carbon dioxide pre-combustion capture. *Energy* 85: 261–279.
19 Yin, Z., Chong, Z.R., Tan, H.K., and Linga, P. (2016). Review of gas hydrate dissociation kinetic models for energy recovery. *Journal of Natural Gas Science and Engineering* 35: 1362–1387.
20 Maini, B.B. and Bishnoi, P.R. (1981). Experimental investigation of hydrate formation behaviour of a natural gas bubble in a simulated deep sea environment. *Chemical Engineering Science* 36 (1): 183–189.
21 Khurana, M., Yin, Z., and Linga, P. (2017). A review of clathrate hydrate nucleation. *ACS Sustainable Chemistry & Engineering* 5 (12): 11176–11203.
22 Linga, P., Daraboina, N., Ripmeester, J.A., and Englezos, P. (2012). Enhanced rate of gas hydrate formation in a fixed bed column filled with sand compared to a stirred vessel. *Chemical Engineering Science* 68: 617–623. doi: 10.1016/j.ces.2011.10.030.
23 Babu, P., Nambiar, A., He, T., Karimi, I.A., Lee, J.D., Englezos, P., and Linga, P. (2018). A review of clathrate hydrate based desalination to strengthen energy–water nexus. *ACS Sustainable Chemistry & Engineering* 6 (7): 8093–8107.
24 Nallakukkala, S. and Lal, B. (2021). Seawater and produced water treatment via gas hydrate. *Journal of Environmental Chemical Engineering* 9 (2). doi: 10.1016/j.jece.2021.1050539 (2).
25 Li, D. and Liang, D. (2012). Seawater desalination plant and seawater desalination method. Patent CN102351255A.
26 Wang, Y., Li, X.S., Li, G., Zhang, Y., Li, B., and Chen, Z.Y. (2013). Experimental investigation into methane hydrate production during three-dimensional thermal stimulation with five-spot well system. *Applied Energy* 110: 90–97.
27 Li, G., Li, X.S., Tang, L.G., and Zhang, Y. (2007). Experimental investigation of production behavior of methane hydrate under ethylene glycol injection in unconsolidated sediment. *Energy & Fuels* 21 (6): 3388–3393.
28 Yang, S.H.B., Babu, P., Chua, S.F.S., and Linga, P. (2016). Carbon dioxide hydrate kinetics in porous media with and without salts. *Applied Energy* 162: 1131–1140. doi: 10.1016/j.apenergy.2014.11.052.

2

Technologies in Desalination

Jai Krishna Sahith and Bhajan Lal

2.1 Introduction

The growing need for freshwater is one of the world's great challenges associated with rapid population growth and climate change. As discussed in Chapter 1, freshwater accounts for just 3% of freshwater resources found in the poles, groundwater, reservoirs, and rivers of the earth, while saline water makes up the remaining 97%. Desalination may be a promising and practical option for generating freshwater from seawater or brackish water in rural and urban areas. Around 1% of freshwater consumed globally was created by desalination in 2014. In 150 countries, nearly 17 000 desalination plants serve over 300 million people worldwide. The cost of desalinated freshwater has decreased dramatically over the past three decades through a combination of clean energy incorporation, development of energy recovery devices, and improvements in membrane technology.

2.2 Conventional Desalination Methods

2.2.1 Multistage Flash Desalination

The desalination method for Multistage Flash (MSF) is based on flash evaporation. Seawater is evaporated by decreasing the pressure as opposed to raising the temperature. The operating concept of MSF desalination is based on regenerative heating, where the seawater being flashed in each chamber gives up a portion of its heat energy to the seawater being flashed in the subsequent phases. The heat emitted by the vapor is referred to as condensation heat, and the absorption of this heat energy induces condensation of the water vapor. The condensation heat released by the water vapor raises the temperature of

Gas Hydrate in Water Treatment: Technological, Economic, and Industrial Aspects,
First Edition. Bhajan Lal and Sirisha Nallakukkala.

the incoming seawater steadily. The general setup for carrying out MSF desalination is shown in Figure 2.1.

Usually, an MSF plant consists of heat input, heat recovery, and heat rejection. Seawater heating is carried out by steam at low pressure in a brine heater. The seawater entering the brine heater follows the heat exchanger side of the tube situated in the upper section of the evaporator. Evaporators consist of multiple successive periods (chambers) sustained from the previous stage at reducing pressures. The seawater flows into the heat exchanger tubes, where the temperature of the condensation heat emitted by flashed water vapor is steadily elevated from one chamber to another.

Usually, the heat exchangers are distributed around the evaporator's diameter. In modern large MSF desalination plants, the evaporator is made up of multiple stages, usually 19–28. In general, depending on the scale control system used, the MSF plants run at a top brine temperature in the range of 90–120°C [1]. Granted, the operation of the plant at higher temperatures increases the efficiency of the process, but it also tends to increase the potential for scale formation [2, 3] and corrosion of the metal surfaces in contact with seawater. A high-temperature additive or even an acid dosage may be used to combat scale.

Pressure is held below the corresponding saturation temperature of the seawater flowing in at each point. Flickering allows the seawater injected into the

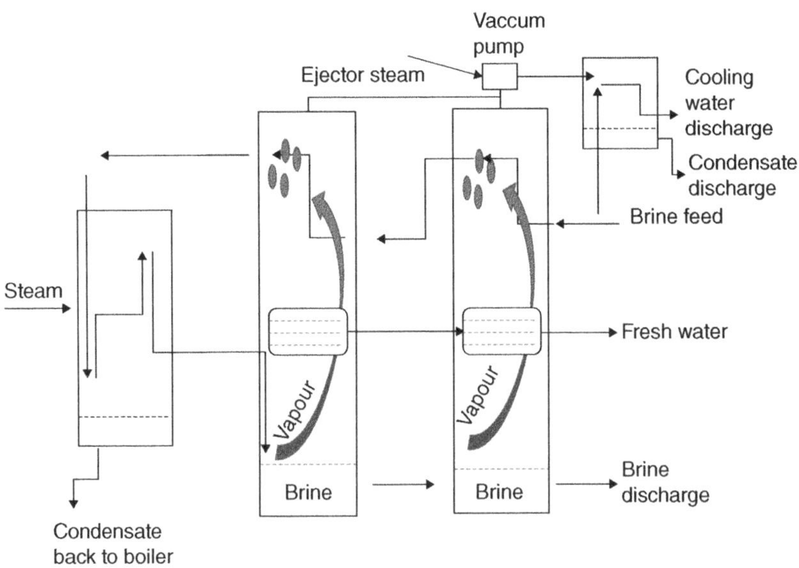

Figure 2.1 Multistage flash desalination.

chamber to boil quickly and intensely [4]. Orifices and baffles mounted between phases minimize the pressure of the brine to that of the equilibrium vapor pressure needed for boiling. Boiling continues until the temperature of the seawater exceeds the chamber's boiling point through the creation of water vapor. Flash distillation occurs in stages while simultaneously reducing the pressure of the boiling seawater.

To increase flickering at lower temperatures, unflashed brine is passed from one step to the next. This results in more seawater flashing without the need for additional thermal energy. Demisters are mounted at each point in the evaporator to decrease the carryover of brine droplets. A decarbonator and vacuum de-aerator are mounted in the evaporator to extract dissolved gases from the brine. The decarbonator eliminates carbon dioxide (CO_2) formed from the reaction between seawater bicarbonates, which is also responsible for producing alkaline scales and acids such as sulfuric acid. The vacuum deaerator extracts non-condensable gases, e.g. O_2 and N_2.

After flashing, cooler seawater flowing into condenser tubes cool and condense the flashed water vapor to create freshwater distillate. The latent heat emitted from the cooled water vapor is then used for heating the incoming brine in the tubes. From one stage to another, the distillate produced from each consecutive stage is cascaded and pumped into a storage tank. This water, however, is not yet ready for use. Usually, distillates from the MSF phase produces dissolved solids of 2–10 ppm. Remineralization by potabilization (post-treatment) is necessary [5].

For the MSF process, the vapor formation depends on the pressure maintained in each phase. The temperature drop range is 2–5°C per stage in a typical process. As seawater temperatures drop, the distillate output rises due to the flash range rising with a decrease in seawater temperature (the average flash range is 50–75°C). The economics of the plant often play an important role in controlling the production rate. The larger the number of stages of an MSF plant, the greater the region for heat transfer. Thus, process and plant productivity are maximized, but with higher initial development costs and higher annual costs for servicing and maintenance.

The energy supply of the brine heater is reused by the cooling water streaming in the heat rejection portion, which is normally composed of two to four phases [5]. This step is referred to as recirculation. A part of the warmed cooling seawater leaving the segment of heat rejection is transferred to the plant to be used as makeup water [6]. This is done to replace the portion of the recirculating brine that is lost to vapor formation. The brine is then heated and flashed through all the processes again.

With a blowdown stream utilized for scale management purposes, the main portion of the cooling seawater used in the heat rejection component is returned to the sea. The blowdown stream is needed to stop the flash brine from over saturating, which will induce a rise in the boiling point.

The seawater system typically consists of an open-intake channel or submarine drain, a pumphouse, sodium hypochlorite generators, and the delivery and return piping/channels that receive the feed water and cooling seawater supply. The feedwater first passes through garbage rakes and traveling screens locate in the pumphouse to isolate debris mechanically. Hot spent brine is discharged back into the ocean from the heat rejection portion via an outfall channel that extends into the sea.

Much progress has been made in terms of improving the MSF process over the past three decades. Improvements cover the specification and configuration of equipment, material collection and structural aspects, methods for installation and shipping, and thermodynamic design. The incremental progression of the MSF method itself entails the general design of the plant, the use of long-tube versus cross-tube, two decks versus a single deck, implementation of vertical MSF, the chemical treatment, the transport and balance of the brine, the heat transfer, the material and techniques of construction, the control and instrumentation, the pumps and the computer role [6].

Large-scale desalination processes emerged as early as 1960 due to the increasing demand for water in the arid regions of the Middle East. However, the process upgrades from prototype to industrial-scale without a complete understanding of the process design parameters caused problems to arise. In the early sixties, the equilibration principle was not thoroughly understood, which manifested in the form of larger, inefficient ranges between brine and related vapor pressures, particularly at lower temperatures. The technology has now shifted to larger units and production plants, and these large-scale units provide significant economies of scale, resulting in lower costs.

Entraining the brine in the vapor stream at certain stages causes the distillate produced to be unsatisfactory in quality. Demisters were installed to combat this problem, but the ultimate solution was to select effective antifoaming agents to decrease foam levels, enabling sufficient height of disengagement before the demisters. In addition, very little care was given to the design of the demister and its location within the evaporator in earlier plant revisions. Seeing as how demisters play an important role in maintaining distillate quality, modifications were proposed for the demister profile and carried out for more efficient performance.

By eliminating boiling, scale formation on heat transfer surfaces during the MSF process has been reduced. The capacity of scale formation, however, was not fully offset. Chemical additives to control and modify scale formation have been developed [6, 7]. However, this deterred effective heat transfer, which, in the long run, hindered successful activity. Another procedure called acid treatment was used to remove scale formation, but, in addition to being costly, it posed corrosion risks. After substantial growth, another technique was developed to battle the problem of scale. This process, known as online ball cleaning, made it possible for MSF plants to work effectively over both short and long-term performances. High-temperature additives to be used along with the online ball cleaning system were introduced further down the line, which allowed operating temperatures to go up to 115°C and ended the use of acid methods.

Carbon steel (CS) was used in MSF plant construction prior to the twenty-first century, especially for shells and interiors. However, CS corrodes in the presence of seawater. To compensate for the corrosion, the thickness of the construction materials had to be increased. This inadvertently increased the weight and size, resulting in higher costs. Construction then switched to stainless steel (SS) and duplex SS to combat this problem. The use of SS allowed for a lower metal thickness requirement in different evaporator components, which reduced the weight and size of the components and, in turn, directly influenced the cost of water production. In addition to process efficiency improvements, such as using titanium tubes in the ejector condenser to improve the system's heat and mass transfer, a better understanding of the equipment materials led to lower costs and less construction time.

Major redundancies were also identified in the plant setup, including the makeup strainer, high conductivity condensate flash tank, cooling water recirculation pump, water box and pump area cranes, ejector condensate extraction pump, and vacuum system ejector standby. The elimination of these has contributed substantially to lowering water rates, simplifying the structure of the plant, and reducing repair costs without adversely affecting the functionality of the plant.

Today, the process's dynamics are very well understood, and traditional control panels have been swapped for monitors that display a wide range of process information. Numerous control, modeling, and simulation studies have been conducted and constantly aid in optimizing the process and reducing costs.

2.2.2 Multi-effect Desalination

Multi-effect desalination (MED) is one of the oldest desalination methods [8] and is thermodynamically efficient [9]. The process takes place in a series of reactors which are called effects, as shown in Figure 2.2, and makes use of the principle of reducing ambient pressure in the various effects. In this process, the feedwater undergoes boiling multiple times without additional heat being supplied after the first effect. When the seawater enters the first effect, its temperature is raised to

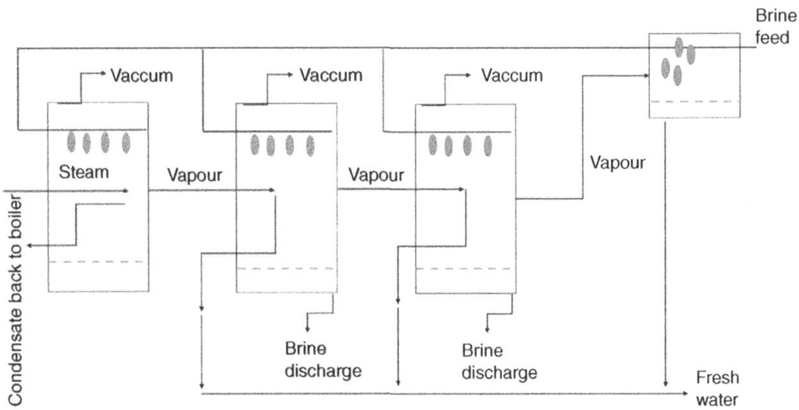

Figure 2.2 Multi-effect desalination.

the boiling point after undergoing preheating in tubes. Seawater is then sprayed onto the surface of evaporator tubes heated by steam supplied externally from what is usually a dual-purpose plant, promoting rapid evaporation. The hot steam condenses due to the release of latent heat on the opposite side of the tubes and the condensate is recycled as boiler feedwater in the power plant.

The steam economy of any MED plant is dependent on the number of effects installed in the plant. The total number of effects that can be installed is dependent on the total available temperature range and the minimum allowable temperature difference between one effect and the next.

Not all the seawater applied to the tubes gets evaporated. On the contrary, only a portion of the applied feedwater gets evaporated, while the remaining is fed to the second effect. The second effect undergoes a process similar to that of the first effect: the only difference being that the tubes in the second effect are heated by the vapors created in the first effect. Upon releasing heat to vaporize the feedwater, the produced vapor in the tubes condenses, forming the desired freshwater product. The evaporation and condensation processes are carried out continuously from one effect to the next, each at a lower temperature and pressure than the previous effect. This process can be optimized for many effects ranging from 4 to 21, with typical performance ratio ratios ranging from 10 to 18 in a large MED plant [10].

As is the case with MSF, scaling can be a potential problem in MED. However, MED plants have been built to operate with a top brine temperature (TBT) of 70°C, thus reducing the risk of scale occurrence. Heat transfer at this lower operating temperature, however, requires additional surface area in the tubes. In terms of power consumption, a MED plant has significantly lower power consumption than an MSF plant, all while having a performance ratio that exceeds that of the latter. Hence, a MED plant is thermodynamically more efficient than an MSF plant.

MED plants can have horizontal, vertical, or even submerged tubes. MED plants with horizontal tubes, as shown in Figure 2.2, have enjoyed successful operation for over three decades. Most of the recent applications of MED plants have been in the Middle East, and although relatively smaller than MSF plants, their numbers have been steadily rising.

MED offers the potential for water cost reduction. The specific power consumption per volume of distillate for a MED plant is 1.8 kWh/m^3 compared to 4 kWh/m^3 for MSF. Additionally, the low TBT conditions used in MED plants have substantially reduced scaling and corrosion rates to manageable levels, overcoming one of the most significant issues plaguing MSF plants.

2.2.3 Reverse Osmosis

Reverse osmosis (RO) is the most popular membrane method for seawater desalination [11, 12], with the highest market share globally, accounting for about 60% of the global desalination capacity. Figure 2.3 shows the working principle behind the RO process. In this process, the osmotic pressure is overcome by adding external pressure to the seawater, which causes the water molecules to flow against their natural flow from regions of low solute concentration to regions of higher solute concentrations. When this happens, dissolved salts in the seawater are unable to pass through the membrane and are left behind, creating a very concentrated salt solution. Water molecules pass through the membrane, enabling the production of freshwater.

Unlike MSF and MED, RO requires no heating or phase separation. The energy input required is primarily for the pressurization of the seawater feed.

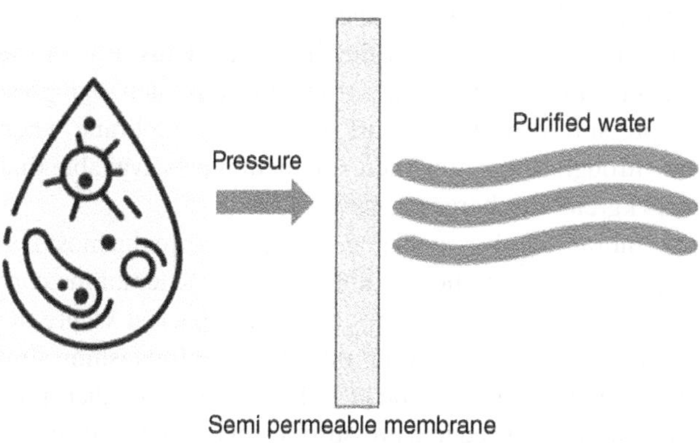

Figure 2.3 Reverse osmosis.

A typical RO plant consists of four major sections: (i) pretreatment of feedwater, (ii) high-pressure pumping, (iii) membrane separation, and (iv) post-treatment of permeate.

Pretreatment is a multistage process. Seawater is first flowed through trash racks and screens to remove any large debris via mechanical separation. The seawater is further cleaned in a multimedia gravity filter which uses media such as anthracite, sand, silica, and granite to remove suspended solids. Next, seawater flows through a cartridge filter that removes particles larger than 10 microns. The purpose of carrying out such extensive filtration is to protect the high-pressure pumps and the membranes from damage that would otherwise occur due to the abrasive nature of particulate matter. Another very important reason for pretreatment is the need to eliminate undesirable constituents in seawater that could potentially cause membrane fouling [13, 14]. Besides multimedia and cartridge filtration, other pretreatment methods include chlorination, dechlorination, coagulation, and acid addition. The type of pretreatment process administered generally depends on the feedwater quality, the type of membrane used, recovery ratio, and desired water quality. Some of the chemicals added to seawater are (i) sodium hypochlorite for the prevention of microorganism growth, (ii) ferric chloride as a flocculant, (iii) sulfuric acid for pH adjustment, and (iv) sodium bisulfite for dechlorination.

The next step entails pressurizing the pretreated seawater to attain the pressure required for the membrane to separate the dissolved salt molecules from water molecules. Typically this is done using high-pressure SS centrifugal pumps. As feed water flows through the membrane, the passage of dissolved salts is restricted while water molecules can flow through. The selectively sized pores prevent the movement of dissolved salts across the membrane, allowing for freshwater to be harvested. The concentrated brine obtained from restricting the motion of dissolved salts across the membrane is then discharged back into the sea. The pressures that the membranes are exposed to range from 50–80 bar, depending on the salt content of the feedwater. Thus, it is of the utmost importance that the membrane installed can withstand such high pressures, including the pressure drop that occurs across the membrane when water molecules flow through. Currently, there are membranes available that can withstand up to 84 kg/cm^2 discharge pressure.

Of the numerous membrane choices available, two options are the most feasible commercially: spiral wound and hollow fine fiber (HFF) membrane configurations [13, 14]. Spiral wound membranes are tightly packed where the permeable membrane is wrapped around a central core in a spiral fashion similar to a roll of fabric and made of polyamide [15]. The hollow fine fiber however is a U-shaped fiber bundle housed in a pressure vessel usually made out of cellulose triacetate.

Once the feedwater passes through the membrane, the procured freshwater will still contain trace amounts of dissolved salts. Residual salts, however, do not pose much of a problem as water quality is usually adjusted in the post-treatment phase. The post-treatment phase consists of pH adjustment, lime addition, removal of dissolved gases (if any) such as hydrogen sulfide (H_2S) and CO_2, and disinfection.

In designing and operating a RO plant, the major design considerations to account for are the flux quantity, recovery ratio, salinity of permeate, shelf life of the membrane, power consumption, and feedwater temperature. Compared to MSF, scaling up RO processes is significantly less problematic due to the ambient temperature operating conditions. This enables reduced use of metal alloys for equipment construction and encourages the utilization of polymeric materials. Over the years, many developments have been made that have helped further reduce the costs required to operate a RO plant.

Two innovations that stand out the most are the advances in membrane technology and the usage of energy recovery devices [16]. The latter makes use of the slight drop in pressure of the concentrated brine, which is in the region of 1–4 bar relative to the applied pressure, to generate rotating energy by using some form of a turbine, pump, or other types of equipment that can convert pressure drop to rotating energy, allowing for the depressurization of one stream to pressurize another stream. Breakthroughs in membrane technology have yielded membranes that boast more compression resistance, a longer shelf life, a better recovery ratio, improved flux, and improved salt passage. Together, these improvements have allowed RO processes to be more economically feasible in terms of operating and capital costs.

Over time, there has been a gradual increase in the RO train size, reaching around 9048–13 626 m^3/day; although this is still very far off from the train size of MSF, which can operate a unit size of 56 775–75 700 m^3/day. Currently, the largest seawater RO plant has a capacity of 624 000 m^3/day, with a recovery rate ranging from 40 to 60%. An RO plant has an energy consumption rate of 6–8 kWh/m^3 without energy recovery and 4–5 kWh/m^3 with appropriate energy recovery systems in place.

2.2.4 Other Desalination Methods

Other processes developed for water desalination have yet to achieve the same level of commercial success as MSF, MED, or RO, but, under certain circumstances and with more development, may prove to show great value. These include freeze and thaw and solar desalination.

The freeze and thaw process, as depicted in Figure 2.4, freezes seawater and then thaws it to obtain freshwater. During freezing, dissolved salts are excluded

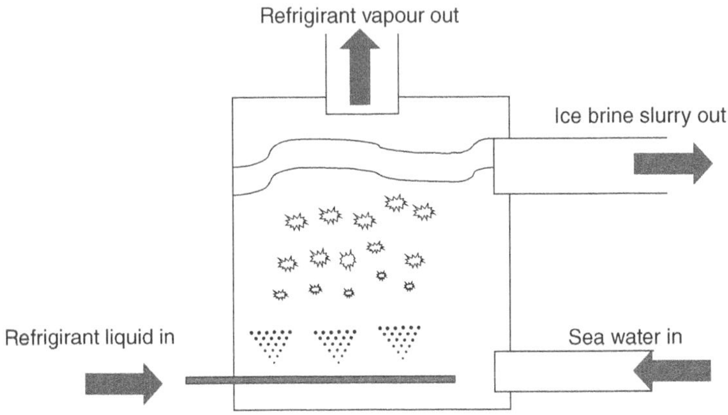

Figure 2.4 Freeze and thaw.

from the formation of ice crystals. Before the entire mass of water has solidified, the mixture is usually rinsed or washed to remove the remaining salts adhering to the surface of the ice. Next, the ice is allowed to thaw to produce freshwater. Hence, the freeze and thaw process consists of cooling the feedwater, partial crystallization of ice, separation of ice from seawater, melting of ice, refrigeration, and finally, heat rejection.

Several processes, including the triple point, hydrate process, eutectic process, secondary refrigerant, and indirect processes, have been developed to pilot plant status [16]. The main advantage of the freeze and thaw process over other processes is the low theoretical energy requirement. Besides that, the risk of scaling, corrosion, and precipitation is minimal because the operating conditions approach the freezing point of seawater. However, the handling of ice and water mixtures poses mechanical complications in terms of moving and processing. This prevents the freeze and thaw method from being as widespread as MSF, MED, or RO. Over the years, a small number of plants have been built, but none have ever been commercialized successfully to desalinate water for municipal purposes.

Solar evaporation is a traditional desalination process dating back thousands of years and has undergone extensive research. This process is similar, in part, to the natural hydrological cycle, where seawater is naturally heated by solar energy to produce water vapor, as shown in Figure 2.5. The water vapor is then condensed on a cool surface, and the resulting condensate is collected as freshwater. An example of this process is the greenhouse solar still, in which saline water is heated in a basin on the floor, and the produced water vapor is condensed on a sloping roof covering the basin [17].

The main reason this process has not been commercialized successfully is that it requires a large land area per unit of daily water production (25 ha/km^3).

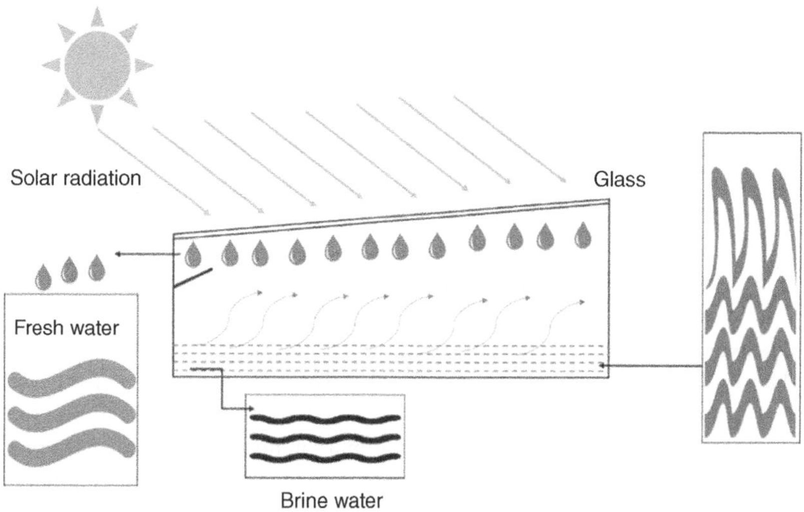

Figure 2.5 Solar evaporation.

This translates to high capital costs and vulnerability to weather-related damages. Additionally, although solar energy is free, additional energy inputs are required to pump water to and from the facility. These inputs contribute further to the high operating costs.

Another process that comes to mind under seawater desalination is potabilization. Potabilization is not technically a seawater desalination process, rather a desalination post-treatment process. For example, desalinated water from MSF plants is incredibly pure with a very small amount of dissolved salts and minerals. This causes the water to be very aggressive and corrosive to the materials commonly used in water distribution systems, such as metals and concrete. Additionally, the pure, mineral-less water has inferior taste and nutrition, making it unsuitable for municipal and commercial uses. Hence, potabilization plays an important role in overcoming the aggressive and corrosive nature of the desalinated water while also ensuring the water is safe for consumption [18].

Some of the processes carried out during potabilization are chlorination with or without aeration, injection of CO_2 and hydrated lime, and passing of the carbonated water through a limestone bed filter. These treatment methods play a vital role in establishing a calcium carbonate equilibrium and forming corrosion inhibiting layers. Chlorination is carried out by injecting chlorine gas, sodium, or calcium hypochlorite to curb and eliminate bacterial growth. Aeration is carried out to increase the oxygen content of the water, which improves taste. Through the processes of liming and carbonation, i.e. the injection of hydrated lime and CO_2, water devoid of minerals can be remineralized. Besides increasing dissolved mineral content, these treatment processes also aid in raising the hardness and alkalinity of the water. For the RO process, potabilization usually includes pH adjustment, the addition of lime, disinfection using chlorination or calcium hypochlorite, and removal of dissolved gases such as CO_2 and H_2S.

2.3 Gas Hydrate-based Desalination

Gas hydrates are ice-like crystalline solids consisting of hydrogen-bonded water (host) and gas (guest/former) molecules [19]. The hydrate former and seawater flow into a reactor, as shown in Figure 2.6. Low-temperature and high-pressure conditions are simulated to provide the optimal conditions for hydrate formation [20–22]. Upon completion of hydrate formation, concentrated brine is removed while the hydrates are allowed to dissociate, providing potable water. It has been observed that the dissociation of 1 m^3 of hydrates can produce 164 m^3 of gas and 0.8 m^3 of pure water at standard temperature and pressure. From the amount of water produced per m^3 of hydrates, gas hydrate-based desalination seems to be a promising alternative to expensive conventional desalination methods. Gas hydrates are non-stoichiometric compounds made up of approximately 86% water and 14% gas molecules on a molar basis. As a technology, gas hydrates have successfully been used for several industrial applications, including gas separation, CO_2 sequestration, gas storage and transportation, energy sourcing, refrigeration, and desalination of saltwater [23–26].

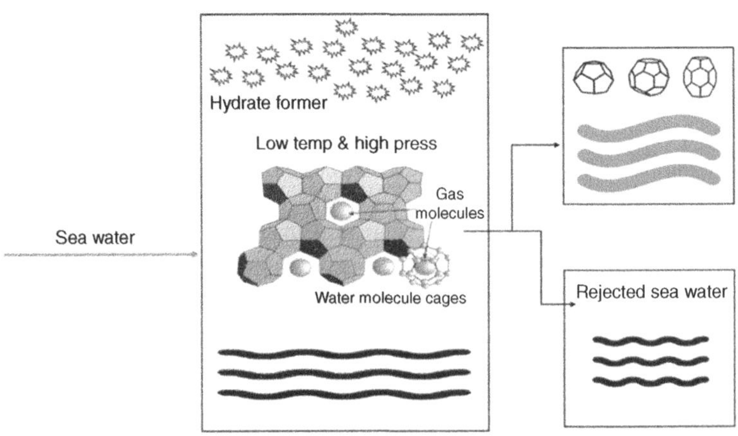

Figure 2.6 Gas hydrate-based desalination.

Gas hydrate-based desalination technology began development in 1940 and gained attention in 1970 due to the development of a desalination process by Sweet Water Development Company and Koppers and Company. Researchers began investigating the kinetics and separation of minerals using hydrate technology, and a pilot-scale plant was developed soon after [27]. Concentrated brines, such as seawater which primarily consists of NaCl, inhibit hydrate formation. However, the salinity in a hydrate once completely formed is almost zero, and this can be utilized to separate dissolved salts from seawater.

Currently, two of the most commercially significant desalination technologies in use today are the MSF and RO methods. Distillation as a desalination technology is a popular method, but it incurs high costs due to large amounts of phase transition energy required, the heat of vaporization of water. RO can compete against conventional desalination methods due to its lower costs. Collet et al. stated that hydrate-based desalination technology could be cheaper than conventional thermal and membrane processes [28].

Researchers have studied the use of various refrigerants for hydrate formation [29]. The use of refrigerants as hydrate formers were considered particularly appropriate for desalination purposes due to these formers being atmospheric hydrate formers that negated the need to work at high pressures. However, their impact on the environment has caused the usage of refrigerants, especially Freons, to be restricted.

During the early industrialization phase, the main obstacle faced was the separation of formed hydrates from the concentrated brine liquid, which led to an uneconomical operation of the process. However, recent studies have shown that gas hydrate-based desalination has the potential to be an economically competitive desalination process compared to conventional desalination methods. Karamoddin et al. [29] investigated the rate of recovery from a 2.5 wt% NaCl solution using the hydrate of propane in a moderately operated vessel in which hydrate nucleation, growth, separation, and dissociation occurred. The average reduction in the salt content of the recovered water from the NaCl solutions was found to be 31%. Cha et al. measured the removal efficiency of high salinity (9 wt%) NaCl solution using CO_2 and insoluble hydrocarbons to form soluble hydrates in a stirring reactor, achieving 90% removal efficiency of salts [30]. Cha et al. also studied the desalination effects using CO_2 hydrate with different concentrations of NaCl solutions, suggesting the remnant salinity should be controlled below 4 wt% [30]. A multigrade hydrate-based desalination process using CO_2 with of R141b (CH_3CCl_2F) additive at a ratio of 1:70 vol. ratio with seawater improved the removal efficiency by as much as three times that reported in Karamoddin et al. [29], enabling removal rates to reach 98.4%.

Laboratory scale experiments have gone a long way in proving the feasibility of hydrate-based desalination. However, two quandaries remain. The first is the choice for hydrate former, in which the phase state, solubility, safety, price, recycling, and the operating conditions are the parameters to consider for selection. The other is in further separation and purification, which will dictate the removal efficiency of freshwater. Gases are easily separated from the liquid and should be considered the best choice as hydrate formers for desalination. Potential hydrate formers that have been investigated for decades are CO_2 and propane for their moderate pressure and temperature requirements for hydrate formation. The formation characteristics of CO_2 hydrates have already been investigated extensively in salt solutions of different concentrations [31]. Compared to CO_2, it is easier for propane to form hydrates at several atmospheric pressures. However, propane is rarely used in hydrate formation due to its flammable characteristic, is thus relegated to co-form hydrates with other gases such as CO_2 [32].

References

1 Bahar, R. and Hawlader, M.N.A. (2013) Desalination: conversion of seawater to freshwater. 2nd International Conference on Mechanical, Automotive and Aerospace Engineering (ICMAAE 2013), 2-4 July 2013, Kuala Lumpur, Malaysia.

2 Alsadaie, S.M. and Mujtaba, I.M. (2016). Generic model control (GMC) in multistage flash (MSF) desalination. *Journal of Process Control* 44: 92–105. doi: 10.1016/j.jprocont.2016.05.006.

3 Said, S.A., Emtir, M., and Mujtaba, I.M. (2013). Flexible design and operation of Multi-stage flash (MSF) desalination process subject to variable fouling and variable freshwater demand. *Processes* 1: 279–295. doi: 10.3390/pr1030279.

4 Khawaji, A.D., Kutubkhanah, I.K., and Wie, J.M. (2008). Advances in seawater desalination technologies. *Desalination* 221: 47–69. doi: 10.1016/j.desal.2007.01.067.

5 Al Ghamdi, M.F., Hughes, C.H., and Kotake, S. (1987). The Makkah-Taif MSF desalination plant. *Desalination* 66: 3–10. doi: 10.1016/0011-9164(87)90190-1.

6 Hamed, O. and Mustafa, G. (2001). Prospects of improving energy consumption of the multi-stage flash distillation process. Proceedings of the Fourth Annual Workshop on Water Conservation in Dhahran the Kingdom of Saudi Arabia, 1–14.

7 Kizhisseri, M.I., Mohamed, M.M., and Hamouda, M.A. (2020). Prediction of capital cost of ro based desalination plants using machine learning approach. *E3S Web of Conferences* 158: 1–5. doi: 10.1051/e3sconf/202015806001.

8 Al-Shammiri, M. and Safar, M. (1999). Multi-effect distillation plants: state of the art. *Desalination* 126: 45–59. doi: 10.1016/S0011-9164(99)00154-X.

9 Mrij, A., Rachid, A., Kouhila, M., and Nouh, F.A. (2017). Comparison of reverse osmosis and multiple effects distillation performances for sea water desalination coupled with solar energy systems. *World Applied Sciences Journal* 35: 1114–1118. doi: 10.5829/idosi.wasj.2017.1114.1118.

10 Nair, M. and Kumar, D. (2013). Water desalination and challenges: The Middle East perspective: a review. *Desalination and Water Treatment* 51: 2030–2040. doi: 10.1080/19443994.2013.734483.

11 Karagiannis, I.C. and Soldatos, P.G. (2008). Water desalination cost literature: review and assessment. *Desalination* 223: 448–456. doi: 10.1016/j.desal.2007.02.071.

12 Loutatidou, S., Chalermthai, B., Marpu, P.R., and Arafat, H.A. (2014). Capital cost estimation of RO plants: GCC countries versus southern Europe. *Desalination* 347: 103–1. doi: 10.1016/j.desal.2014.05.033.

13 Abushaban, A., Salinas-Rodriguez, S.G., Dhakal, N., Schippers, J.C., and Kennedy, M.D. (2019). Assessing pretreatment and seawater reverse osmosis performance using an ATP-based bacterial growth potential method. *Desalination* 467: 210–218. doi: 10.1016/j.desal.2019.06.001.

14 Khanzada, N.K., Khan, S.J., and Davies, P.A. (2017). Performance evaluation of reverse osmosis (RO) pre-treatment technologies for in-land brackish water treatment. *Desalination* 406: 44–50. doi: 10.1016/j.desal.2016.06.030.

15 Balster, J. (2015). Encyclopedia of membranes. *Encyclopedia of Membranes* 3–5. doi: 10.1007/978-3-642-40872-4.

16 Stover, R.L. (2007). Seawater reverse osmosis with isobaric energy recovery devices. *Desalination* 203: 168–175. doi: 10.1016/j.desal.2006.03.528.

17 Buros, O.K. (2000). The ABCs of desalting. Topsfield, MA, USA: International Desalination Association.

18 Bajahlan, A.S. and Wie, J.M. (2013). Post-treatment of desalinated water and water quality characteristics in Yanbu industrial city. *Desalination and Water Treatment* 51: 1790–1803. doi: 10.1080/19443994.2012.714450.

19 Sayani, J.K.S., Pedapati, S.R., Kassim, Z., and Lal, B. (2021). Investigation on thermodynamic equilibrium conditions of methane hydrates in multiphase gas-dominant pipelines. *ACS Omega* 6: 2505–2512. doi: 10.1021/acsomega.0c04204.

20 Krishna, J., Sayani, S., Pedapati, S.R., and Lal, B. (2020). Phase behavior study on gas hydrates formation in gas dominant multiphase pipelines with crude oil and highCO_2 mixed gas. *Scientific Reports*, 1–12. doi: 10.1038/s41598-020-71509-6.

21 Sahith, S.J.K., Pedapati, S.R., and Lal, B. (2020). Investigation on gas hydrates formation and dissociation in multiphase gas dominant transmission pipelines. *Applied Sciences* 10: 5052. doi: 10.3390/app10155052.

22 Sahith, S.J.K., Sivabalan, V., Rao, S., and Bhajan, P. (2020). Investigation of CO_2 hydrate formation in the presence of gasoline. *Advances in Research Engineering* 200: 125–131.

23 Ai Krishna Sahith, S.J., Pedapati, S.R., and Lal, B. (2019). Application of artificial neural networks on measurement of gas hydrates in pipelines. *Test Engineering and Management* 81: 5769–5774.

24 Lal, B. and Nashed, O. (2020). *Chemical Additives for Gas Hydrates*. Switzerland AG: Springer Nature. n.d. doi: 10.1007/978-3-030-30750-9.

25 Kaur, M., Sahith, J.K., Sivabalan, V., and Lal, B. (2020). CO2 transportation via hydrates and liquefaction: technical analysis and comparison. *Proceedings of the Third International Conference on Separation Technology 2020 (ICoST 2020)*. Atlantic Press. https://doi.org/10.2991/aer.k.201229.020.

26 Sivabalan, V., Sahith, J.K., and Lal, B. (2020). Deep eutectic solvents as the new norm for oil and gas industry: a mini review. *Proceedings of the Third International Conference on Separation Technology 2020 (ICoST 2020)*. Atlantic Press. doi: 10.2991/aer.k.201229.017.

27 Sangwai, J.S., Patel, R.S., Mekala, P., Mech, D., and Busch, M. (2013). Desalination of seawater using gas hydrate technology – current status and future direction. Proceedings of XVIII Conference on Hydraulics, Water Resources, Coastal and Environmental Engineering (HYDRO 2013 INTERNATIONAL) December 4–6, 2013, IIT Madras, Chennai INDIA.

28 Collett, T.S., Johnson, A.H., Knapp, C.C., and Boswell, R. (2009). Natural gas hydrates: A review. *AAPG Memoir* 89: 146–219. doi: 10.1306/13201101M891602.

29 Karamoddin, M. and Varaminian, F. (2014). Water desalination using R141b gas hydrate formation. *Desalination and Water Treatment* 52: 2450–2456. doi: 10.1080/19443994.2013.798840.

30 Cha, J.H. and Seol, Y. (2013). Increasing gas hydrate formation temperature for desalination of high salinity produced water with secondary guests. *ACS Sustainable Chemistry & Engineering* 1: 1218–1224. doi: 10.1021/sc400160u.

31 Dholabhai, P.D., Parent, J.S., and Bishnoi, P.R. (1997). Equilibrium conditions for hydrate formation from binary mixtures of methane and carbon dioxide in the presence of electrolytes, methanol and ethylene glycol. *Fluid Phase Equilibria* 141: 235–246. doi: 10.1016/s0378-3812(97)00214-8.

32 Babu, P., Kumar, R., and Linga, P. (2014). Unusual behavior of propane as a co-guest during hydrate formation in silica sand: potential application to seawater desalination and carbon dioxide capture. *Chemical Engineering Science* 117: 342–351. doi: 10.1016/j.ces.2014.06.044.

3

Prospectives on Gas Hydrates-based Desalination

Jesa Singh and Bhajan Lal

3.1 Introduction

The utilization of gas hydrate formation as a concept for desalination has recently gained much attention due to growing acknowledgment of the water shortage crisis and the pursuit of a more cost-effective desalination process relative to the conventional methods. Gas hydrates are ice-like, non-stoichiometric compounds formed in the presence of a host (water) molecule and a guest molecule (hydrate former) under high-pressure and low-temperature conditions [1]. The host and guest molecules are bonded by van der Waals forces to form hydrates, which typically consist of 85% water molecules and 15% gas molecules. It is a common misconception that the guest molecules used to form hydrates can only be gas molecules. The guest molecule used to form hydrates can be either gaseous or liquid, as exhibited in [2], where the authors use R141b to form hydrate crystals.

While easily accessible freshwater reservoirs continue to be depleted, harvesting alternative freshwater sources trapped in glaciers, icecaps, or deep underground is not considered economically feasible nor environmentally sound [3]. Compounded with population growth, the finite nature of these dwindling supplies poses a very serious problem. The rate of water consumption has been increasing at more than twice that of population growth in the last century. By 2025, approximately 1.8 billion people will be living in arid regions with absolute water scarcity, while two-thirds of the remaining population will be living under the looming threat of water scarcity [4].

This is where conventional desalination processes have played a significant role in converting seawater, a resource unsuitable for human consumption or even industrial or agricultural use, into usable water. In layman's terms, desalination can be defined as a process that removes salts and other minerals such as sodium, magnesium, calcium, and carbonate from saline or reclaimed waters to produce potable water that can be utilized both industrially and

Gas Hydrate in Water Treatment: Technological, Economic, and Industrial Aspects,
First Edition. Bhajan Lal and Sirisha Nallakukkala.

domestically. Desalination has emerged as one of the most important processes to ensure the continuity of the survival of the human species. Various desalination processes, including multistage flash processes and reverse osmosis, have been developed over the last several decades. Chapter 2 provides an overview of the processes of conventional technologies.

More recently, gas hydrate-based desalination processes have been touted as being capable of reinventing the general desalination process. Capital and operating costs for gas hydrate-based desalination are significantly lower than those of conventional processes. These cost savings translate to reduced freshwater costs for various segments of societies around the globe that depend on desalinated water. In this chapter, we look at some of the more prominent designs of hydrate-based desalination setups. The information that follows is obtained from various journals, all of which are intended for the sole purpose of enlightenment on the topic is gas hydrate-based desalination.

3.2 General Proposed Gas Hydrate-based Desalination Design

3.2.1 Design 1

Figure 3.1 shows the schematic of an apparatus for water desalination with seawater based on the formation and dissociation of gas hydrates [6]. Before flowing seawater into the hydrate reactor, the temperature of the seawater is decreased using a heat exchanger. The reactor is connected to a gas storage vessel from which hydrate forming gas is supplied into the reactor. During the hydrate formation process, the gas within the reactor can be used up, leading to a pressure drop. As a control mechanism, the reactor is connected to a gas cylinder that recharges the pressure as needed, as well as replaces any gaseous guest molecules lost.

The seawater and the gaseous guest species, along with small quantities of hydrate formation promoters [7, 8], are agitated and blended within the reactor using a stirrer at a desirable rpm under requisite pressure and temperature conditions. Low-temperature and high-pressure conditions are maintained in the reactor throughout the gas hydrate formation process. Hydrate promoters play a vital role in significantly lowering the interfacial tension between the water and gaseous phases, thereby enhancing hydrate formation [6].

The hydrate slurry formed in the reactor is sent to a crystallizer using a hydrate slurry pump. There, the hydrate slurry is converted to a crystalline solid hydrate structure, producing a residue of concentrated brine solution. Upon completion, the crystalline solid hydrate structures are sent to a hydrate decomposer, where heat is used to dissociate the hydrates into gas and desalinated water [6]. The gas released during hydrate dissociation is sent to a storage tank through the upper

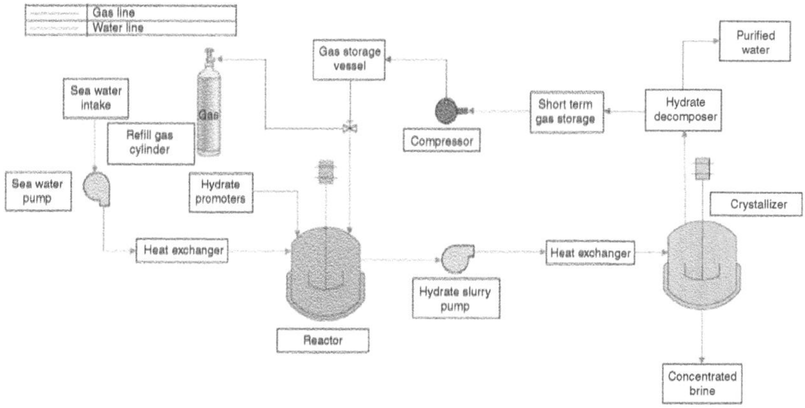

Figure 3.1 Schematic of apparatus used for gas hydrate-based desalination.

part of the hydrate decomposer and compressed before being sent back to the gas storage vessel to be reused. The desalinated water is gathered at the unit's base is ready to be utilized in industrial and domestic zones for daily use.

Studies have shown recovery rates in the range of 60% to 80% over a one-hour run at operating pressures of fewer than 60 bars (6MPa) using a unique sand bed reactor [9], which will be discussed later in this chapter. The utilization of propane (C_3H_8) as a co-guest at various concentrations, alongside other constituents of the gas mixture, also increases water recovery. The use of hydrate promoters such as acetone can also decrease the operating pressure required for hydrate-based desalination [10, 11].

Hydrate formation temperatures can also be increased [12], in which authors found the use of a co-guest such as cyclohexane (CH) or cyclopentane (CP) could increase the temperature of carbon dioxide (CO_2) hydrate formation from -2°C to 6°C or 7°C. Kinetic studies have revealed that mixed gases also increase the rate of hydrate formation for dual hydrates through the use of CH and CP co-guests and are shown to be 16 and 22 times greater than utilizing CO_2 to form pure CO_2 hydrates with a 90% purity [12].

3.2.2 Jacketed Reactor Designs

3.2.2.1 Design 1

Figure 3.2 shows the schematic of the desalination apparatus for experiments carried out by Park et al. [13] Detailed information regarding the setup is presented in Table 3.1. The main equipment of the setup, the reactor, is a 316 stainless steel duplex jacketed hydrate reactor, within which the water-ethylene glycol coolant mixture is circulated. The dual cylinder unit consists of pistons that move vertically in a connection pipe to pelletize the formed gas hydrates.

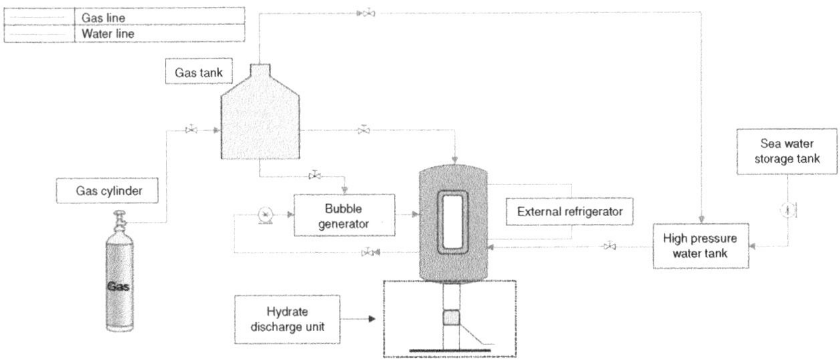

Figure 3.2 Apparatus used for gas hydrate-based desalination with dual cylinder unit.

Table 3.1 A detailed breakdown of the desalination apparatus.

Reactor type	1800 cm^3 duplex jacket type hydrate reactor (316 stainless steel) with dual cylinder unit, dual polycarbonate viewing windows, liquid level sensor, and drain valve
Components of setup	Duplex jacketed hydrate reactor with a dual cylinder unit, high-pressure circulation pump, century-tube-type bubble generator, hydrate discharge unit
Materials used	Water-ethylene glycol solution as refrigerant fluid, carbon dioxide (99.8%) as hydrate former, seawater samples from the southeast coast of Korea (35° 5′ 1″ north, 128° 47′ 11″ east)
Setup efficiency	78.1%

Pressure transmitters and copper-constantan thermocouples measure temperature and pressure within the reactor and both the gas and feedwater tanks. The gas and feedwater storage tanks are kept at a higher pressure than that of the hydrate reactor. The flow of the gas and feedwater from these pressurized tanks is controlled by control valves and a proportional-integral-derivative (PID) controller [13]. The reactor contents are mixed using a high-pressure circulation pump and a Venturi tube-type bubbler generating bubbles of approximately 10 μm as measured using an optical microscope.

The seawater used in this setup was retrieved from the southeast coast of Korea. Carbon dioxide was used as the guest due to the benign presence of dissolved CO_2 in drinking water upon the completion of desalination. Once the entire system reached the set temperature of 280 K, the hydrate forming reactor and gas tank were flushed at least thrice with CO_2 to remove any residual air. Next, 1000 mL of seawater was added to the reactor and CO_2 until the desired pressure of 2.9 MPa was achieved [13].

The hydrate formation reaction proceeded for a total of 150 min, during which time the CO_2 was consumed. Additional gas was automatically added from the tank using the PID controller and a control valve to maintain the set pressure. A constant level of water was also maintained by means of a water level sensor and control valve.

Figure 3.3 shows the cross-sectional view of the hydrate reactor and the dual cylinder unit. The connection pipe that houses the pistons has multiple holes of different sizes and shapes for fluid to pass through, allowing hydrate slurry to flow from the reactor into the connection pipe. When there is a sufficient hydrate slurry in the connection pipe, the upper piston exerts a downward motion to squeeze the slurry. The residual water forced out of the hydrate slurry during compression flows out through the smaller-sized holes. In Park et al. [13] the pistons in the connection pipe were operated at hydraulic pressures of 50–150 kg/cm^2. The squeezing motion is repeated a few times until the pelletized hydrate reaches a certain thickness. After compression, both pistons move downward at a similar rate to

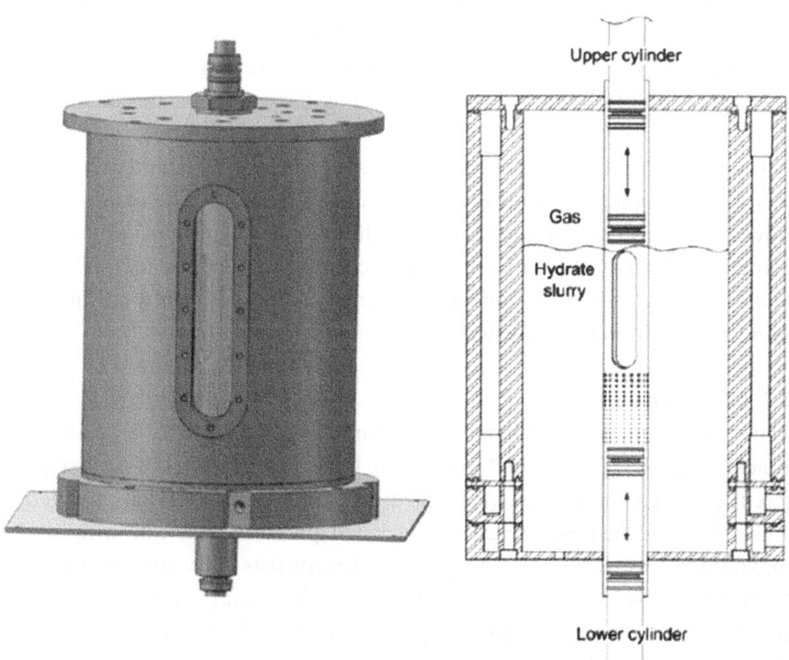

Figure 3.3 Cross-sectional view (front and rear) of the hydrate reactor (left) and dual cylinder unit (right) [13].

Figure 3.4 Actual extracted pelletized gas hydrate. The squeezed pellet in the left image exhibits gas bubbles on the surface due to dissociation at atmospheric conditions [13], Figure 03 (p.93) / with permission of ELSEVIER.

enable the pellet to be extracted, and then. Once the pelletized gas hydrate is recovered, the pistons return to their initial positions to begin the process again. Figure 3.4 shows an actual pellet made up of gas hydrates that have been extracted.

Once extracted, the pellet is immediately weighed before hydrate dissociation. As seen in Figure 3.4, the pellet in the left image exhibits gas bubbles on the surface [13]. This is attributed to dissociation during exposure to atmospheric conditions. The dissociated water is also weighed to calculate the hydrate conversion.

The concentration of salt in the water is analyzed using a digital conductivity meter, carrying a maximum uncertainty of less than 5%. The content of dissolved minerals (calcium, potassium, sodium, and magnesium) are analyzed using Inductively Coupled Plasma Atomic Emission Spectroscopy (ICP-AES), and the removal efficiency for each mineral is calculated based on the initial mineral content in the seawater.

3.2.2.2 Design 2

Figure 3.5 depicts the schematic diagram of the apparatus assembled by Bradshaw et al. [2] to conduct gas hydrate formation experiments utilizing different guest gas molecules. A detailed account of equipment is provided in Table 3.2. The installed reactor, a jacketed pressure cell, can withstand a maximum pressure of 1500 psi [2]. The jacket consists of coolant circulating in from a thermostatic circulation bath and provides temperature control. The cell is also fitted with a magnetically coupled stirrer shaft and a controller for the stirrer motor [2].

The molecules used to form gas hydrates can be directed to the pressure cell in either liquid or gaseous form, with the option of an external reservoir if not

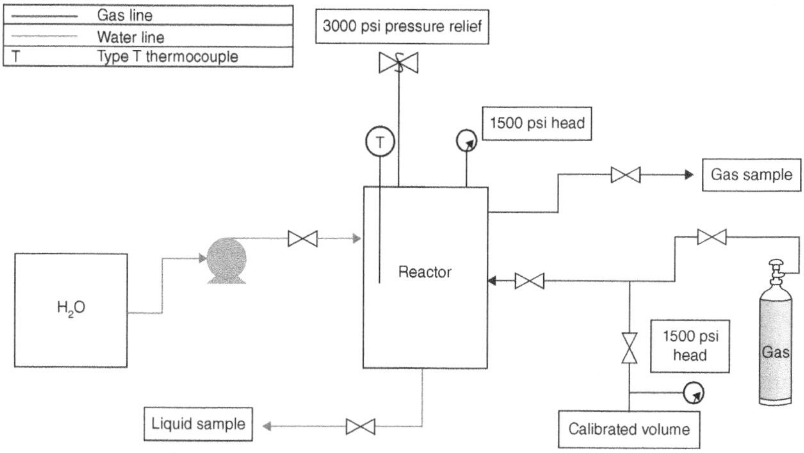

Figure 3.5 Schematic of apparatus for gas hydrate-based desalination.

Table 3.2 Detailed breakdown of the desalination apparatus.

Reactor type	500 mL Parr Model 4562 jacketed stirred pressure cell, stirrer shaft (magnetic couple), stirring motor speed controller
Components of setup	500 mL Parr Model 4562 jacketed, stirred pressure cell, thermostatic circulating bath, feed system for guest gas, injection pump for water, Type T and Type K thermocouples, Pressure transducers, Data acquisition software
Materials used	**Guest molecules**: R141b (1-fluoro, 1,1-dichloroethane,), Ethylene (99.5%), HFC-32 (difluoromethane), and R152a (difluoroethane, 99%), Cyclopentane – cooperative guest molecules when primary guest molecules such as HFC-32 or R152a **Saline solution**: Using deionized water (18 meg-ohm specific resistivity) and Sodium Chloride Heat exchange fluid (inert) – Fluorinert FC3283

initially present in the pressure cell. The majority of the conducted experiments were carried out in simple batch mode.

Agitation is an important aspect of hydrate formation as it enables the host and guest molecules of different phases to properly mix, ensuring efficient hydrate formation. In this setup, the impeller was a four-blade Parr propeller rotated at a constant speed of 600 rpm in almost all the cases studied by the authors. In a liquid host and guest species system, the impeller is located in the region of the denser liquid layer, where it can produce sufficient agitation to enable the less dense liquid to be dispersed in the denser liquid [14]. With gaseous guest species, the impeller is partially immersed at the liquid layer to enhance contact between the gas and liquid molecules [2].

Thermocouples and pressure transducers connected to a data acquisition system were used to measure the temperature and pressure. Data is acquired at 6 s intervals, providing sufficient time resolutions for the carried out hydrate formation experiments.

Hydrate formation experiments carried out using refrigerant-141b (R141b) are performed by first adding weighed amounts of chilled R141b to the cell, within which a measured amount of saline solution is already present and cooled to the required temperature. Upon the addition of the chilled R141b into the cell, the two liquids are then stirred at a low rpm until the temperature within the cell has stabilized at the desired value, as indicated by the thermocouples [2]. Next, seed crystals are added to the cell and the cell head is secured. The presence of air does not affect the formation of hydrates, so the cell is not flushed after being secured. The stirrer speed is then increased to 600 rpm [2], at which point the hydrate formation and growth are allowed to commence. Hydrate formation is indicated by a sudden increase in the temperature of the cell contents, as measured by the two inserted thermocouples.

Approximately 0.1 gm of R141b seed crystals, grown separately and stored in a refrigerator, are then added to the chilled contents of the cell. The resultant secondary nucleation leads to sustained growth of R141b hydrates with reduced induction times compared to primary or homogenous nucleation, processes that could lead to long, variable induction times prior to hydrate growth [15]. The operating environment may vary at larger scales of production, with a significantly higher amount of solid hydrate material.

Recent environmental regulations have restricted the use of hydrofluorocarbon compounds such as R141b due to their potential for ozone depletion, which has severe environmental effects. Thus, they are no longer used for hydrate-based desalination processes. Safer compounds have been explored and eventually identified as suitable replacements for R141b, including difluoromethane and difluoroethane. Both alternatives form hydrates at relatively high temperatures and moderate pressures [2].

Hydrate formation experiments using gaseous guest species are carried out primarily by adding a weighed amount of saline solution to the cell and cooling to the required temperature while simultaneously stirring at low rpm to attain the required steady-state values. The cell head is then secured, and the internal gaseous space is flushed with argon several times before admitting the gaseous guest species, after which the speed of the stirrer is increased to 600 rpm [2]. During the last argon flush, but before admitting the gaseous guest species, the pressure in the cell is maintained at above 100 psi while sealed to check for leaks and ensure the pressure within the cell remains constant [2].

Homogenous nucleation takes place when gaseous guest species are used as seed material. However, using a seed material in this instance is impractical and unnecessary, as high pressures would be needed to stabilize the seed crystals [2]. The formation of hydrates is indicated by a sustained increase in the temperature of the cell contents and a continuous decrease in cell pressure due to the guest gas species being used up by the process of hydrate formation.

When the hydrate formation process is complete, the formed hydrate samples are extracted at atmospheric conditions. The hydrate materials are extracted using a spatula that has been chilled and placed in a pre-chilled vial before storing it in a freezer for subsequent analysis if required. Otherwise, the hydrate samples are allowed to dissociate, and the salt content is determined by measuring the refractive index. The obtained refractive index data is converted to the corresponding concentration of NaCl in the solution [2].

3.2.3 Silica Sand Bed Crystallizer Reactor Design

3.2.3.1 Design 1

Table 3.3 presents the details for the apparatus and materials used in the setup displayed [16] in Figure 3.6. The crystallizer has ports through which four thermocouples can be inserted at various heights radially. Thermocouples T1 through T3 are placed in the bed for temperature measurements, while thermocouple T4 is placed at the bottom of the bed. Gaseous feed and vent lines are installed for gas to enter and exit the reactor. A water bath maintains the temperature of the contents within the crystallizer at 277.2 K [16]. A PID controller is equipped at the exit valve of the reactor, connected to the gas reservoir to help maintain a constant pressure.

Silica sand (645.16 g) is placed in the crystallizer to form a silica sand bed with a height of 5 cm. Seawater or pure water is added to obtain 100% or 75% silica saturation, 140 mL or 105 mL, respectively. The sand and water are alternately layered in five to seven stages to avoid the formation of any air pockets [17, 18]. Once the silica sand bed is formed, the crystallizer is closed and the thermocouples are inserted in their respective positions.

The process begins by immersing the crystallizer in the water bath and connecting the crystallizer to the gaseous inlet, reservoir cylinder, vent lines, and online data acquisition system. Methane (CH_4) is injected into the crystallizer until the pressure reaches 1 MPa and then flushed thrice to remove any atmospheric air that may have remained lingering in the crystallizer. Once the temperature and pressure are set at the required conditions, the formation profiles of pressure and temperature are recorded at 20-second intervals until the

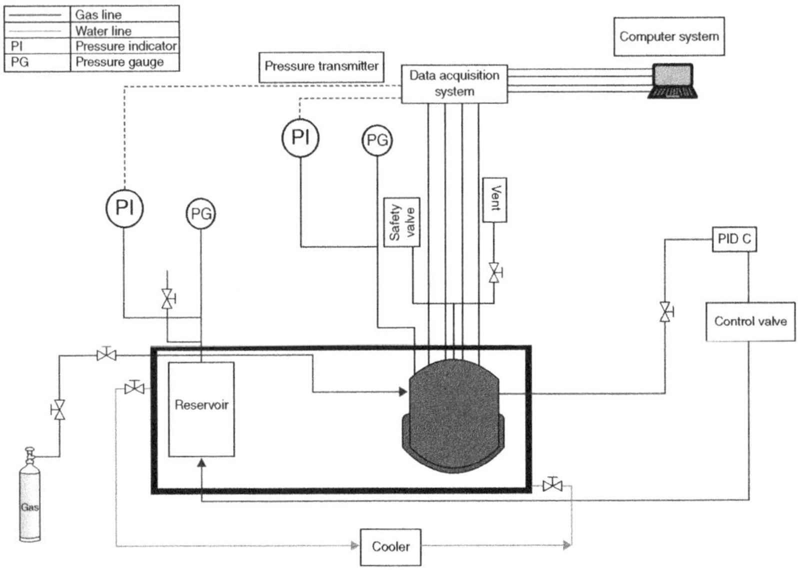

Figure 3.6 Experimental setup for hydrate formation and dissociation studies in porous media.

Table 3.3 Detailed breakdown of the desalination apparatus.

Reactor type	1240 cm^3 (SS-316) Stainless Steel crystallizer with an inner diameter of 10.2 cm and height of 15.2 cm
Components of setup	Stainless steel crystallizer,4 thermocouples, pressure transducer, water bath, PID controller, a gas reservoir
Materials used	Seawater (3.03 wt % salinity), Silica sand (0.34 porosity, 100 μm – 500 μm size, 0.217 cm^3/g pore volume, and 1.56 g/cm^3 density), Deionized water, methane (99.9% purity)

crystallizer steadily reaches the experimental temperature and pressure conditions. The temperature rise is eventually followed by a drop in pressure within the crystallizer, indicating the formation of gas hydrates. Formation is complete when the pressure drop recorded in the crystallizer is considered negligible (around 1–2 kPa/h).

The formation process is followed by hydrate dissociation in porous media via thermal stimulation. Dissociation begins as the temperature in the silica sand bed approaches the equilibrium temperature and is indicated by an increase in the crystallizer's internal pressure. The excess gas released during the dissociation process increases the pressure to above 4.8 MPa [16]. At this point, the PID controller causes the crystallizer valve to open, directing the excess gas from the crystallizer into the reservoir. As the hydrates dissociate, the temperature is maintained at 277.2 K, but the pressure of the crystallizer is slowly vented off to 4.8 MPa [19]. Dissociation was studied for two incremental temperature rises of $\Delta T = 10$ and $\Delta T = 20$ in a silica sand bed. The temperature rise in the crystallizer is at a rate of 0.1– 0.2 K/min [16].

3.2.3.2 Design 2

One of the conceptual schematics for hydrate-based desalination is shown in Figure 3.7. The concept illustrated [9] harnesses the ability of a C_3H_8 co-guest to draw water from within the silica sand bed to address the slow kinetics and advocate the effective separation of hydrate crystals from the residual concentrated brine.

Seawater flows into the crystallizer and fills the interstitial pore spaces available between the silica particles. The crystallizer is then chilled to the experimental temperature required for hydrate formation to occur. This process is carried out by harnessing the cold energy released during the re-gasification of liquefied natural gas or a traditional cooling system such as one that is ammonium-based. When C_3H_8 is used as a co-guest, water is drawn up and out from within the interstitial pore spaces of the silica, allowing hydrates to form in the hydrate formation zone above the sand bed [9].

The salts initially present in the seawater are preferably excluded from the hydrates within the silica sand bed, with the residual concentrated brine solution remaining below the retainer, as shown in Figure 3.6. However, a percentage of the initial salts might end up adsorbed into the hydrates by

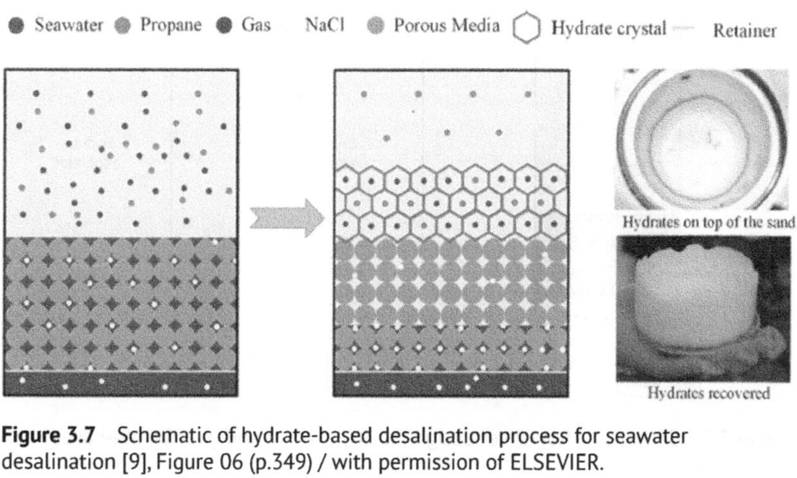

Figure 3.7 Schematic of hydrate-based desalination process for seawater desalination [9], Figure 06 (p.349) / with permission of ELSEVIER.

moving along with the water. Still, According to [13], up to 80% removal of dissolved minerals can be achieved with a well-agitated system. Higher removal efficiencies are experimentally possible for this system but require further investigation for consistent results [9].

Once the hydrate formation process is complete, the formed hydrate crystals are removed from the hydrate formation zone above the silica sand bed and dissociated to yield salt-free water. The location of the formation zone directly above the silica sand bed does pose operational problems; however, this can be dealt with through more novel reactor designs.

The water conversion rates achieved from this can design match those of conventional desalination methods. Moreover, ranges of 60–80% recovery within an hour of operating below 6 MPa [9] suggest the possibility for significantly higher recovery rates. Opportunities for increasing these rates while simultaneously reducing the operating pressure, depending on the concentration of C_3H_8 co-guest and other gas mixture constituents, are currently under investigation at a bench scale.

3.2.3.3 Design 3

The ability for C_3H_8 to draw water and enhance formation kinetics in the presence of 2.5% C_3H_8 in the fuel gas mixture is demonstrated in a study [9] for simultaneous CO_2 desalination and CO_2 capture, as shown in Figure 3.8.

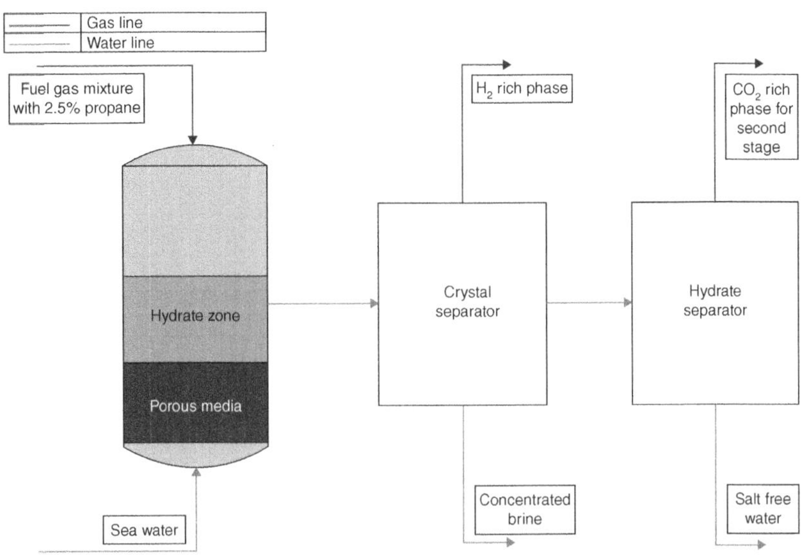

Figure 3.8 Conceptual flowsheet design for CO_2 capture and gas hydrate-based desalination.

Hydrates are formed in the hydrate crystallizer from seawater and a flue gas mixture containing 2.5% C_3H_8. Formation begins in the silica sand bed, where hydrate nucleation occurs in the interstitial pore spaces of the silica, which spurs hydrate growth in the gas phase above the sand bed. To sustain the growth of hydrates, water is continuously drawn from the silica sand bed and into the region where hydrate growth is taking place, using C_3H_8 as a co-guest. Once hydrate formation has occurred, the hydrogen-rich residual gas stream is sent to a gas turbine to be repurposed for energy generation, while the formed hydrates are recovered and allowed to dissociate [9]. The CO_2 obtained upon the completion of dissociation can be further purified or sequestered and stored for usage. The concentrated brine which remains in the crystallizer can be flushed out by washing the silica sand bed with seawater.

This process can significantly reduce the overall cost and is an attractive option from an energy and environmental point of view, as it combines two energy-intensive processes into one and recycles energy sources. However, it is noted that several unseen engineering and process challenges, alongside issues in scaling up the process, may arise and require proper mitigation.

3.2.4 Stirred Reactor Design

3.2.4.1 Design 1

The setup shown in Figure 3.9 and detailed in Table 3.4 comprises a stirred hydrate reactor with an approximate volume of 300 cm^3. It has a circulatory system that utilizes a cooling medium to control the temperature, a storage cylinder for injecting gas into the reactor, and a stirrer located within the reactor to ensure proper mixing of the contents within the hydrate reactor [20].

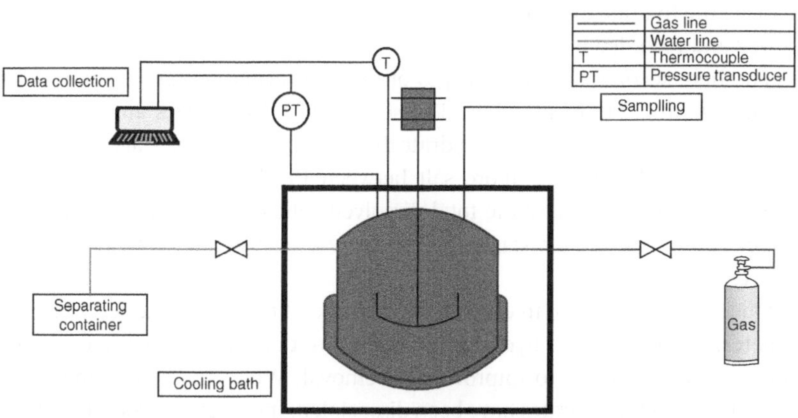

Figure 3.9 Schematic of the experimental setup for hydrate formation.

Table 3.4 A detailed breakdown of the desalination apparatus.

Reactor type	300 cm^3 hydrate reactor with stirrer
Components of setup	Cooling medium circulatory system, gas cylinder, pressure transducer with a scale of 0.5 psi (approximately 0.5% accuracy), thermocouple with a scale of 0.1 K (approximately 0.4% accuracy), separating container
Materials used	Compressed natural gas, synthesized produced water, sodium chloride (>99.5%), potassium chloride (>99.5%), calcium carbonate (98%), magnesium chloride hexahydrate (>99.5%), sodium sulfate anhydrous (99%), hydrogen chloride 1 N Solution
Setup efficiency	Up to 81.7%

Pressure and temperature are measured by a pressure transducer (0.5 psi increments, accuracy up to 0.5%) and thermocouple (0.1 K increments, accuracy up to 0.4%). Once the hydrate formation and dissociation begin, the produced concentrated brine is drained into a separate container.

The pressure in the reactor is increased to 1600 psi (110 bar) using an inert gas (N_2) and is left for 24 h at a temperature of 20°C (293.2 K) to ensure that the system has no leaks [20]. The reactor is then washed using distilled water, and 100 cm^3 of synthesized produced water is charged into the reactor. Next, the reactor and its content are chilled to the required temperature, which in this case is 274.2 K. Once the temperature stabilizes, the reactor is pressurized to 1380 psi (95 bar) using compressed natural gas, and the stirrer is turned on with a rotation speed of 360 rpm to initiate hydrate formation. The stirrer is later reduced to 220 rpm for the second and third stages of hydrate formation. During the hydrate formation process, the pressure and temperature are monitored and recorded every 20 s.

The first, second, and third stages of hydrate formation are monitored for 20 h, 10 h, and 2 h, respectively. The second and third stages are performed using water produced from the hydrate dissociation of the first stage, with each successive cycle trapping more salt between the hydrate crystals. Hydrate memory is improved, and the total dissolved solids (TDS) in the desalinated water is reduced at each stage, thereby increasing the rate of formation and process efficiency.

Once formation is complete, the concentrated brine is discharged from the reactor, and the formed hydrates are filtered out and washed with a suitable amount of freshwater to improve salt removal. When the brine removal is complete, and the hydrates have been dissociated, the electrical conductivity of the initial (before hydrate formation) and final (after hydrate formation) brine

Table 3.5 Efficiency and electrical conductivity of desalinated water after each stage [20].

Stage	EC (mS/cm)	η (%)
1	74.6	43
2	34.3	74
3	23.7	81.7

solution is measured using a conductivity meter. Finally, the content of dissolved mineral components such as potassium, magnesium, sodium, calcium, sulfate, bicarbonate, and TDS is analyzed for the initial and final brine solutions using titration, atomic absorption spectroscopy, and gravimetric methods [20]. The process efficiency increases at each stage are shown in Table 3.5 by the decreasing electrical conductivity. By the end of the third stage, an efficiency of 81.7% was obtained associated with an electrical conductivity of 23.7 mS/cm.

3.2.4.2 Design 2

Figure 3.10 shows the schematic for a CP-based hydrate formation process [21]. The apparatus used in this setup is a 763 cm^3 stainless steel reactor equipped with a speed adjustable electromagnetic stirrer, a platinum resistance thermometer with an uncertainty of ± 0.1 K placed at the bottom of the reactor, and an air bath with a view window. The temperature of the air bath is kept stable with a deviation of ± 0.1 K. A vacuum filter is connected to a vacuum pump to remove and filter the hydrate slurry, and a conductivity meter is employed to measure the conductivity of the desalinated water to determine the salinity of the dissociated water. Table 3.6 provides a detailed breakdown of the equipment and materials used.

Table 3.6 A detailed breakdown of the desalination apparatus.

Reactor type	763 cm^3 Stainless steel reactor with a speed adjustable electromagnetic stirrer
Components of setup	Stainless steel reactor, platinum resistance thermometer, air bath with a view window, vent, vacuum filter connected to a vacuum pump, conductivity meter, Buchner funnel
Materials used	Deionized water (electrical conductivity < 10^{-4} $S.m^{-1}$), cyclopentane (96%), sodium chloride (99.5%)
Setup efficiency	80%

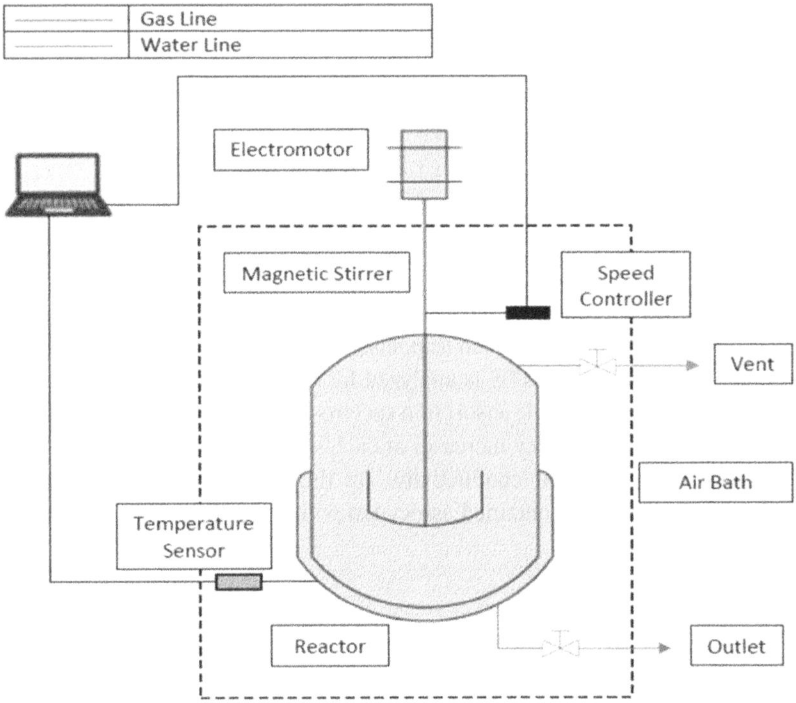

Figure 3.10 Schematic of the experimental setup for cyclopentane (CP) hydrate formation process.

The process for this setup is carried out in batch mode. First, deionized water is used to clean the reactor three times [21]. The temperature in the reactor is lowered to the desired value, and a 400 cm^3 water–oil dispersion is prepared immediately after. The prepared dispersion is first refrigerated to facilitate hydrate nucleation and, as soon as hydrate particles are observed, is promptly charged into the reactor. The electromagnetic stirrer is preset to a constant speed of 300 rpm. This step is carried out to negate variations in induction time as well as for better process reproducibility. This point is considered time zero for hydrate formation, and a data acquisition unit is used to record the temperatures at 60 s intervals.

Once hydrate formation has occurred for 8 h, the process is halted. The hydrate slurry that has formed is then extracted and transferred into the chilled vacuum filter and pump set up to separate the formed hydrates from the concentrated brine. After completing the initial filtration, the entire Buchner funnel with the filter cake of hydrates is weighed.

The amount of chilled deionized water required to wash the hydrates is determined based on the weight of the filter cake, the ratio of washing water to dissociated water (0.5 g/g is used for this case), and a constant empirical coefficient. The filter cake of hydrates is further filtered when washing water is flushed through the filter cake, from top to the bottom of the Buchner funnel. After this stage of filtration with washing, the hydrates will experience a loss in weight, believed to be due to the removal of residual brine with washing water.

The produced hydrates are then weighed and allowed to dissociate at room temperature (298 K) for over 10 h. At this point, the cyclopentane produced via dissociation is gradually removed using evaporation. The salinity and weight of the produced desalinated water are determined when the total mass of desalinated water is no longer decreasing, indicating negligible residual cyclopentane. The removal efficiency reported for this setup was 80% [21].

3.2.4.3 Design 3

Figure 3.11 displays an experimental schematic used by [12] to conduct hydrate-based desalination. The reactor is a 120 cm^3 type-316 stainless steel round-bottom cup unit set with a magnetically-driven mechanical stirrer. Three access ports for gas input, output, and a thermocouple are fitted to a stainless steel cap placed on top of the unit. Temperature regulation is achieved by submerging the reactor in a water-ethylene glycol bath. Temperature and pressure readings in the reactor are measured using a K-type thermocouple with a resolution of 0.1 K and a pressure transducer with a digital readout.

For the hydrate formation experiments, synthesized produced water is prepared using the chemicals listed in Table 3.9, based on the average composition of produced water presented in Table 3.8 [12]. Organic constituents usually present in produced water are not included in the formulation of synthesized produced water because the focus of the experiment was the removal of mineral salts in the desalination process. This exclusion is also convenient due to

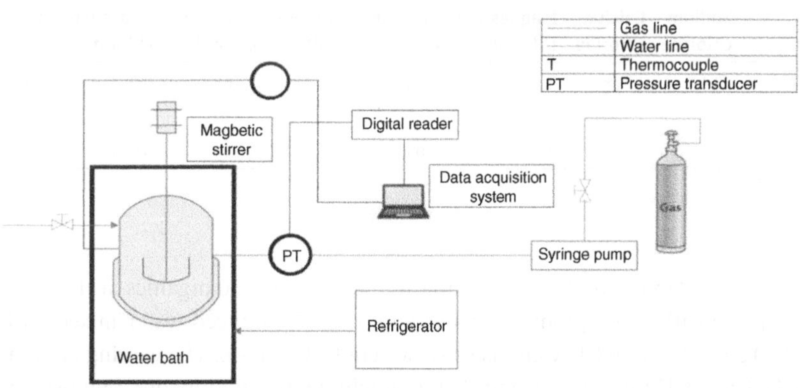

Figure 3.11 Experimental setup for hydrate formation reaction.

Table 3.7 A detailed breakdown of the desalination apparatus.

Reactor type	120 cm^3 316 stainless steel round-bottom cup reactor with a magnetically-driven mechanical stirrer
Components of setup	316 stainless steel round-bottom cup reactor with a magnetically-driven mechanical stirrer, thermocouple, water-ethylene glycol pool for cooling, K-type thermocouple, pressure transducer with digital readout, syringe pump, vacuum suction
Materials used	**For synthesized produced water** sodium chloride (> 99.0%), potassium chloride (> 99 %), calcium carbonate (> 99.0%), calcium chloride (96%), magnesium chloride hexahydrate (> 99.0%), sodium sulfate anhydrous (99.3%), and hydrogen chloride 1 N solution **For hydrate formation** Cyclopentane (99 %), cyclohexane (> 99 %), and carbon dioxide (99.999 %)
Setup efficiency	90%

Table 3.8 Average concentrations of cations found in produced water [12].

	Calcium	Chlorine	Sodium	$Bicarbonate^-$	Magnesium	Sulfate	TDS	pH
Avg. Conc (mg/L)	4874.56	53 620.85	24 609.11	656.05	1028.05	1131.56	89 253.91	7.14

Table 3.9 The concentration of mineral salts utilized in the production of synthesized produced water [12].

	Sodium chloride	Calcium chloride	Magnesium chloride	Calcium carbonate	Sodium sulfate	Potassium chloride	Hydrogen chloride	Total
Conc. (g/L)	60.13	12.18	8.43	1.06	1.64	11.68	0.74	95.80
Wt %	5.62	1.13	0.79	0.099	0.15	1.09	0.069	8.95

the unavailability of detailed compositional data on the organics at the time. 35 g of synthesized produced water and 35 g of produced water mixed with 7.30 g CP + 8.77 g CH were used as two controlled tests. The varying amount of CP and CH used corresponds to the stoichiometric amount of 5.56 mol % for 100% sII hydrate conversion from water, maximizing the amount of hydrate formed from brine. The amount of produced water used, 35 g, is the minimum amount of water that would allow for the submersion of the impeller in the liquid phase, as uniform mixing is important in expediting the hydrate forming process [12].

Prior to starting the hydrate forming process, CO_2 is used to pressurize the reactor below the expected equilibrium pressure at a designated temperature and is slowly purged for ten minutes at atmospheric pressure. This process is repeated thrice to remove air in the head volume. The reactor is charged with CO_2 gas until the partial pressure reaches 3.1 MPa to initiate the hydrate forming process. A syringe pump is constantly in operation to maintain the pressure within the reactor throughout this period. As hydrate formation starts taking place, the volume in the syringe pump decreases, and the variation in volume in the pump is recorded as a function of time at 30 s intervals.

Once the reaction runs its course, the hydrates are extracted from the reactor, crushed, and then filtered by vacuum suction for 4 m to remove any interstitial brine remaining between hydrate crystals. Without this last step, the higher salt concentration of the remaining brine can reduce the efficiency of the desalination process [12]. The hydrates are then dissociated at ambient conditions, and the water immiscibility of CP/CH is used to physically separate the water. Figure 3.12 briefly shows what happens to the hydrates as soon as they leave the reactor and how desalinated water is extracted from said hydrates.

By utilizing secondary guests such as CP and CH alongside the CO_2 primary hydrate former, this double hydrate desalination process achieves 90% salt removal efficiency compared to the 70% removal seen in desalinated water produced from simple hydrates. However, the conversion rate of double hydrates remains the same as that of simple hydrates. The secondary guests serve to

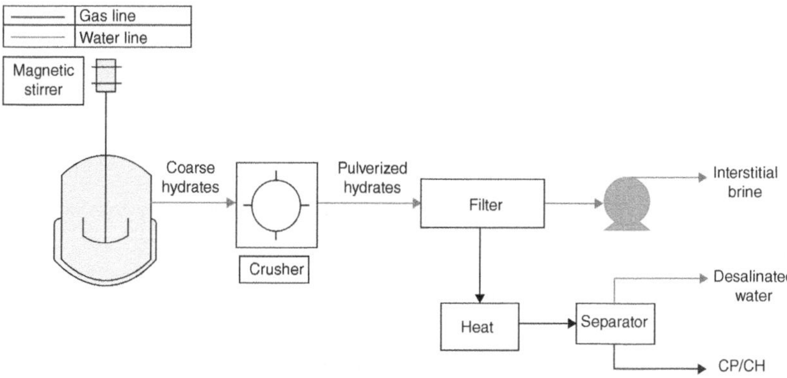

Figure 3.12 Schematic process of obtaining desalinated water from gas hydrates using water-immiscible hydrate formers.

increase the temperature from −2°C to 16°C for CP and to 7°C, for CH, thereby increasing the rate of hydrate formation 22 and 16 times, respectively, relative to that reported in CO_2 formers systems at the upper-temperature limit [12].

3.2.5 Novel Reactor Design

The system depicted in Figure 3.13 and detailed in Table 3.10 comprises a gas supply unit, a saline water supply unit, a reactor, and a drain unit. The gas supply unit provides a controlled supply of gas into the reactor, which is contacted with the saline water supplied by the saline water supply unit through a high-pressure injection pump. At appropriate pressure and temperature, this will initiate gas hydrate formation in the reactor.

The drain unit then separates the gas hydrate crystals from the remaining liquid (brine). The supply gas is provided through a gas cylinder connected to the gas inlet. The gases that could be used include, but are not limited to, CO_2, CH_4, ethane, C_3H_8, or any other gas that can be used to form hydrates. The pressure within the reactor is controlled by a series of valves, while the pressure is maintained throughout the process using a gas booster.

A sieve is submerged in the saline water region within the reactor, and the formation of gas hydrates takes place above this sieve. A stirrer is located right

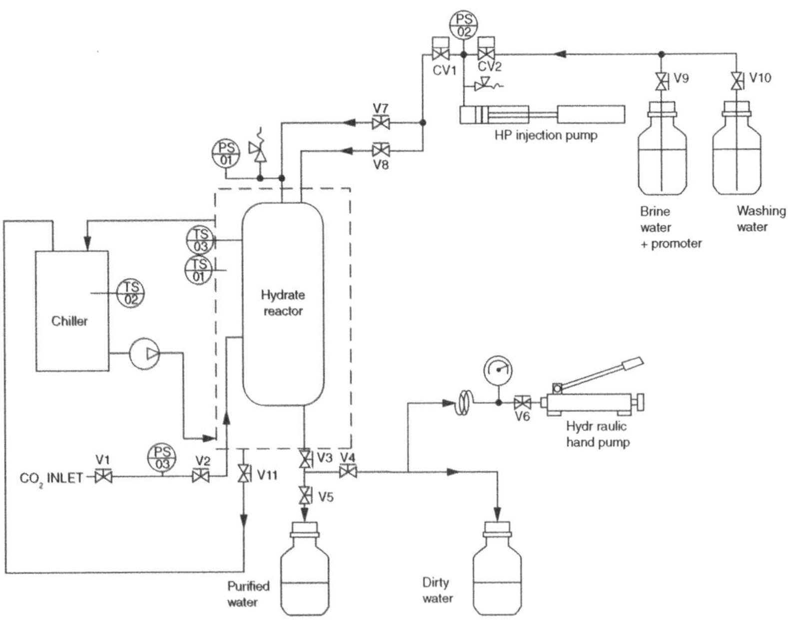

Figure 3.13 Experimental setup for hydrate formation reaction.

Table 3.10 A detailed breakdown of the desalination apparatus.

Reactor type	2 L hydrate reactor with a stirrer and sieve
Components of setup	Reactor; containers for pure, brine, saline, and washing liquid; pressure sensors; temperature sensors; pneumatic control valves; high-pressure injection pump; hydraulic hand Pump; gas booster; chiller; drain unit
Materials used	saline solution, carbon dioxide
Setup efficiency	50%

beneath the sieve. The purpose of this stirrer is to force the liquid content in an upward motion and contact the gas phase to promote hydrate formation and growth. The use of the stirrer also prevents any formation of gas hydrates beneath the sieve while maximizing the number of water molecules that can form gas hydrates.

Forcing the liquid through the sieve creates microbubbles, which serve to increase the rate of gas hydrate formation. Moreover, the stirrer creates a marginal cyclonic separation effect when it is switched on, forming a vortex in the liquid. Larger particles in the rotating stream having higher inertia do not follow the tight curve formed by the vortex, and as such, these particles will move toward the sides of the reactor and eventually end up at the bottom of the reactor leaving only pure water molecules to form gas hydrates.

A washing unit is connected to the reactor and is used to introduce a washing liquid to remove any traces of salt that may remain on the hydrate crystals. The now contaminated washing liquid settles at the bottom of the reactor and is removed by the drain unit, the same way that brine solution was removed earlier. Once the washing liquid has been collected, the temperature and pressure of the enclosed space are controlled to allow gas hydrate crystals to dissociate, forming pure water. A hydraulic handpump alongside a combination of valves is used to regulate pressure release from the reactor due to the release of hydrate forming gas from dissociation, the brine solution, and the contaminated washing liquid to avoid a sudden pressure drop occurring in the reactor.

The valves in the setup are closed before turning on the power for the equipment and accessing the Desal application. A ramp input temperature, 274.15–281.15 K for CO_2-based hydrate formation, is set. Once the system has stabilized, the sample is injected into the reactor via the corresponding valves.

Pressurization of the system entails setting the desired pressure value via the Desal application, which in the case of CO_2-based hydrate formation is in the range of 20–40 bars while turning on the gas booster and the corresponding valve. As soon as the gas is charged into the reactor, the stirrer is turned on as well. Once the desired pressure is obtained, the valve is closed, and the operation of the gas booster is halted.

Once hydrates have formed, the residual solution is drained by opening the corresponding valves. The formed hydrates are then washed using spraying clean water. The wash water is then collected. The system is then depressurized, allowing the formed hydrates to dissociate while the pure water is collected by opening the corresponding valves. Then the residual freshwater and initial brine sample concentrations are analyzed for their final and initial concentrations, metal removal efficiencies., electrical conductivity, and pH values. Water recovery and yield are analyzed based on the volume of water obtained.

References

1 Lal, B. and Nashed, O. (2020). Chemical additives for gas hydrates. 2020, 19–21. doi: 10.1007/978-3-030-30750-9.

2 Bradshaw, R.W., Greathouse, J.A., Cygan, R.T., Simmons, B.A., Dedrick, D.E., and Majzoub, E.H. (2008). Desalination Utilizing Clathrate Hydrates (LDRD Final Report). (No. SAND2007-6565). Sandia National Laboratories.

3 Khawaji, A.D., Kutubkhanah, I.K., and Wie, J.M. (2008). Advances in seawater desalination technologies. *Desalination* 221 (1–3): 47–69. doi: 10.1016/j.desal.2007.01.067.

4 Khawas, V. (2006). Human Development Report 26, Social Change, 245–246.

5 Zheng, J., Cheng, F., Li, Y., Lü, X., and Yang, M. (2019). Progress and trends in hydrate based desalination (HBD) technology: a review. *Chinese Journal of Chemical Engineering* 27(9): 2037–2043. doi: 10.1016/j.cjche.2019.02.017.

6 Khan, M.S., Lal, B., Sabil, K.M., and Ahmed, I. (2019). Desalination of seawater through gas hydrate process: an overview. *Journal of Advanced Research in Fluid* 1(1): 65–73.

7 Aliev, A.M., Yusifov, R.Y., Kuliev, A.R., and Yusifov, Y.G. (2008). Method of gas hydrate formation for evaluation of water desalination. *Russian Journal of Applied Chemistry* 81(4): 588–591. doi: 10.1134/S1070427208040034.

8 Miller, J.E. (2003). Review of water resources and desalination techniques. Albuquerque, NM: Sandia National Laboratories. (49, 2003-0800).

9 Babu, P., Kumar, R., and Linga, P. (2014). Unusual behavior of propane as a co-guest during hydrate formation in silica sand: potential application to seawater desalination and carbon dioxide capture. *Chemical Engineering Science* 117: 342–351. doi: 10.1016/j.ces.2014.06.044.

10 Partoon, B., Sabil, K.M., Roslan, H., Lal, B., and Keong, L.K. (2016). Impact of acetone on phase boundary of methane and carbon dioxide mixed hydrates. *Fluid Phase Equilibria* 412: 51–56. doi: 10.1016/j.fluid.2015.12.027.

11 Khan, M.S., Lal, B., Partoon, B., Keong, L.K., Bustam, A.B., and Mellon, N.B. (2016). Experimental evaluation of a novel thermodynamic inhibitor for CH_4 and CO_2 hydrates. *Procedia Engineering* 148 (December): 932–940. doi: 10.1016/j.proeng.2016.06.433.

12 Cha, J.H. and Seol, Y. (2013). Increasing gas hydrate formation temperature for desalination of high salinity produced water with secondary guests. *ACS Sustainable Chemistry & Engineering* 1(10): 1218–1224. doi: 10.1021/sc400160u.

13 Park, K.N., Hong, S.Y., Lee, J.W., Kang, K.C., Lee, Y.C., Ha, M.G. et al. (2011). A new apparatus for seawater desalination by gas hydrate process and removal characteristics of dissolved minerals (Na^+, Mg^{2+}, Ca^{2+}, K^+, B^{3+}). *Desalination* 274 (1–3): 91–96. doi: 10.1016/j.desal.2011.01.084.

14 Skelland, A.H.P. and Ramsay, G.G. (1987). Minimum agitator speeds for complete liquid—liquid dispersion. *Industrial & Engineering Chemistry Research* 26(1): 77–81. doi: 10.1021/ie00061a014.

15 Ohmura, R., Ogawa, M., Yasuoka, K., and Mori, Y.H. (2003). Statistical study of clathrate-hydrate nucleation in a water/hydrochlorofluorocarbon system: search for the nature of the memory effect. *The Journal of Physical Chemistry A* 107(22): 5289–5293. doi: 10.1021/jp027094e.

16 Mekala, P., Babu, P., Sangwai, J.S., and Linga, P. (2014). Formation and dissociation kinetics of methane hydrates in seawater and silica sand. *Energy and Fuels* 28(4): 2708–2716. doi: 10.1021/ef402445k.

17 Haligva, C., Linga, P., Ripmeester, J.A., and Englezos, P. (2010). Recovery of methane from a variable-volume bed of silica sand/hydrate by depressurization. *Energy and Fuels* 24(5): 2947–2955. doi: 10.1021/ef901220m.

18 Linga, P., Haligva, C., Nam, S.C., Ripmeester, J.A., and Englezos, P. (2009). Gas hydrate formation in a variable volume bed of silica sand particles. *Energy and Fuels* 23(11): 5496–5507. doi: 10.1021/ef900542m.

19 Loh, M., Falser, S., Babu, P., Linga, P., Palmer, A., and Tan, T.S. (2012). Dissociation of fresh- and seawater hydrates along the phase boundaries between 2.3 and 17 MPa. *Energy and Fuels* 26(10): 6240–6246. doi: 10.1021/ef3008954.

20 Fakharian, H., Ganji, H., and Naderifar, A. (2017). Desalination of high salinity produced water using natural gas hydrate. *Journal of the Taiwan Institute of Chemical Engineers* 72: 157–162. doi: 10.1016/j.jtice.2017.01.025.

21 Lv, Y.N., Wang, S.S., Sun, C.Y., Gong, J., and Chen, G.J. (2017). Desalination by forming hydrate from brine in cyclopentane dispersion system. *Desalination* 413: 217–222. doi: 10.1016/j.desal.2017.03.025.

22 Corak, D., Barth, T., Høiland, S., Skodvin, T., Larsen, R., and Skjetne, T. (2011). Effect of subcooling and amount of hydrate former on formation of cyclopentane hydrates in brine. *Desalination* 278(1–3): 268–274. doi: 10.1016/j.desal.2011.05.035.

4

Hydrate Promoters in Gas Hydrate-based Desalination

Sirisha Nallakukkala and Bhajan Lal

4.1 Chemical Additives in Desalination

Chemical treatment began as one of the most prominent methods for treating water, starting with the use of disinfectants to prevent waterborne sicknesses [1]. Chemical treatment gained even more recognition upon the discovery of other additives, such as coagulants that assist in filtration methods. Chemical additives are indispensable for preventing corrosion, water softening, and numerous other applications. Typically, the factors considered when planning a chemical treatment are the order in which chemicals are added to the system and the dosage. Table 4.1 shows the various chemical additives commonly implemented in water treatment systems worldwide.

Disinfectants are the most commonly used synthetic chemicals. In no particular order, other heavily used chemical additives are precipitation softeners, coagulants and flocculants, algaecides, antiscalants, water purifying biocides like amines, chlorinated phenolic compounds, and salts of copper. Strong inorganic acids and bases, such as chelating agents, resin cleansers, and fouling cleansers, are also sometimes used [2].

The decline in feed quality and stringent water standards have led to an increase in the use of chemicals both for water treatment and equipment safeguarding. However, the present technology in chemicals is still sufficient in tackling most water treatment requirements. Research has recently shifted to the pursuit of retrieving drinking water from more challenging sources. The long-term objective is to explore further how freshwater can be obtained from locations with poor water quality while also understanding the general environmental and health consequences entailed as the resources are claimed and processed.

Gas Hydrate in Water Treatment: Technological, Economic, and Industrial Aspects,
First Edition. Bhajan Lal and Sirisha Nallakukkala.

Table 4.1 Chemicals used in water treatment to improve quality of water.

Type	Chemicals	Purpose/Description
Disinfectants and oxidants	Oxygen, chlorine, sodium hypochlorite ($NaOCl$), calcium hypochlorite ($CaOCl_2$), potassium permanganate ($KMnO_4$), hydrogen peroxide (H_2O_2), ammonia	Remove heavy metals and oxidize organics
Chemical conditioning	Ferric chloride ($FeCl_3$), lime (CaO), alum, and organic polymers	Decrease the moisture content in sludge
Scale inhibitor	Calcium carbonate ($CaCO_3$), barium sulfate ($BaSO_4$), calcium sulfate ($CaSO_4$), strontium sulfate ($SrSO_4$), antiscalant	Delay or prevent scale formation
Corrosion inhibitors		
Passive	Oxidizing (chromate, nitrite, nitrate) and non-oxidizing anions (phosphate and molybdate)	Cause a shift of the corrosion potential, forcing metallic surface into passive range
Cathodic	Ions like calcium, zinc, magnesium, and compounds of arsenic, antimony	Generally, precipitate as oxides to form a protective layer on the metal.
Volatile corrosion inhibitors	Morpholine and hydrazine and volatile salts like dicyclohexylamine, cyclohexylamine, and hexamethylene-amine	They get in contact with the metal surface, the vapor of these salts condenses and is hydrolyzed by moisture to liberate protective ions
Antiscalant/Depressant carbonate/sulfate	$CaCO_3$, $BaSO_4$, $CaSO_4$, $SrSO_4$, and depressants for silt and metal oxides	Control hardness, scaling, and membrane fouling
Antioxidant	Removes chlorine as reducing agent	React with chlorine as pretreatment
Coagulants & Flocculants	Aluminum sulfate ($Al_2SO_4)_3$, ferric sulfate $Fe_2(SO_4)_3$, ($FeCl_3$), polyaluminum chloride, sodium aluminates, alum, $Ca(OH)_2$, polyelectrolytes	Remove solids, clarify water, softening, sludge thickening, and solids dewatering
Neutralizing agents	$NaOH$, $CaCO_3$, CaO, dilute H_2SO_4, and dilute HCl	Increase/decrease pH
Precipitation and softening	CaO, $Ca(OH)_2$, $NaOH$, CO_2, Na_2CO_3, and $NaCl$	Reduce hardness, alkalinity, silica

(Continued)

Table 4.1 (Continued)

Type	Chemicals	Purpose/Description
Algaecides	$CuSO_4$, iron salts, rosin amine salts, and benzalkonium chloride	Control algal growth
Biocides	Acrolein, amines, chlorinated phenolics, copper salts, organo-sulfur compounds, quaternary ammonium salts	Kill bacteria
Chlorine reducing agent	Sodium metabisulfite (SMBS)	Remove free chlorine and as a biostatic
Antifoam	Oils coalesce with slight quantities of silica	Prevent the formation of foam or break a foam
Off-line chemicals for the treatment of equipment		
Chelating agents	Citric acid, EDTA, phosphates	Bind and remove ions from solutions
Fouling cleaners	Inorganic acids and bases, oxidizers	
Resin cleaners	NaCl, potassium chloride (KCl), citric acid, and chlorine dioxide (ClO_2)	Regenerate ion exchange resin
Others	Fluoride compounds Activated carbon	

4.2 Overview of Gas Hydrate Additives in the Desalination Process

Gas hydrate-based desalination is projected to become a prospective technology in treating seawater [3–6]. Chemical additives can be used to encourage the formation of hydrates at milder conditions, increase the rate of gas intake and hydrate formation, and enhance the selectivity of the hydrate [4–6, 8]. Hydrate formers are the guest molecules essential for the formation of hydrates. Promoters are chemicals that affect the essential characteristics of hydrate formation and separation. Promoters either increase the temperature or decrease the pressure, both of which can improve the hydrate formation process. Hydrates formed by various formers are dissimilar in terms of induction time, formation conditions, density, and crystal structure.

Hydrate crystals exclude most of the salts and ions that were dissolved in the formation cycle [7]. A close analog to the hydrate-based technique is the freeze-and-thaw desalination method. Hydrate-based desalination, however, requires higher pressures and lower temperatures to deliver enough driving force. Dissociation of the hydrate produces freshwater water and the guest gas (former) molecules, which can be recycled in subsequent desalination runs.

4.3 Favorable Conditions Used to Determine Suitable Hydrating Agents

The criteria used in determining the suitable hydrate formers are the miscibility of said hydrate former or guest molecule in water, cost, and ease of access. The miscibility of the hydrate former in water can assist in deciding which kind of partition step should be implemented to recover the former from the liquid phase when the hydrate is dissociated. The cost of the guest molecule and the purity of the recovered water will dictate the extent to which the hydrate former must be recovered. The economic feasibility of the process depends largely on the cost and availability of the selected formers. An overview of the selection criteria for pure gases is presented in Table 4.2.

Several properties to be considered in the selection of hydrate forming agents are the following:

(i) *Environmentally acceptable*. The former must have low ozone depletion and the greenhouse effect.
(ii) *Non-toxic*. The guest molecule should have low toxicity and should be non-carcinogenic.
(iii) *Non-flammability*. The hydrate former should have a high flash point to reduce the risk of fire.
(iv) *Chemical stability*. The former should have low chemical activity and react very slowly with chemicals.
(v) *Low cost*. Hydrate formers should have low operating costs.
(vi) *Availability*. Commercially available.
(vii) *Water solubility*. The former should have low water solubility. This will enhance the degree of retrieval of the guest molecules from water.
(viii) *Structure of unit cell*. Should be capable of forming either sI, sII, sH, or semiclathrate cell structure.
(ix) *Suitable operating temperature and pressure*. Formers should have operating conditions close to ambient temperature and atmospheric pressure to reduce operating costs.
(x) *Quadruple point*. Operation near the upper quadruple point is recommended for the energy intensity of the process.
(xi) *Critical point*. The heat of vaporization should be considerable. The upper quadruple point should be, to the extent possible, far from the former's critical point.

4.4 Formers and Promoters in Hydrate-based Desalination

Early studies were based on the usage of a single guest gas molecule. More recently, research has focused on multi-gas and gas–liquid mixtures. The

Table 4.2 Specific criteria were used for choosing the most suitable common formers for hydrate formation.

Chemical name	Methane	Ethane	Propane	Carbon dioxide
Chemical formula	CH_4	C_2H_6	C_3H_8	CO_2
Ecological effect	Ozone depletion potential (ODP) = No Data Available Global Warming Potential (GWP) = 25	ODP = 0 GWP = 6	ODP = 0 GWP = 3	ODP = 0 GWP = 1
Toxicity effect	Asphyxiation at high concent-rations. Contact with gas or vaporizing liquid may cause burns or frostbite.	Asphyxiation at high concent-rations. Contact with gas or vaporizing liquid may cause burns or frostbite.	Asphyxiation at high concentrations. Contact with gas or vaporizing liquid may cause burns or frostbite.	Asphyxi ation at high concent-rations. Contact with gas or vaporizing liquid may cause burns or frostbite.
Flammability	Flammable	Flammable	Flammable	Non-flammable
Explosive limits % by volume	Lower level = 1.8 % Upper level = 8.4 %	Lower level = 3 % Upper level = 12.5 %	Lower level = 2.2% Upper level = 9.5%	None
Chemical stability	Stable under normal conditions. Unstable at high temperatures outside explosive limits.	Stable under normal conditions. May decompose to form hydrogen at high T and low P.	Stable under normal conditions. Unstable at high temperatures outside explosive limits.	Stable under normal conditions.
Compatibility	Incompatible with oxidizing agents.	Incompatible with oxidizing agents.	Incompatible with oxidizing agents.	No reaction with any common materials in wet or dry conditions.

benefit of using gaseous guest molecules is the ease of separation; thus, gases are expected to remain key formers in mixtures using either liquid or solid co-guest molecules. A generalized list of various guest gas molecules is listed in Table 4.3, mostly in terms of their phase behavior and miscibility levels for easier post-separation.

The small guest molecules that engage the cavities inside the hydrate crystal are known as hydrate formers, and hydrate formation cannot take place without their presence. The structure of hydrates is stabilized through intermolecular exchanges between the hydrate former and the water molecule. Hydrate formers exist in either liquid or gaseous form, are either organic or inorganic, are usually hydrophobic, and have various degrees of toxicity and flammability.

The most important characteristics of the formers are their chemical nature, size, shape, structure, and stability during hydrate formation. Semiclathrates are a type of clathrate in which a portion of the water cage is broken [14]. Semiclathrate hydrates have similar properties as conventional gas hydrates, but the guest gas molecule in a semiclathrate engages the hydrate cavity and contributes to the structure of the lattice [15].

Common hydrate formers found in the literature [16], such as methane (CH_4), ethane (C_2H_6), propane (C_3H_8), and carbon dioxide (CO_2), are initially among the guest gas molecules or formers nominated for hydrate-based desalination. The hydrate forming environments of these potential formers, in pure liquid water and in the existence of salts, have been measured extensively in the literature. C_3H_8 is a hydrate former with flammable characteristics and can

Table 4.3 Potential hydrate formers for hydrate-based desalination.

Hydrate Former (Guest)	Advantage	Disadvantage	Examples	Features
Gaseous	Easy separation	Higher cost for pressurizing	CH_4, CO_2,[8–10]	Non-toxic and accessible
Liquid			C_3H_8 [11, 12]	Flammable
Gaseous + liquid	Easier to separate from water Low cost for pressurizing	High cost for pressurizing Essential to separate further	CO_2 + CP (cyclopentane)* [3, 13] CO_2 + R141b** [5]	Insoluble in water and improve hydrate formation Inexpensive with better efficiency

combust if not managed correctly. For this reason, C_3H_8 is not selected as a suitable former despite its extensive use in the industry. CO_2 is more miscible than the other hydrate formers but is a known greenhouse gas and requires pressures of up to 4.5 MPa to form hydrates. It is, however, non-flammable and can form hydrates at 282 K. It is noted that Malaysia has high amounts of CO_2-rich natural gas, making it easily accessible as a hydrate former, thus decreasing the operational cost. Fluorinated hydrocarbons form hydrates at substantially lower pressures compared to more common hydrocarbon formers. Compared to the ranges of 277.5–282.5 K and 2.048–4.020 MPa for hydrocarbon formers [17–19], formation conditions with an R134a former are between 277.1–282.6 K and 0.114–0.382 MPa.

Promoters and other additives are used to improve performance and efficiency in both conventional and hydrate-based technologies. Chemicals commonly used in desalination fall into one of three general categories: (i) ionic liquid (IL), (ii) amino acid, or (iii) surfactant. Table 4.4 lists the chemicals commonly found in the literature under each category.

ILs are thermally stable and non-flammable and are, therefore, regarded as global solvents for industrial applications [20–22]. ILs act as non-volatile liquids below 100°C. In desalination, ILs are employed to thin the aromatic polyamide reverse membranes, thus improving the properties of the surface and its efficiency. In a 2017 study [23], RO membranes are dipped in 1,3-dimethyl imidazolium dimethyl phosphate ([MMIM]DMP), 1-octyl-3-methylimidazolium chloride ([OMIM]Cl), 1-hexyl-3-methylimidazolium chloride ([HMIM]Cl), and 1-butyl-3-methylimidazolium chloride ([BMIM]Cl). [BMIM]Cl exhibited the maximum thinning effect of the layer, thus enhancing its hydrophilic nature by increasing activation temperatures and reducing membrane roughness for anti-smudging properties. Application of [MMIM][DMP] showed similar effects.

ILs have also been used for decomposing polymers [24]. Imidazolium-based ILs have demonstrated different water solubilities to various anions [25]. An innovative method to alter the efficiency of polyamide membranes was done by exposing them to ILs like [BMIM][Cl], 1-Methyl-3-Octylimidazolium Chloride ([C8MIM][Cl]), and 1-Butyl-3-Methylimidazolium Bromide ([BMIM][Br]) is reported [26]. Other significant ILs to note are quaternary ammonium salts [27].

Amino acids, such as cyclodextrin-containing composites, have similarly attracted extensive recognition as a group of adsorbents aimed at the exclusion of dye from wastewater, owing to their unusual physical and chemical properties [28–31].

Surfactants are usually implemented as hydrating and emulsifying agents or detergents, though some, such as quaternary ammonium compounds (QACs),

Table 4.4 Various additives/promoters used in desalination.

Ionic Liquid	Application	Reference
1-butyl-3-methylimidazolium hexafluorophosphate [BMIM][PF6] 1-butyl-3-methylimidazolium bis(trifluoro methyl sulfonyl)imide [BMIM][NTf2] 1-butyl-3-methylimidazolium tris(pentafluoroethyl)trifluoro phosphate [BMIM][FAP]	Applied on the membrane for phenol removal and shows hydrophilic nature	[37]
1-Butyl-3-Methylimidazolium Chloride [BMIM][Cl] 1-Methyl-3-Octylimidazolium Chloride [C_8MIM][Cl] 1-Butyl-3-Methylimidazolium Bromide [BMIM][Br]	Used to vary the performance of membrane by decreasing contact angle and surface roughness, and increased hydrophilicity	[26]
([BMIM]Cl) 1-hexyl-3-methylimidazolium chloride ([HMIM]Cl) 1-octyl-3-methylimidazolium chloride ([OMIM]Cl)	Exhibited an increase in surface hydrophilicity and electronegativity through the decreased roughness and elevated anti-fouling properties.	[24]
Amino Acids		
L-cysteine	Fouling resistance by surface modification of membrane. Improved membrane displayed a smoother and further hydrophilic surface with enhanced exclusion of salt.	[38]
L-lysine	Improved membrane performance and improvements in water flux and removal efficiency.	[39]
Cyclodextrin-based composites	Dye removal from wastewater by adsorption	[28]
Arginine with $ZnCl_2$	Increase osmotic pressure and water flux	[40]
3-(3,4-dihydroxyphenyl)-L-alanine (L-DOPA)	Create a zwitterionic surface that resists membrane fouling	[41]

(Continued)

Table 4.4 (Continued)

Ionic Liquid	Application	Reference
Surfactants		
Anionic		
Linear alkyl benzene sulfonates	Nitrogen, phosphorus, and chemical oxygen demand removal	[33]
Sodium dodecylbenzene sulfonate (SDBS)		
Secondary alkane sulfonates		
Alcohol ether sulfates (alkyl ethoxy) sulfates		
Cationic		
Quaternary ammonium-based compounds	Used as emulsifiers, fabric softeners, disinfectants, pesticides, corrosion inhibitors	
Alkyl trimethyl ammonium halides & alkyl dimethyl ammonium halides	Cr, Phosphate, Ammonium, Bacteria, BTEX, Arsenic	
Hexadecyltrimethylammonium (HDTMA)	Cr, Phosphate, Ammonium, Bacteria, BTEX, Arsenic	
Hexadecylpyridinium (HDP)	Phosphate	
Cetylpyridinium bromide (CPB)	Nitrates	
Hexadecyltrimethylammonium bromide (HTAB)	Textile wastewater	
Cetyltrimethylammonium bromide (CTAB)	Reactive black dye	
Nonionic		
Alkylphenol ethoxylates	In nanofiltration: adsorption of nonionic surfactants can cause an increase and decrease in flux, reliant on surface hydrophilicity	
Alcohol ethoxylates		

have remained well-known for their ability to harm the cell membranes of bacteria, leading to its use as a disinfectant [32]. Several researchers have stated that surfactants can increase the solubility of organic hydrocarbon compounds in water or can be sorbed into soils and sediments where it can cause harm to agricultural crops. In both instances, the hydrophobic content and organic compounds end up degraded [33]. Benzalkonium chlorides (BACs) can enhance microbial resistance to antibiotics [34]. Surfactant Modified Zeolites (SMZs) utilize surfactants like hexadecyltrimethylammonium (HDTMA) to modify zeolite surface chemistry [35]. This charge modification permits the adsorption of contaminants [36] and leads to the utilization of these HDTMA-SMZs for eliminating pollutants.

The industrialization of gas hydrate technology is hindered by the slow kinetics of formation and the energy costs associated with the required low-temperature, high-pressure conditions. Promoters, used in conjunction with formers, enable hydrates to form at reduced pressures or increased temperatures. This is due to the stabilization effect caused by the promoter on the hydrate structure [42]. The stabilization arises from the promoter occupying small or large cages that the hydrate former could not fill [43], thereby reducing mass transfer and kinetic challenges during hydrate formation.

4.5 Hydrate Formers Investigation

Investigations of various hydrate formers have yielded positive results in reducing hydrate nucleation time, improving the rate of hydrate growth, and increasing water-to-hydrate conversion through the use of additives and promoters. A more comprehensive review of various hydrate formers and their application has been discussed in detail in [44]. Additives generally reduce the energy required for mixing while also enhancing the rate of hydrate formation. Thermodynamic promoters (THPs) are mainly implemented to decrease the hydrate formation pressure or increase the temperature. The most commonly explored THPs are cyclic ethers, like tetrahydrofuran (THF), 1,4-dioxane, propylene oxide, cyclopentane (CP), C_3H_8, neohexene, acetone, and methylcyclohexane [45]. The gases released with the use of THF promoters are volatile and require further separation.

The foremost problem that has mired practical application of hydrates is its slow kinetics. Therefore, scientists have vested considerable interest in emerging kinetic hydrate promoters (KHPs) to accomplish faster hydrate formation. KHPs are preferred in that they do not engage with the water cages and help to increase uptake. Moreover, KHPs require smaller dosage strengths (< 10 000 ppm) to reduce the surface tension between the gas and liquid phases.

4.5.1 Gaseous Hydrate Formers

Gases, in general, can be isolated from water without much difficulty. Not all gas molecules that can form hydrates, however, are necessarily selected as guest molecules. Gas hydrate formers require phase equilibrium conditions to determine whether the gas hydrate is stabilized. Probable guest molecules could be CO_2 and C_3H_8 for their mild hydrate formation pressure and temperature settings. Studies on CO_2 hydrates alongside their phase equilibrium in aqueous solutions at various strengths are shown in Table 4.5 [10, 46–54]. It must be noted that CO_2 as a former is used along with additional guest molecules, else the pressure–temperature conditions can be difficult to achieve.

Compared to CO_2 gas hydrates, C_3H_8 gas hydrates are relatively simpler to form at numerous atmospheric pressures [62]. Though C_3H_8 gas is not often used for hydrates due to its flammability, it has seen use as a co-former [11, 63]. Yang et al. [12] utilized the combination of CO_2 and C_3H_8 to analytically prove the thermodynamic enhancement of C_3H_8 by quite a few degrees at similar pressures. C_3H_8 is also used with other gas combinations to form gas hydrates for CO_2 or natural gas capture and hydrogen storage [64–70]. The density of C_3H_8 hydrate is less than that of water, whereas CO_2 hydrates have a higher density than water [71]. However, the most preferred gaseous formers are usually taken from mixtures involving C_3H_8.

4.5.2 Liquid Hydrate Formers

The immense energy and material (gas) cost associated with the use of liquid hydrate-based formers is often overlooked. Usually, fluid-based guest molecules comprise THF, HCFC (hydrochlorofluorocarbons), and liquid hydrocarbons [72, 73]. Due to its immiscibility in water, THF is rarely utilized for desalination,

Table 4.5 Measured hydrate–liquid–vapor (HLV) equilibrium points data from the literature.

Formers	Aqueous solution	Measured HLV Equilibrium data,	Reference
CO_2	NaCl	273–282 K, 2.1–4.2 MPa	Sun et al. [54]
CO_2	NaCl	269–282 K, 1–5 MPa	Yang et al. [55]
CO_2	NaCl, KCl, $CaCl_2$	259–281 K, 0.9–4.1 MPa	Dholabhai et al. [56]
CO_2	Acetone	269–281 K, 1–4 MPa	Maekawa et al. [57]
CO_2	Glycerol	270–280 K, 1.5–3.3 MPa	Breland et al. [58]
$CO_2 + C_3H_8$	NaCl	271–284 K, 0.8–3.6 MPa	Yang et al. [12]
CO_2 + CP	NaCl	280–286 K, 0.5–3.6 MPa	Zheng et al. [59]
CO_2 + CP	—	287–293 K, 0.9–3.5 MPa	Zhang et al. [60]
CO_2 + CP derivatives	—	274–292 K, 0.1–4.9 MPa	Matsumoto et al. [61]

as it will incur additional expenses. The most recommended fluorinated hydrate formers are R134a, R410a, and R507, all of which can easily be separated by centrifuging. However, studies of HCFC-141b hydrate efficiency are promising [74, 75]. Of the liquid hydrocarbons, CP has gained considerable attention as a co-former in hydrate-based desalination [76–79], alongside its derivatives [3, 59, 80, 81]. Liquid formers are advantageous in reducing costs associated with pressurization, though much of the technology is not mature enough for commercial-scale desalination. Hence, for both gaseous and liquid formers, further development is needed before they can be implemented successfully at larger scales [82].

4.5.3 Functional Additives

Functional additives are synthetic chemicals that promote the formation of hydrates [83–86]. The addition of these surfactants, of either amphipathic or amphiphilic types, can diffuse both polar and nonpolar constituents. They can adjust the interfacial tension and alter the contact angle between the phases, consequently charging their surface properties and viscosity. At appropriate concentration strengths, the molecules of surfactants agglomerate to arrange into several varieties of structures, having diverse shapes and positionings [87].

Surfactants used in promoting hydrate formation include sodium dodecyl sulfonate (SDS) [88], sodium dodecylbenzene sulfate (SDBS) [89], alkyl polyglycosides (APG), and linear alkyl sodium. Kang et al. reported that SDS shows promoting effects at an optimum concentration, though higher than optimal concentrations can constrain hydrate growth [71]. Nonionic surfactants like Tween80 are superior to cationic surfactants [90]. Zhong et al. determined the effect of CP and SDS on CH_4 isolation from a small-powered coal gas mine [91]. The gas intake and rate of hydrate formation were found to rely on SDS strength; however, SDS presence did not show any impact on CH_4 recovery. Larger compounds like THF [84, 92], tetra-butyl ammonium bromide (TBAB) [93–97], and CP [81, 98] have seen use as chemical additives. When used in smaller amounts, these chemical additives can significantly lessen the pressure required for hydrate formation and thus contribute to building the structure of the hydrate crystal. Nevertheless, surfactants are immiscible in water and are not considered the best choice for hydrate-based desalination.

Amino acids comprise an amino group ($-NH_2$), acidic carboxyl group ($-COOH$), and an organic R group. Proponents for amino acids as potential hydrate promoters highlight their environmental friendliness, inexpensiveness, and stability from frothing as some of the reasons for candidacy in being used for hydrate-based processes [99]. Liu et al. [100] examined natural amino acids as promoters for CH_4 hydrate, from low strengths up to 1 wt%. In their investigation, leucine demonstrated maximum promotion of CH_4 hydrate in contrast to tryptophan, phenylalanine, methionine, glutamic acid, histidine, and arginine at 0.5 wt%. However, leucine displayed an inhibition impact in C_2H_6 and THF hydrates [101, 102]. The innate nature of amino acids alters within systems of CO_2 or hydrocarbons. Moreover, histidine displayed kinetic promotion with CH_4 hydrate [103] while hindering CO_2 hydrates, indicating that the kinetic promotion or inhibition is dependent on the guest molecule [104]. Remarkably, tryptophan and methionine both encourage CO_2 and CH_4 hydrates [105]. The surface activity and adsorption behavior by capillary action of amino acids may explain their effectiveness as hydrate promoters [100, 105].

The production of amino acids (through breaking down of proteins) expressly for the application of promoting and inhibiting hydrate formation has been investigated for a reasonable period [106]. In summarizing previous studies, it is evident that the same amino acid will exert different effects on different guest formers. The amino acids side chain, extending from a nonpolar alkyl chain to a charged or uncharged polar chain, plays a vital part in their physicochemical properties [107]. Among the various amino acids, leucine has shown the best promoter effect for a CH_4 hydrate, whereas, for the CO_2 system, L-methionine has shown the best results.

For CO_2 hydrates, 0.2 wt% concentrations of the following amino acids, ordered from highest effect to lowest, have shown kinetic promoting effects: L-methionine > L-norleucine > L-tryptophan > L-norvaline > n-hexylamine [105]. At higher concentrations (0.5 wt%), L-methionine still shows the highest promotion effect but is followed by: L-cysteine > L-valine > L-threonine > L-phenylalanine [108]. Most researchers deliberate that the promotion effect is due to the surface activity induced by the amino acids. Amino acids with an aromatic-sided chain and hydrophobic properties are the types primarily exhibiting a promoting effect [106].

The use of common hydrate promoters like SDS and sodium tetradecyl sulfate may still not be suitable for water purification as these surfactants are considered pollutants. Several researchers have studied novel bio-additives to enhance hydrate kinetics [109]. Sun et al. investigated the use of L-arginine and iso-octyl glucoside in CH_4 hydrate formation and determined that these additives enhanced CH_4 gas consumption [110]. Leucine has also been studied for use as a promoter with CH_4 gas [73]. More recently, it was reported that the kinetics of hydrate formation could be improved by using the edible surfactant lecithin with natural gas and CP. The addition of lecithin improved the kinetics of hydrate formation, subsequently resulting in more efficient desalination [111].

4.6 Conclusion

Amino acids have shown promise as gas hydrate promoters that are environmentally safe, stable, and inexpensiveness. The use of surfactants as additives, in general, has enhanced hydrate formation rates with high water-to-hydrate conversion ratios for various former molecules. However, it remains unclear whether the concentration dependency would be affected by reactor configuration. Hence, subsequent research on the efficiency of separation and ion exclusion is suggested to ensure a more substantial hydrate-based desalination technology.

References

1 Wattie, E. and Butterfield, C.T. (1944). Relative resistance of Escherichia coli and Eberthella typhosa to chlorine and chloramines. *Public Health Reports* 59 (52): 1661–1671.

2 Shi, X., Tal, G., Hankins, N.P., and Gitis, V. (2014). Fouling and cleaning of ultrafiltration membranes: A review. *Journal of Water Process Engineering* 1: 121–138.

3 Cha, J.H. and Seol, Y. (2013). Increasing gas hydrate formation temperature for desalination of high salinity produced water with secondary guests. *ACS Sustainable Chemistry & Engineering* 1: 1218–1224.

4 Yu, Z., Qi, Y., Ji, L., Xing, Y., and Liu, Y. (2013). Experimental study of effects on hydrate seawater desalination by CO_2. *Low Temperature and Specialty Gases* 31: 21–25.

5 Liu, C., Ren, H., Meng, Q., and Sun, S. (2013). An experimental study of CO_2 hydrate-based seawater desalination with the R141b as an accelerant. *Natural Gas Industry* 33: 90–95.

6 Kang, K.C., Hong, S.Y., Cho, S.J., Kim, D.H., and Lee, J.D. (2017). Evaluation of desalination by nanostructured hydrate formation and pellet production process. *Journal of Nanoscience and Nanotechnology* 17: 4059–4062.

7 Babu, P., Nambiar, A., He, T., Karimi, I.A., Lee, J.D., Englezos, P., and Linga, P. (2018). A review of clathrate hydrate-based desalination to strengthen energy-water nexus. *ACS Sustainable Chemistry & Engineering* 6: 8093–8107.

8 Kang, K.C., Linga, P., Park, K.N., Choi, S.J., and Lee, J.D. (2014). Seawater desalination by gas hydrate process and removal characteristics of dissolved ions (Na^+, K^+, Mg^{2+}, Ca^{2+}, B_3^+, Cl^-, SO_4^{2-}). *Desalination* 353: 84–90.

9 Yang, M.J., Song, Y.C., Jiang, L.L., Liu, W.G., Dou, B.L., and Jing, W. (2014). Effects of operating mode and pressure on hydrate-based desalination and CO_2 capture in porous media. *Applied Energy* 135: 504–511.

10 Sarshar, M. and Sharafi, A. (2011). Simultaneous water desalination and CO_2 capturing by hydrate formation. *Desalination and Water Treatment* 28: 59–64.

11 Babu, P., Kumar, R., and Linga, P. (2014). Unusual behavior of propane as a co-guest during hydrate formation in silica sand: Potential application to seawater desalination and carbon dioxide capture. *Chemical Engineering Science* 117: 342–351.

12 Yang, M., Zheng, J., Liu, W., Liu, Y., and Song, Y. (2015). Effects of C_3H_8 on hydrate formation and dissociation for integrated CO_2 capture and desalination technology. *Energy* 93: 1971–1979.

13 Li, S., Fan, S., Wang, J., Lang, X., and Wang, Y. (2010). Clathrate hydrate capture of CO_2 from simulated flue gas with cyclopentane/water emulsion. *Chinese Journal of Chemical Engineering* 18: 202–206.

14 Jeffrey, G. (1984). Hydrate inclusion compounds. *Journal Of Inclusion Phenomena* 1: 211–222.

15 Bouchemoua, A., Brantuas, P., and Herri, J.-M. (2011). Equilibrium data of CO_2-based semi-clathrates from quaternary ammonium solutions. *7th International Conference on Gas Hydrates* 439. (ICGH 2011)

16 Sloan, E.D. and Koh, C. (2013). Clathrate Hydrates of Natural Gases. New York: CRC Press.

17 Mohammadi, A.H., Anderson, R., and Tohidi, B. (2005). Carbon monoxide clathrate hydrates: Equilibrium data and thermodynamic modeling. *AIChE Journal* 51: 2825–2833.

18 Ngema, P.T., Petticrew, C., Naidoo, P., Mohammadi, A.H., and Ramjugernath, D. (2014). Experimental measurements and thermodynamic modeling of the dissociation conditions of clathrate hydrates for (refrigerant + Nacl + water) systems. *Journal of Chemical & Engineering Data* 59: 466–475.

19 Mooijer-Van Den Heuvel, M.M. (2004). Phase behaviour and structural aspects of ternary clathrate hydrate systems, the role of additives. Tu Delft, Delft University Of Technology.

20 Sas, O.G., Domínguez, I., González, B., and Domínguez, Á. (2018). Liquid-liquid extraction of phenolic compounds from water using ionic liquids: Literature review and new experimental data using [C_2mim]FSI. *Journal of Environmental Management* 228: 475–482.

21 Yan, X., Anguille, S., Bendahan, M., and Moulin, P. (2019). Ionic liquids combined with membrane separation processes: A review. *Separation and Purification Technology* 222: 230–253.

22 Zhu, G., Cheng, G., Lu, T., Cao, Z., Wang, L., Li, Q., and Fan, J. (2019). An ionic liquid functionalized polymer for simultaneous removal of four phenolic pollutants in real environmental samples. *Journal of Hazardous Materials* 373: 347–358.

23 Zhang, J., Qin, Z., Yang, L., Guo, H., and Han, S. (2017). Activation promoted ionic liquid modification of reverse osmosis membrane towards enhanced permeability for desalination. *Journal of the Taiwan Institute of Chemical Engineers* 80: 25–33.

24 Meng, H., Gong, B., Geng, T., and Li, C. (2014). Thinning of reverse osmosis membranes by ionic liquids. *Applied Surface Science* 292: 638–644.

25 Fredlake, C.P., Crosthwaite, J.M., Hert, D.G., Aki, S.N.V.K., and Brennecke, J.F. (2004). Thermophysical properties of imidazolium-based ionic liquids. *Journal of Chemical & Engineering Data* 49 (4): 954–964.

26 Raval, H., Mehta, B., Joshi, R., and Kumar, A. (2018). A novel thin film composite reverse osmosis membrane modified BY IONIC liquid. *Brazilian Journal of Chemical Engineering* 35 (4): 1249–1256.

27 Isosaari, P., Srivastava, V., and Sillanpää, M. (2019). Ionic liquid-based water treatment technologies for organic pollutants: Current status and future prospects of ionic liquid mediated technologies. *Science of the Total Environment* 690: 604–619.

28 Berber, M.R. (2020). Current advances of polymer composites for water treatment and desalination. *Journal of Chemistry*. doi: 10.1155/2020/7608423

29 Zhou, Y., Hu, Y., Huang, W., Cheng, G., Cui, C., and Lu, J. (2018). A novel amphoteric β-cyclodextrin-based adsorbent for simultaneous removal of cationic/anionic dyes and bisphenol a. *Chemical Engineering Journal* 341: 47–57.

30 Huang, W., Hu, Y., Li, Y. et al. (2018). Citric acid-crosslinked β-cyclodextrin for simultaneous removal of bisphenol a, methylene blue and copper: The roles of cavity and surface functional groups. *Journal of the Taiwan Institute of Chemical Engineers* 82: 189–197.

31 Li, Y., Zhou, Y., Zhou, Y., Lei, J., and Pu, S. (2019). Cyclodextrin modified filter paper for removal of cationic dyes/Cu ions from aqueous solutions. *Water Science and Technology* 78 (12): 2553–2563.

32 McDonnell, G. and Russell, A.D. (1999). Antiseptics and disinfectants: Activity, action and resistance. *Clinical Microbiology Reviews* 12 (1): 147–179.

33 Ying, G.G. (2006). Fate, behavior and effects of surfactants and their degradation products in the environment. *Environment International* 32 (3): 417–431.

34 Tandukar, M., Oh, S., Tezel, U., Konstantinidis, K., and Spyros, G. (2013). Long-term exposure to benzalkonium chloride disinfectants results in change of microbial community structure and increase antimicrobial resistance. *Environmental Science & Technology* 47 (17): 9730–9738.

35 Naghash, A. Nezamzadeh-Ejhieh (2015). Comparison of the efficiency of modified clinoptilolite with HDTMA and HDP surfactants for the removal of phosphate in aqueous solutions. *Journal of Industrial & Engineering Chemistry Research* 31: 185–191.

36 Bowman, R.S. (2003). Applications of surfactant-modified zeolites to environmental remediation. *Microporous and Mesoporous Materials* 61: 1–3.

37 Ng, Y.S., Jayakumar, N.S., and Hashim, M.A. (2011). Behavior of hydrophobic ionic liquids as liquid membranes on phenol removal: Experimental study and optimization. *Desalination* 278 (1–3): 250–258.

38 Azari, S. and Zou, L. (2013). Fouling resistant zwitterionic surface modification of reverse osmosis membranes using amino acid l-cysteine. *Desalination* 324: 79–86.

39 Liu, T., Chen, D., Yang, F., Chen, J., Cao, Y., Xiang, M., ... Xu, R. (2019). Enhancing the permeability and anti-fouling properties of a polyamide thin-film composite reverse osmosis membrane via surface grafting of l-lysine. *RSC Advances* 9 (35): 20044–20052.

40 Nematzadeh, M., Shokrollahzadeh, S., Samimi, A., and Mohebbi-Kalhori, D. (2020). Synergistic effect of amino-acids and metal salts as draw solutions to enhance the performance of fertilizer-drawn forward osmosis. *Environmental Science: Water Research & Technology* 6 (11): 3121–3131.

41 Azari, S. and Zou, L. (2012). Using zwitterionic amino acid l-DOPA to modify the surface of thin film composite polyamide reverse osmosis membranes to increase their fouling resistance. *Journal of Membrane Science* 401: 68–75.

42 Kuo, P.-C., Chen, L.-J., Lin, S.-T., and Chen, Y.-P. (2010). Measurement for the dissociation conditions of methane hydrate in the presence of 2-methyl-2-propanol. *Journal of Chemical & Engineering Data* 55: 5036–5039.

43 Mandal, A. and Laik, S. (2008). Effect of the promoter on gas hydrate formation and dissociation. *Energy & Fuels* 22: 2527–2532.

44 Nallakukkala, S. and Lal, B. (2021). Seawater and produced water treatment via gas hydrate: Review. *Journal of Environmental Chemical Engineering* 9 (2): 105053. doi: 10.1016/j.jece.2021.105053.

45 Veluswamy, H.P., Kumar, A., Seo, Y., Lee, J.D., and Linga, P. (2018). A review of solidified natural gas (SNG) technology for gas storage via clathrate hydrates. *Applied Energy* 216: 262–285.

46 Xu, C.G. and Li, X.S. (2014). Research progress of hydrate-based CO_2 separation and capture from gas mixtures. *RSC Advances* 4: 18301–18316.

47 Wendland, M., Hasse, H., and Maurer, G. (1999). Experimental pressure-temperature data on three- and four-phase equilibria of fluid, hydrate, and ice phases in the system carbon dioxide-water. *Journal of Chemical & Engineering Data* 44: 901–906.

48 Yang, S.O., Yang, I.M., Kim, Y.S., and Lee, C.S. (2000). Measurement and prediction of phase equilibria for water + CO_2 in hydrate forming conditions. *Fluid Phase Equilibria* 175: 75–89.

49 Li, X., Xu, C., Chen, Z., Wu, H., and Cai, J. (2011). Effect of temperature fluctuation on hydrate-based CO_2 separation from fuel gas. *Journal of Natural Gas Chemistry* 20: 647–653.

50 Dholabhai, P.D., Parent, J.S., and Bishnoi, P.R. (1996). "Carbon dioxide hydrate equilibrium conditions in aqueous solutions containing electrolytes and methanol using a new apparatus, Ind. *Industrial & Engineering Chemistry Research* 35: 819–823.

51 Dholabhai, P.D., Parent, J.S., and Bishnoi, P.R. (1997). Equilibrium conditions for hydrate formation from binary mixtures of methane and carbon dioxide in the presence of electrolytes, methanol and ethylene glycol. *Fluid Phase Equilibria* 141: 235–246.

52 Kang, S.P., Chun, M.K., and Lee, H. (1998). Phase equilibria of methane and carbon dioxide hydrates in the aqueous $MgCl_2$ solutions. *Fluid Phase Equilibria* 147: 229–238.

53 Mohammadi, A.H., Afzal, W., and Richo, D. (2008). Gas hydrates of methane, ethane, propane and carbon dioxide in the presence of single NaCl, KCl and $CaCl_2$ aqueous solutions: Experimental measurements and predictions of dissociation conditions. *Journal of Chemical Thermodynamics* 40: 1693–1697.

54 Sun, S., Liu, C., and Ye, Y. (2013). Phase equilibrium condition of marine carbon dioxide hydrate. *Journal of Chemical Thermodynamics* 57: 256–260.

55 Yang, M.J., Song, Y.C., Liu, Y., Lam, W.H., and Li, Q.P. (2011). Equilibrium conditions for CO_2 hydrate in porous medium. *Journal of Chemical Thermodynamics* 43: 334–338.

56 Dholabhai, P.D., Kalogerakis, N., and Bishnoi, P.R. (1993). Equilibrium conditions for carbon dioxide hydrate formation in aqueous electrolyte solutions. *Journal of Chemical & Engineering Data* 38: 650–654.

57 Maekawa, T. (2011). Equilibrium conditions of clathrate hydrates formed from carbon dioxide and aqueous acetone solutions. *Fluid Phase Equilibria* 303: 76–79.

58 Breland, E. and Englezos, P. (1996). Equilibrium hydrate formation data for carbon dioxide in aqueous glycerol solutions. *Journal of Chemical & Engineering Data* 41: 11–13.

59 Zheng, J.N., Yang, M.J., Liu, Y., Wang, D.Y., and Song, Y.C. (2017). Effects of cyclopentane on CO_2 hydrate formation and dissociation as a co-guest molecule for desalination. *Journal of Chemical Thermodynamics* 104: 9–15.

60 Zhang, J.S. and Lee, J.W. (2009). Equilibrium of hydrogen plus cyclopentane and carbon dioxide plus cyclopentane binary hydrates. *Journal of Chemical & Engineering Data* 54: 659–661.

61 Matsumoto, Y., Makino, T., Sugahara, T., and Ohgaki, K. (2014). Phase equilibrium relations for binary mixed hydrate systems composed of carbon dioxide and cyclopentane derivatives. *Fluid Phase Equilibria* 362: 379–382.

62 Bishnoi, P.R. and Dholabhai, P.D. (1993). Experimental study on propane hydrate equilibrium conditions in aqueous electrolyte solutions. *Fluid Phase Equilibria* 83: 455–462.

63 Robinson, D.B. and Mehta, B.R. (1971). Hydrates in the propane-carbon dioxide-water system. *Journal of Canadian Petroleum Technology* 10: 33–35.

64 Babu, P., Yang, T., Veluswamy, H.P., Kumar, R., and Linga, P. (2013). Hydrate phase equilibrium of ternary gas mixtures containing carbon dioxide, hydrogen and propane. *Journal of Chemical Thermodynamics* 61: 58–63.

65 Veluswamy, H.P., Chen, J.Y., and Linga, P. (2015). Surfactant effect on the kinetics of mixed hydrogen/propane hydrate formation for hydrogen storage as clathrates. *Chemical Engineering Science* 126: 488–499.

66 Kumar, R. and Hui-jiewu, P.E. (2006). Incipient hydrate phase equilibrium for gas mixtures containing hydrogen, carbon dioxide and propane. *Fluid Phase Equilibria* 244: 167–171.

67 Kumar, R., Linga, P., Moudrakovski, I., Ripmeester, J.A., and Englezos, P. (2008). Structure and kinetics of gas hydrates from methane/ethane/propane mixtures relevant to the design of natural gas hydrate storage and transport facilities. *AIChE Journal. American Institute of Chemical Engineers* 54: 2132–2144.

68 Yoslim, J., Linga, P., and Englezos, P. (2010). Enhanced growth of methane-propane clathrate hydrate crystals with sodium dodecyl sulfate, sodiumtetradecyl sulfate, and sodium hexadecyl sulfate surfactants. *Journal of Crystal Growth* 313: 68–80.

69 Lee, H., Lee, J.W., Kim, D.Y., Park, J., Seo, Y.T., Zeng, H., Moudrakovski, I.L., Ratcliffe, C.I., and Ripmeester, J.A. (2005). Tuning clathrate hydrates for hydrogen storage. *Nature* 434: 743–746.

70 Florusse, L.J., Peters, C.J., Schoonman, J., Hester, K.C., Koh, C.A., Dec, S.F., Marsh, K.N., and Sloan, E.D. (2004). Stable low-pressure hydrogen clusters stored in a binary clathrate hydrate. *Science* 306: 469–471.

71 Sloan, E.D. and Koh, C. (2007). Molecular structures and similarities to ice. In: Clathrate Hydrates of Natural Gases (T.F. Group ed.), 46–90. CRC Press.

72 Veluswamy, H.P., Kumar, A., Premasinghe, K., and Linga, P. (2017). Effect of guest gas on the mixed tetrahydrofuran hydrate kinetics in a quiescent system. *Applied Energy* 207: 573–583.

73 Veluswamy, H.P., Kumar, A., Kumar, R., and Linga, P. (2017). An innovative approach to enhance methane hydrate formation kinetics with leucine for energy storage application. *Applied Energy* 188: 190–199.

74 Yu, Z., Qi, Y., Liu, Y., and Zhang, H. (2012). Experimental research on HCFC-141b hydrate seawater desalination. *Chinese Journal of Refrigeration Technology* 32: 46–48.

75 Song, Y., Dong, H., Yang, L., Yang, M., Li, Y., Ling, Z., and Zhao, J. (2016). Hydrate-based heavy metal separation from aqueous solution. *Scientific Reports* 6: 21389.

76 Xu, H., Khan, M.N., Peters, C.J., Sloan, E.D., and Koh, C.A. (2018). Hydrate-based desalination using cyclopentane hydrates at atmospheric pressure. *Journal of Chemical & Engineering Data* 63: 1081–1087.

77 Li, F., Chen, Z., Dong, H., Shi, C., Wang, B., Yang, L., and Ling, Z. (2018). Promotion effect of graphite on cyclopentane hydrate based desalination. *Desalination* 445: 197–203.

78 Han, S., Rhee, Y., and Kang, S. (2017). Investigation of salt removal using cyclopentane hydrate formation and washing treatment for seawater desalination. *Desalination* 404: 132–137.

79 Corak, D., Barth, T., Høiland, S., Skodvin, T., Larsen, R., and Skjetne, T. (2001). Effect of subcooling and amount of hydrate former on formation of cyclopentane hydrates in brine. *Desalination* 278: 268–274.

80 Lim, Y.A., Babu, P., Kumar, R., and Linga, P. (2013). Morphology of carbon dioxide-hydrogen cyclopentane hydrates with or without sodium dodecyl sulfate. *Crystal Growth & Design* 13: 2047–2059.

81 Zheng, J., Zhang, B., Wu, Q., and Linga, P. (2018). Kinetic evaluation of cyclopentane as a promoter for CO_2 capture via a clathrate process employing different contact modes. *ACS Sustainable Chemistry & Engineering* 6: 11913–11921.

82 Zheng, J., Cheng, F., Li, Y., Lü, X., and Yang, M. (2019). Progress and trends in hydrate based desalination (HBD) technology: A review. *Chinese Journal of Chemical Engineering* 27 (9): 2037–2043. doi: 10.1016/j.cjche.2019.02.017.

83 Mohammadi, A.H., Eslamimanesh, A., Belandria, V., and Richon, D. (2011). Phase equilibria of semiclathrate hydrates of CO_2, N_2, CH_4, or H_2+ tetra-n-butylammonium bromide aqueous solution. *Journal of Chemical & Engineering Data* 56: 3855–3865.

84 Kang, S.P., Lee, H., Lee, C.S., and Sung, W.M. (2001). Hydrate phase equilibria of the guest mixtures containing CO_2, N_2 and tetrahydrofuran. *Fluid Phase Equilibria* 185: 101–109.

85 Mohammadi, A.H. and Richon, D. (2010). Phase equilibria of semi-clathrate hydrates of tetra nbutylammonium bromide + hydrogen sulfide and tetra-n-butylammonium bromide + methane. *Journal of Chemical & Engineering Data* 55: 982–984.

86 Mohammadi, A.H. and Richon, D. (2011). Phase equilibria of binary clathrate hydrates of nitrogen +cyclopentane/cyclohexane/methylcyclohexane and ethane +cyclopentane/cyclohexane/methyl cyclohexane. *Chemical Engineering Science* 66: 4936–4940.

87 Nashed, O., Lal, B., Shariff, A.M., and Sabil, K.M. (2020). Gas hydrate promoters. In: Chemical *Additives for Gas Hydrates. Green Energy and Technology* Cham: Springer Nature Switzerland AG.

88 Alberti, M., Pirani, F., and Lagana, A. (2013). Carbon dioxide clathrate hydrates: Selective role of intermolecular interactions and action of the SDS catalyst. *The Journal of Physical Chemistry A* 117: 6991–7000.

89 Li, Y., Zhu, C., and Wang, W. (2012). Promoting effects of surfactants on carbon dioxide hydrate formation and the kinetics. *Petrochemical Technology* 41: 699–703.

90 Nallakukkala, S., Sivabalan, V., Lal, B., and Mokhtar Che Ismail, N.D.N. (2019). Nonionic surfactants as corrosion inhibitors for carbon steel in hydrochloric acid medium. *TEST Engineering & Management* 81: 5830–5835.

91 Zhong, Y. and Rogers, R.E. (2000). Surfactant effects on gas hydrate formation. *Chemical Engineering Science* 5: 4175–4187.

92 Komatsu, H., Hayasaka, A., Ota, M., Sato, Y., Watanabe, M., and Smith, R.L. (2013). Measurement of pure hydrogen and pure carbon dioxide adsorption equilibria for THF clathrate hydrate and tetra-n-butyl ammonium bromide semi-clathrate hydrate. *Fluid Phase Equilibria* 357: 80–85.

93 Mohammadi, A.H., Eslamimanesh, A., Belandria, V., Richon, D., Naidoo, P., and Ramjugernath, D. (2012). Phase equilibrium measurements for semi-clathrate hydrates of the ($CO_2 + N_2 +$ tetra-n-butylammonium bromide) aqueous solution system. *Journal of Chemical Thermodynamics* 46: 57–61.

94 Duc, N.H., Chauvy, F., and Herri, J.-M. (2007). CO_2 capture by hydrate crystallization A potential solution for gas emission of steelmaking industry. *Energy Conversion and Management* 48: 1313–1322.

95 Li, X.S., Xu, C.G., Chen, Z.Y., and Wu, H.J. (2010). Tetra-n-butyl ammonium bromide semi-clathrate hydrate process for post-combustion capture of carbon dioxide in the presence of dodecyl trimethyl ammonium chloride. *Energy* 35: 3902–3908.

96 Li, X.S., Xu, C.G., Chen, Z.Y., and Cai, J. (2012). Synergic effect of cyclopentane and tetra-n-butyl ammonium bromide on hydrate-based carbon dioxide separation from fuel gas mixture by measurements of gas uptake and X-ray diffraction patterns. *International Journal of Hydrogen Energy* 37: 720–727.

97 Xu, C.G., Zhang, S.H., Cai, J., Chen, Z.Y., and Li, X.S. (2013). CO_2 (carbon dioxide) separation from CO_2–H_2 (hydrogen) gas mixtures by gas hydrates in TBAB (tetra-n-butylammonium bromide) solution and Raman spectroscopic analysis. *Energy* 59: 719–725.

98 Li, X., Xu, C., Chen, Z., and Wu, H. (2011). Hydrate-based pre-combustion carbon dioxide capture process in the system with tetra-n-butyl ammonium bromide solution in the presence of cyclopentane. *Energy* 36: 1394–1403.

99 Veluswamy, H.P., Hong, Q.W., and Linga, P. (2016). Morphology study of methane hydrate formation and dissociation in the presence of amino acid. *Crystal Growth & Design* 16: 5932–5945.

100 Liu, Y., Chen, B., Chen, Y., Zhang, S., Guo, W., Cai, Y., and Wang, W. (2015). Methane storage in a hydrated form as promoted by leucines for possible application to natural gas transportation and storage. *Energy Technology* 3: 815–819.

101 Abbasian Rad, S., Rostami Khodaverdiloo, K., Karamoddin, M., Varaminian, F., and Peyvandi, K. (2015). Kinetic study of amino acids inhibition potential of glycine and l-leucine on the ethane hydrate formation. *Journal of Natural Gas Science and Engineering* 26: 819–826.

102 Naeiji, P., Mottahedin, M., and Varaminian, F. (2014). Separation of methane–ethane gas mixtures via gas hydrate formation. *Separation and Purification Technology* 123: 139–144.

103 Bhattacharjee, G., Choudhary, N., Kumar, A., Chakrabarty, S., and Kumar, R. (2016). Effect of the amino acid l-histidine on methane hydrate growth kinetics. *Journal of Natural Gas Science and Engineering* 35: 1453–1462.

104 Roosta, H., Dashti, A., Mazloumi, S.H., and Varaminian, F. (2016). Inhibition properties of new amino acids for prevention of hydrate formation in carbon dioxide–water system: Experimental and modelling investigations. *Journal of Molecular Liquids* 215: 656–663.

105 Cai, Y., Chen, Y., Li, Q., Li, L., Huang, H., Wang, S., and Wang, W. (2017). CO_2 hydrate formation promoted by a natural amino acid l–methionine for possible application to CO_2 capture and storage. *Energy Technology* 5: 1195–1199.

106 Bavoh, C.B., Lal, B., Osei, H., Sabil, K.M., and Mukhtar, H. (2019). A review on the role of amino acids in gas hydrate inhibition, CO_2 capture and sequestration, and natural gas storage. *Journal of Natural Gas Science and Engineering* 64: 52–71. doi: 10.1016/j.jngse.2019.01.020.

107 Madeira, P.P., Bessa, A., Álvares-Ribeiro, L., Aires-Barros, M.R., Rodrigues, A.E., Uversky, V.N. et al. (2014). Amino acid/water interactions study: A new amino acid scale. *Journal of Biomolecular Structure & Dynamics* 32: 959–968. doi: 10.1080/07391102.2013.800994.

108 Prasad, P.S.R. and Kiran, B.S. (2018). Are the amino acids thermodynamic inhibitors or kinetic promoters for carbon dioxide hydrates. *Journal of Natural Gas Science and Engineering* 52: 461–466. doi: 10.1016/j.jngse.2018.02.001.

109 Arora, A., Cameotra, S.S., Kumar, R., Balomajumder, C., Singh, A.K., Santhakumari, B., Kumar, P., and Laik, S. (2016). Biosurfactant as a promoter of methane hydrate formation: Thermodynamic and kinetic studies. *Scientific Reports* 6 (1): 20893.

110 Sun, Q., Chen, B., Li, X., Guo, X., and Yang, L. (2017). The investigation of phase equilibria and kinetics of CH4 hydrate in the presence of bioadditives. *Fluid Phase Equilibria* 452: 143–147.

111 Gaikwad, N., Nakka, R., Khavala, V., Bhadani, A., Mamane, H., and Kumar, R. (2020). Gas hydrate-based process for desalination of heavy metal ions from an aqueous solution: Kinetics and rate of recovery. *ACS ES&T Water.* 1(1): 134–144.continue

5

Modeling of Seawater Desalination by Gas Hydrate Method

Sirisha Nallakukkala and Bhajan Lal

5.1 Introduction

Industrial implementation of hydrate desalination has seen much support through continued research and development. The chemicals recommended for use as hydrate formers in the literature are designated for low-temperature and high-pressure conditions, which impedes their commercial implementation [1]. Therefore, choosing a hydrate-forming agent and optimizing the desalination procedure is necessary. The suggested steps in this optimization are learning the process conditions, modeling the hydrate desalination, and calculating the design values per standards. Numerous desalination models have been proposed and fall into three classes: (i) thermodynamic, (ii) kinetic, and (iii) machine learning approaches.

Thermodynamic-based approaches were developed in the 1950s and are mainly based on the forces driven by chemical potentials. Henceforth, any discussions regarding chemical or additive properties are described with their appropriately accessible parameters of chemical potential. Kinetic-based models primarily look at the impact of pressure conditions on the growth rate and are used to build dynamic models of the system [2]. More recently, multiple machine learning algorithms, such as neural networks, decision trees, and support vector machines, have been used to predict formation and dissociation conditions of hydrates.

Gas Hydrate in Water Treatment: Technological, Economic, and Industrial Aspects,
First Edition. Bhajan Lal and Sirisha Nallakukkala.

5.2 Gas Hydrate Thermodynamic and Kinetic Models

Studies on the prediction of hydrate formation are more commonly offered in two classes: thermodynamic and experimental relationships. Classically, thermodynamic and mathematical correlations use simulations and are programmed with a clear relationship among input and output variables. The typical thermodynamic models applied for the formation of hydrates are K value and van der Waals–Plateeuw (vdW–P) [5]. The former is derived from vapor–solid equilibrium constants, and the latter is constructed from numerical thermodynamic analysis.

The foremost disadvantage of the thermodynamic models is their poor prediction capability for generalized pressure and temperature behavior. Another disadvantage is the intricate programming required for a model to be done well [6]. Use of thermodynamic methods as a function of equation of state (EoS) with numerous constituents takes a great deal of time and effort as the intermolecular attractions and appropriate rules of mixing must be assessed by experimentation [3, 7]. For experimental relationships, the variables comprise pressure, temperature, and gas gravity. Despite these drawbacks, thermodynamic models are still considered beneficial due to the specificity of experiments.

Most of the difficulties faced in the industrial use of hydrates are in understanding the dynamics of hydrate formation behavior and dissociation or their kinetic behavior [8]. In contrast to the time-invariant studies on gas hydrate properties carried out since the 1960s, more recent exploration has placed a greater emphasis on understanding the kinetic behavior. A better understanding of hydrate formation kinetics can help identify aspects leading to enhanced hydrate formation rates.

The kinetic behavior of hydrate formation in desalination processes is more intricate, as the formation of hydrate comprises two mechanisms: (i) stochastic nucleation of hydrate, followed by (ii) the growth mechanism [8]. A thorough understanding of hydrate formation behavior requires familiarity with thermodynamics, heat and mass transport phenomena, colloidal science, reaction engineering, as well as mathematical [9] and numerical simulation methods [10]. All these fields have had some hand in creating these specific models: models that can solve the governing equations, using foundational assumptions, to develop further interrelations that can closely describe the behavior of hydrate formation in desalination.

Examination of gas hydrate growth behavior is almost always initiated from laboratory experiments, where hydrates are formed in high-pressure vessels and their formation rate measured [11]. Naturally, any newly proposed numerical or analytical solution based on the mathematical models needs to be substantiated with experimental results. Within this print, kinetic models related to hydrate growth and desalination efficiency are evaluated based on the control mechanism, type of hydrate-forming guest molecule, and reactor configuration. The objective is to communicate to scientists regarding the latest and most updated knowledge on growth kinetics. The salient features of the kinetic models were surveyed based on designs, hypotheses, driving force, and governing equations with their limiting conditions, solution methods, and model-based determinations. It is noted that most of the models listed in Table 5.1 are not established from these principles. Additionally, there is no single model yet available to describe the kinetics of gas hydrate growth, water recovery, and desalination efficiency. The lack of a convenient model that incorporates all these elements shows a clear research gap within this field.

Hydrate growth kinetics usually rely on the molar consumption of gas, which is directly and indirectly measured through:

- Water and gas mass transfer (MT) during hydrate development.
- Transport of exothermal heat generated throughout crystal development.
- The intrinsic kinetics of hydrate growth.

The hydrate structures can be determined depending on these aspects.

5.3 Statistical Thermodynamic Modeling of Hydrate Equilibrium

Based on the understanding of the hydrate's structure, the pioneering phase equilibrium statistical thermodynamic model was perceived in 1959 by Barrer and Stuart [48] in the form of the vdW–P model [49]. The model established equations for sI and sII structures based on the chemical potential of water by implementing a mechanism equivalent to the Langmuir isotherm. The vdW–P model built upon the driving forces of μ_w^H and μ_w^β, as shown in Eq. (5.1).

$$\frac{\Delta\mu_w^H}{RT} = \frac{\mu_w^\beta - \mu_w^H}{RT} = \sum_{m=1}^{2} \upsilon_m \ln\left(1 + \sum_{i=1}^{nc} C_{mi} f_i\right) \tag{5.1}$$

The equation for the chemical potential of water in the aqueous-ice phase has been abridged by [6, 50] and is represented as in Eq. (5.2).

$$\frac{\Delta\mu_w^\alpha}{RT} = \frac{\mu_w^\beta - \mu_w^\alpha}{RT} = \frac{\Delta\mu_w^0}{RT} - \int_{T_0}^{T} \frac{\Delta h_w}{RT^2} dT + \int_{P_0}^{P} \frac{\Delta v_w}{RT} dP - RT\ln\left(a_w^\alpha\right) \tag{5.2}$$

Table 5.1 Kinetic model structures for growth of hydrate.

Rate Control process	Model structures/driving force/rate expression	Reference	Hydrate former mixtures
Reaction kinetics	Arrhenius rate equation $r = Aa_s \exp\left(-\frac{\Delta E_a}{RT}\right)\exp\left(-\frac{a}{\Delta T^b}\right)P^{\gamma}$ First-order reaction rate (mole fraction) $r = K_R * x_G(t)$ Elementary pseudo reactions with rate constants Arrhenius expression through subcooling driving force $r = uk_1 \exp\left(-\frac{k_2}{RT}\right)A_s\left(T_{eq} - T_{sys}\right)$	[12–17]	Methane + glycol Ethane + glycol Carbon dioxide + ethylene glycol Methane + methanol

Mass transfer (MT)	Concentration difference among gas–liquid phase $r = k_L A_g - l^{c_{w0}} \left(x_{int} - x_b \right)$	[18–23]	Methane, ethane
	Fugacity difference of gas phase based upon population balance $r = \sum_{1}^{2} K_j^* A_p \left(f - f_{eq} \right)_j$		Methane + glycol Ethane + glycol
	Methane concentration difference between the oil and equilibrium phase $r = k_{g-0} A_{g-0} \left(C_{CH4,0}^{eq} - C_{CH4,0} \right)$		Carbon dioxide + glycol Methane + methanol
	Concentration difference (advanced nucleation) $r = k_g \left(C_b - C_{eq} \right)$		Methane + glycol
Heat transfer (HT)	One dimensional HT model	[24–27]	Carbon dioxide
	One dimensional HT due to conduction and convection $v_f \delta = \Delta T^{3/2}$		Carbon dioxide
			Methane, ethane,
	One dimensional HT model due to convection $v_f \delta = \psi \Delta T^{5/2}$		Methane + ethane
	Two-dimensional HT model due to conduction		Carbon dioxide

(Continued)

Table 5.1 (Continued)

Rate Control process	Model structures/driving force/rate expression	Reference	Hydrate former mixtures
Model related to kinetics and MT	Change in concentration based on MT and kinetics	[28–38]	Methane + methanol
	$r = K\left(C_{sol} - C_{eq}\right)$		Propane + mono-ethylene glycol
	$\frac{1}{K} = \frac{1}{k_L A_g} + \frac{1}{k_S A_C}$		Methane + water
	Interface MT of liquid-vapor and kinetics		Methane + heavy water
	$r = \frac{1}{A_p}\left(\frac{1}{k_{H-L}} + \frac{1}{k_r}\right) + \frac{1}{k_{\dot{G}-L}}$		carbon dioxide
	Shell and core model with diffusion/integrated through reaction		carbon dioxide
	$(1-\alpha)^{\frac{1}{3}} = \left(-\frac{(2k)^{\frac{1}{2}}}{r_0}\right)\left(t-t^*\right)^{\frac{1}{2}} + \left(1-\alpha^*\right)^{\frac{1}{3}}$		carbon dioxide carbon dioxide
	First-order diffusion of H_2O kinetics		Methane + calcium chloride
	First-order diffusion with formation and dissociation H_2O kinetics		
	First-order diffusion with carbon dioxide formation kinetics		
	Reaction and diffusion restricted arrangements with induction time		

Models related to kinetics and HT	rate integrated with kinetics and HT $\frac{dX}{dt}=\frac{\Delta T}{\rho_h\lambda\left(\frac{1}{k\Delta T^n}+\frac{1}{h_{avg}(t)}\right)}$	[39–45]	Tetrahydrofuran
	Surface kinetic rate combined with HT model $\frac{dX}{dt}=\frac{T_{eq}-T_{bulk}}{\frac{\Delta H}{k_h A_h}+\frac{1}{k_r A_r}}$		Methane, carbon dioxide
	Kinetic rate combined with a heat convection model $\frac{dX}{dt}=\frac{T_{eq}-T_{bulk}}{\rho_h\lambda\left(\frac{1}{k}+\frac{1}{h}\right)}$		Methane + diesel Oil
	Equilibrium in a porous medium (sand)		Methane
Kinetics and fluid heat flow	Rate expression for formation and dissociation of hydrate	[46, 47]	Methane, carbon dioxide
	Rate expression constructed on Kim et al. mechanism		Methane
	Simple Slug flow mechanism united with kinetic and flow of heat		Methane

Within the calculation of the fugacity coefficient, the critical point varies by choice of the EoS and the mixing rules. The Peng–Robinson [51] and Soave–Redlich–Kwong [52] equations are some examples of tried and true methods. For complicated schemes consisting of electrolytes, EoS such as Valderrama–Patel–Teja [53], Nasrifar–Bolland [54], Cubic Plus Association [55], or statistical associating fluid theory (SAFT) [56], can lead to improved estimates. Additionally, the surplus Gibbs energy mixing rules like the improved Huron–Vidal first- and second-order (MHV1 or MHV2) [57, 58] could enhance the precision of fugacity calculations.

The subsequent parameter is the activity of water, which can shift dramatically as part of the equilibrium of hydrates when additives are introduced. This parameter must be precisely modeled to be reliably predicted. The last constant, the Langmuir constant, is vital if thermodynamic promoters are utilized [59].

From a thermodynamic point of view, the equilibrium hydrate temperature decreases due to the intermolecular interactions, which reduce the water molecule's activity. The suppression temperature can be calculated as in Eq. (5.3).

$$\Delta T = -\frac{n_H R T_0^2}{\Delta H^d} \frac{18W}{M(100 - W)} \tag{5.3}$$

Furthermore, the works of [60, 61] improved the enthalpy expression and incorporated ionic solution concentrations to enable implementation for electrolyte systems. Another researcher [62] optimized the model further by increasing the variables of the data sets, culminating into Eq. (5.4).

$$\frac{\Delta H^d}{n_H R} = \frac{e_1 I^{e_2}}{1 + e_3 P + e_4 \ln P} \tag{5.4}$$

5.3.1 Modeling Thermodynamic Equilibrium of Cyclopentane Hydrates in the Presence of Salts

Salts are excluded from hydrate crystals during the crystallization process in hydrate-based desalination. Hydrates can be segregated from the saline solution by filtration and dissociated to recover freshwater. If the hydrate formers are gases, they can be separated from the freshwater and reused in subsequent desalination. This technique has been extensively discussed in the literature yet is not ready for commercial utilization due to economic and other challenges, specifically high energy usage, the difficulty of segregating hydrates from residual water, and slow kinetics [63, 64]. Solutions proposed to overcome such problems include using cold energy liquefied natural gas for the exterior cooler [65] to decrease the specific energy intake consumption and the use of cyclopentane (CP) as a guest molecule.

In CP hydrate-based desalination, the guest and host form sII hydrate structures at 7.1°C under atmospheric pressure [66], reducing process energy inputs. The immiscibility of CP also makes separation from water easier. CP can also be used to optimize the driving force (subcooling temperature) for CP hydrate-based desalination, as seen in the work of [67]. Post-hydrate formation treatment procedures for this process include washing, centrifuging, and sweating to remove residual brine from CP hydrates [68]. As a tradeoff, CP hydrate-based desalination requires a good understanding of its kinetics and thermodynamic properties to optimize the amount of salt entrapped and removal efficiency.

The CP hydrate thermodynamics within salts, like sodium chloride (NaCl), potassium chloride (KCl), NaCl-KCl, or calcium chloride ($CaCl_2$), are readily available as experimental data in the literature [66]. In another source, five thermodynamic methods were applied [69] to simulate them based on: (i) standard freezing point depression (SFPD), (ii) Hu–Lee–Sum relationship, (iii) vdW–P, (iv) Kihara, and (v) activity-based occupancy correlation (ABOC) approaches. Correlation with experimental data found ABOC to be the best method to produce consistent equilibrium data of CP hydrate in electrolytes.

5.3.1.1 Standard Freezing Point Depression Calculation

The brine water activity with CP hydrate is articulated by [70] as in Eq. (5.5):

$$\ln a_w = \frac{\Delta H_{fm}}{R}\frac{(T_f - T)}{T_f T} + \frac{\Delta C_{fm}}{R}\left[\frac{(T_f - T)}{T} - \ln\left(\frac{T_f}{T}\right)\right], \tag{5.5}$$

where ΔC_{fm} is the undetermined parameter and is established by investigating the data in the existence of NaCl using Eq. (5.6).

$$\Delta C_{fm} = F(T) = a * \exp(b * T) \tag{5.6}$$

Equilibrium temperatures of CP hydrates in an electrolyte mixture of sodium sulfate (Na_2SO_4), magnesium chloride ($MgCl_2$), $MgCl_2$ + NaCl, or $MgCl_2$ + NaCl + KCl are estimated by using Eq. (5.5) and Eq. (5.6). Using this approach [69], a large gap was found between the modeling and experimentation data in the existence of $MgCl_2$ (≥ 18 mass%) and $MgCl_2$ + NaCl mixture (≥ 20 mass%). This difference is due to the a_w calculations.

5.3.1.2 Hu–Lee–Sum Correlation

While $\frac{nR}{\Delta H_{diss}}$ is perceived to remain a constant, $\frac{\Delta T}{T_0 T}$ was seen to rely on the effective mole fraction [71, 72]. Therefore, a new relationship based on Eq. (5.5) was established for the suppression temperature ($\Delta T = T_o - T$) and is described in Eq. (5.7) as:

$$\frac{\Delta T}{T_0 T} = -\frac{nR}{\Delta H_{diss}} \ln a_w = C_1 X + C_2 X^2 + C_3 X^3, \tag{5.7}$$

while the effective mole fraction is described as in Eq. (5.8), and the formation of hydrate temperature is given as in Eq. (5.9).

$$X = \sum_{j=salts} \sum_{i=ions} \left| z_{j,i} \right| x_{j,i} \tag{5.8}$$

$$T = T_0 [1 + \left(\frac{\Delta T}{T_0 T} \right) T_0]^{-1} \tag{5.9}$$

Equation. (5.7) was established for sI and sII hydrates obtained for the salt solution. For sII hydrates, a new parameter α was proposed. The simulated results using this method agreed well with experimental data [72]. The parameter is expressed as given in Eqs. (5.10 and 5.11).

$$\alpha = \frac{\left(\frac{\Delta T}{T_0 T}\right)_{SII}}{\left(\frac{\Delta T}{T_0 T}\right)_{SI}}, \tag{5.10}$$

so that $T = T_0 \, [1 + \alpha \left(\frac{\Delta T}{T_0 T} \right) T_0]^{-1}$. (5.11)

5.3.1.3 Kihara Approach

Equilibrium is determined based on μ_w in the liquid-hydrate phase. A reference state is implemented and expressed in Eq. (5.12) as:

$$\Delta\mu_w^{\beta-H} = \Delta\mu_w^{\beta-L}, \tag{5.12}$$

where $\Delta\mu_w^{\beta-H}$ can be estimated by vdW–P, as in Eq. (5.13).

$$\Delta\mu_w^{\beta-H} = -RT \sum_i \nu_i \ln \left(1 - \sum_j \theta_j^i \right) \tag{5.13}$$

Occupancy factor θ_j^i is found by integration of the Kihara potential [60] and the hydrate former fugacity. This approach can reasonably predict the CP formation temperature, as proven by Han et al. [69] The only drawback to this method is the necessity of the integration of Kihara potential.

5.3.1.4 Activity-Based Occupancy Correlation Approach

Overcoming the need for Kihara parameters, a more straightforward approach was discovered through interaction parameters. As an alternative to the

Langmuir method for cavity occupancy, the relationship between the activity of water and occupancy factor, $\theta = F(a_w)$ is stated in Eq. (5.14) as:

$$\theta(a_w) = m * (a_w^2) + n * (a_w) + p, \tag{5.14}$$

where *m*, *n*, and *p* are the experimental constants. The application of this method was tested and verified by Ho-Van et al. [71], where the experimental temperature dissociation was reproduced for CP hydrates in electrolyte solutions of Na_2SO_4 or $MgCl_2$–NaCl or $MgCl_2$–NaCl–KCl with an average absolute deviation of 0.1°C.

5.3.2 Modeling of Thermodynamic Equilibrium of Mixed Cyclopentane/Carbon Dioxide Hydrates

When the guest molecules are gases, they require favorable conditions such as high pressures and low temperatures to form gas hydrates, consuming lots of energy in the process. Thus, alternative hydrate formers with milder hydrate formation conditions are ideal for improving economic efficiency [73]. Hence, the use of promoters has gained interest in hydrate-based desalination as they favor moderate equilibrium pressures for hydrate formation. Tetrahydrofuran (THF) and tetrabutylammonium bromide (TBAB) are among the most widely discussed types of additives [74, 75]. Despite their advantages in increasing the removal efficiency of salt, the solubility of THF and TBAB is a major concern, making CP hydrates an attractive candidate for desalination [67, 68, 73, 76–79].

Cha and Seol [81] examined the upper-temperature boundary of carbon dioxide (CO_2) hydrate formation in the existence of CP within a simulation. The produced water showed 8.95 wt% salinity at isobaric conditions, with a salt removal efficiency 20% higher than usual with the introduction of CP [80]. The phase equilibrium of CP/CO_2 and THF/CO_2 binary hydrates in various concentrations of NaCl solutions was studied based on an isochoric method [81]. CP and THF reduced the equilibrium pressure and enhanced the removal efficiency. Additionally, the stability of CP/CO_2 hydrates is an improvement compared to THF-CO_2 hydrates at lower NaCl mass fractions. The thermodynamic effects of CP, along with its derivatives including cyclopentanone (CP-one) and cyclopentanol (CP-ol), with CO_2 hydrates for seawater desalination application, were analyzed [82], and the thermodynamic stabilities of these binary hydrates were found to be as follows: CP/CO_2 > CP-one/CO_2 > CP-ol/CO_2. However, there is still a need for more phase equilibrium data of CP/CO_2 hydrates in different concentrations of salt solutions to uncover more regarding CP hydrate-based desalination.

Pressure-temperature illustrations from Zhang et al. [83] showed that only sII CP/CO_2 hydrate should be formed in pure water. However, the presence of KCl, and to a lesser extent, NaCl, might lead to the formation of both sI simple CO_2 hydrate and sII binary CP/CO_2 hydrate structures. In this case, the hydrate dissociation could start with the declathration of simple CO_2 hydrate cavities and then present a decomposition of binary CP/CO_2 hydrate. The equilibrium temperature of CP/CO_2 hydrates in the presence of 7 wt% KCl was slightly less than the equilibrium temperature of 7 wt% NaCl. This signifies that the inhibition effect of NaCl is higher than that of KCl and is in agreement with the experimental results [69]. Coulombic interactions in the presence of NaCl are stronger than KCl, which tends to lower temperature equilibrium.

Babakhani et al. modeled a thermodynamic equilibrium using the Kihara approach [83], where the occupancy factor was achieved by integrating the Kihara potential variables. Maximum attractive potential (ε), core and wall cavity distance at zero potential (σ), and hard-core radius (a) CO_2 Kihara parameters were taken from [84]. The values were optimized by reducing the deviation among the predicted and experimental data of CP/CO_2 binary hydrates. It was observed that the deviation is huge when (σ) is < 2.55 or > 2.74 irrespective of the value of ε/k_b (k_b being the Boltzmann constant), with a good capability of reproducing the dissociation temperatures of mixed CP/CO_2 hydrate in pure water and the presence of salts.

5.4 Kinetic Models for Hydrate Formation

Comprehensive studies on the growth models for hydrate formation, listed in Table 5.2, are numerous in the literature. The majority of this section consists of kinetic expressions that encompass a variety of mixtures concerning the development of hydrates. Remarkably, among the various models considered, none of them focus on the physical behavior of hydrate formation or, more specifically, water recovery in hydrate-based desalination. Thus, there is a clear necessity for a universal kinetic growth model. Though many models have progressed enough to be utilized industrially, there is still a lack of accuracy when scaling up the equipment. Hence, the development of a new universal model could also close the gap between the application of the evolved kinetic expressions and the industrial scaling up of the processes.

5.4.1 Mathematical Model for Seawater Desalination

Shi et al. [88] studied the effect of various reaction sections on the throughput of the hydrate-based desalination process for a Freon R-142-NaCl system to optimize the system per the mathematical expression model. An optimal

Table 5.2 Comprehensive growth models for hydrate growth.

Researcher	Features of model	Model for Hydrate growth
[39]	Model based on kinetics and heat transfer (HT)	$\lambda_H \rho_H \frac{dX}{dt} \left\{ \frac{1}{\frac{1}{[k_0 \exp\left(\frac{E_a}{RT_{eq})}\right]} + \frac{1}{h}} \right\} (T_{eq} - T_{bulk})$
[23]	Model based on kinetics and mass transfer (MT)	$t = \frac{\rho_h r_d^2 \theta_{g-h}}{6D_{CH_4,h} C_{CH_4,o}} [1 - 3(1 - \frac{n_{CH_4,h}}{\frac{4\prod r_d^3 \rho_h \theta_{g-h}}{3}})^{\frac{2}{3}} + 2\left(1 - \frac{n_{CH\$,h}}{\frac{4\prod r_d^3 \rho_h \theta_{g-h}}{3}}\right)]$

(Continued)

Table 5.2 (Continued)

Researcher	Features of model	Model for Hydrate growth
[85]	Integrated kinetics with MT and HT	$D_{eff}\frac{\partial y}{\partial x}=\frac{\psi K_f H}{C_{w0}}\left(C^*-C_{eq}\right)$ $k_w\frac{\partial T}{\partial x}=\frac{\psi\lambda K_f H}{C_{wo}}\left(C^*-C_{eq}\right)$
[86]	Intrinsic growth kinetic model, MT and HT	$\frac{dR_c}{dt}=\frac{\frac{M_w\lambda D_{eff}\left(C_s-C_{eq}\right)}{\rho_w}}{\left(\frac{1}{R_c}-\frac{1}{R_s}\right)R_c^2+\frac{D_{eff}C_{w0}}{K_f H}}$ $\frac{dR_s}{dt}=\frac{\frac{M_w\lambda D_{eff}\left(C_s-C_{eq}\right)}{\rho_w}}{\left(\frac{1}{R_c}-\frac{1}{R_s}\right)R_s^2+\frac{D_{eff}C_{w0}}{K_f H}}$

[87]	Model based on thermodynamics, MT, and HT	$\frac{dr_{in}}{dt} = \frac{\beta M_w}{\rho_w} \sum_{N}^{i=1} \left[\frac{\Omega_i C_{\frac{H}{0},i} - \Omega_{eq,i} C_{eq,i}}{\frac{1}{K_i^*} + \Omega_i \left(r_{in}^{\Delta t-1} \right)^2 \left(\frac{1}{r_{in}^{\Delta t-1}} - \frac{1}{r_{out}^{\Delta t-1}} \right) / D_{f,i}} \right]$ $V^{w,\frac{H}{0}} = \frac{\prod \varepsilon_H \sigma \Delta t}{4\mu_{water} \left(r_{out}^{\Delta t-1} - r_{in}^{\Delta t^*} \right)}$ $\frac{\lambda_w}{\lambda_w + \lambda_L} \frac{Q_{heat}^{\Delta t}}{4\prod \Delta t (r_{in}^{\Delta t})} = \lambda_w \frac{dT}{dr}\Big\rvert_{\mathrm{r=rin}} \Delta \mathrm{t}$ $\frac{\lambda_L}{\lambda_w + \lambda_L} \frac{Q_{heat}^{\Delta t}}{4\prod \Delta t (r_{out}^{\Delta t})} = -\lambda_L \frac{dT}{dr}\Big\rvert_{\mathrm{r=r\,out}} \Delta \mathrm{t}$

configuration of the model could serve as a foundation for designing a hydrate-based desalination process. Gas–liquid operations were performed in column reactors where higher rates of the interphase MT are achieved among the gas–liquid regions with perfect mixing with the help of sieve plates. The MT coefficient can be calculated using Eqs. (5.15 and 5.16), while the total volume of the liquid region in the reactor is given by Eq. (5.17), and the hydrate former concentration outlet in the gas phase is described by Eq. (5.18).

$$\beta Sc^{\frac{2}{3}} = 0.42\left[\frac{(\rho_1 - \rho_g)\mu_1 g}{\rho_1}\right]^{\frac{1}{3}} \tag{5.15}$$

$$\mathrm{Sc} = \frac{\mu_1}{\rho_1 D^*} \tag{5.16}$$

$$D^* = 7.4 * 10^{-8} \frac{(FM_1)^{\frac{1}{2}} T}{\mu_1 V_A^{0.6}}, \tag{5.17}$$

$$(C_F)_i = \{A\}_i - [\{A\}_i - (C_F)_{i-1}] e^{\frac{\beta\sigma}{\varepsilon U_g}\Delta h_i}, \tag{5.18}$$

The average hydrate former concentration is calculated according to Eq. (5.19).

$$(C_F)_i^{av} = \frac{1}{h_i - h_{i-1}} \int_{h_{i-1}}^{h_i} (C_F)_i (h) dh = \{A\}_i [\{A\}_i - (C_F)_{i-1}] \frac{\varepsilon U_g}{\beta\sigma\Delta h_i}\left(1 - e^{\frac{-\beta\sigma}{\varepsilon U_r}\Delta h_i}\right) \tag{5.19}$$

By relating this mathematical model, Shi et al. found that the hydrate concentration relies on the number of sections at the rate of feed of guest molecule [88]. Stabilization of the hydrate concentration represents the optimum number of sections for the reactor.

5.4.2 Lattice Boltzmann Model for Hydrate Formation

The lattice Boltzmann model is efficient in handling interfacial dynamics and geometric boundaries [89, 90]. In recent years, it has also been implemented to evaluate the formation of hydrates in an electrolyte solution [91]. The model evaluates the kinetics of mass transfer and hydrate formation by considering the effects of salts while also analyzing the mechanism of hydrate formation and desalination efficiency.

According to Wang et al. [92], the formation of hydrates in clean water can be thought of as a reaction of a single solute, followed by a transport problem in two steps: (i) conveyance of the hydrate former to the hydrate surface, and (ii) reaction at the surface of the hydrate. Assuming that the diffusion of the hydrate former into hydrate is insignificant, the transportation of the hydrate former is analogous to a convection–diffusion model, as in Eq. (5.20).

$$\frac{\partial C_G}{\partial t} + (u.\nabla)C_G = \nabla.(D_G \nabla C_G) \tag{5.20}$$

The above equation can be used to describe the conveyance of both the hydrate former and salt ions in brine. However, since hydrate formation excludes salt ions, the concentration of rejected salt ions in the brine increases. Additionally, changes in the rejected salt concentration can influence the solubility and molecular diffusivity of the guest gas molecule [93]. Thus, the transport of guest and salt molecules by hydrate formation is better expressed by Eqs. (5.21 and 5.22).

$$\frac{\partial C_1}{\partial t} + (u.\nabla)C_1 = \nabla.(D_1 \nabla C_1) + S_{1,con} \tag{5.21}$$

$$\frac{\partial C_G}{\partial t} + (u.\nabla)C_G = \nabla.(D_G \nabla C_G) + S_{G,con} + S_{G,out} \tag{5.22}$$

Equations (5.23 and 5.24) can be used to calculate $S_{1,\,con}$ and $S_{G,\,out}$.

$$S_{1,con} = C_1 \frac{V_{cw}}{V - V_{cw}} \tag{5.23}$$

$$S_{G,out} = C_G \frac{V_{cw}}{V - V_{cw}} \tag{5.24}$$

The hydrate formation rate is calculated using Eq. (5.25) [93].

$$R(t) = k_r A(C_G - C_{Geq}) \tag{5.25}$$

If the effect of salt ions on water activity is considered, then Eq. (5.25) must be modified to Eq. (5.26), and the formation rate of hydrate is expressed by Eq. (5.27).

$$D_G \frac{\partial C_G}{\partial n} = k_r A a_w (C_G - C_{Geq}) \tag{5.26}$$

$$\left(\frac{\partial m_H}{\partial t}\right) = M_H k_r A a_w (C_G - C_{Geq}) \tag{5.27}$$

In Staykova et al. [33], application of the lattice Boltzmann model to numerically simulate crystal growth managed to evaluate water conversion and inspect the kinetics of the reaction and mass transfer during hydrate formation.

The efficiency in directing such complex geometries has also been noted as an advantage of the model by others [94].

Solute transfer during hydrate formation with no agitation is defined by the Bhatnagar–Gross–Krook (BGK) equation, presented in Eq. (5.28) [92].

$$g_{k,a}(x + ce_a \Delta t) - g_{k,a}(x,t) = -\frac{1}{\tau\kappa}\begin{pmatrix} g_{k,a}(x,t) - g_{k,\alpha}^{eq}(x,t) + \\ J_{k,a}\Delta t\left(S_{k,out} + S_{k,con}\right)\end{pmatrix} \tag{5.28}$$

The equilibrium distribution equation is used as in Eq. (5.29) [95].

$$g_{k,\alpha}^{eq} = C_k\left(J_{k,\alpha} + \frac{1}{2}e_\alpha u\right) \tag{5.29}$$

The J_0 value in the range of 0 to 1 is expressed as in Eq. (5.30).

$$J_{k,\alpha} = \begin{cases} J_{k,0} & \alpha = 0 \\ \dfrac{(1 - J_{k,0})}{4} & \alpha = 1,2,3,4 \end{cases} \tag{5.30}$$

The diffusivity in the lattice element is connected to the time of relaxation as in Eq. (5.31):

$$D_k = \frac{1}{2}\left(1 - J_{k,0}\right)\left(\tau_k - 0.5\right), \tag{5.31}$$

where τ_k is the relaxation time of the solute species.

The limiting conditions for the reaction at the hydrate surface are given in Eq. (5.32) as:

$$D_k \frac{C_{Fk} - C_{Rk}}{c} = k_r\left(C_{Fk} - C_{k,eq}\right), \tag{5.32}$$

where C_{Fk} is the concentration of the fluid node in the solute species (mole/m^3) and C_{Rk} is the concentration of the interface node in the solute species (mole/m^3).

For the CP–brine hydrate system, Wang et al. have established a lattice Boltzmann model to simulate the formation of hydrates complete with a quantitative representation of water conversion with kinetics, where the parameters of the numerical simulation are specified elsewhere [91]. The salting-out effect reduces the concentration of the hydrate former near the surface of the hydrate, thereby lowering the rate of hydrate formation. Yet, the rapid increase in the surface area of the hydrate enhances the formation rate compared to its initial state. The low concentration of hydrate former and increased concentration of salt govern the hydrate formation rate and progressively reduce the conversion of water to hydrate. Also, in the presence of electrolytes, the morphology of CP hydrate or hydrate formers notably depends on the hydrate formers' diffusion length. This model can have future applications for analyzing the formation mechanism of oceanic hydrates and predicting desalination efficiencies.

5.5 Machine Learning Models to Predict Desalination Efficiency

Given the complexity of water treatment modeling within the numerous physical constraints [97, 98], using data-driven methods to mimic the desalination process is inevitable [96, 98, 99]. The benefits of machine learning (ML) are in its competence to model complex methods without requiring thorough information [100].

5.5.1 Machine Learning Techniques to Model Hydrate-based Desalination

Examples of ML implementations include an enhanced neural network technique by Wei et al. [98] to estimate the levels of total suspended solids that can deteriorate water quality, and usage of adaptive neuro-fuzzy inference system (ANFIS) and comprehensive linear techniques to create a black-box model of a desalination process by Araromi [101]. So far, there has been no description or work performed in using ML for hydrate-based desalination.

Sadi and Shahrabadi studied the application of ANFIS and support vector machine (SVM) for the identification of a desalination process [99], with optimal parameters determined by a genetic algorithm (GA) coupled with SVM. The experimental salt removal efficiencies for desalination using compressed natural gas and CO_2 as formers were applied for validation.

Data normalization is first implemented to increase the speed and accuracy of data convergence. Commonly used functions and techniques are: (i) sigmoid, (ii) unit norm, (iii) min-max scaling, (iv) soft-max, and (v) decimal scaling. Sadi et al. [101] utilized the min-max normalization, as shown in Eq. (5.33):

$$X_{\text{norm}} = 2 * \frac{X - X_{min}}{X_{max} - X_{min}} - 1, \tag{5.33}$$

where X denotes actual data prior to normalization, X_{max} and X_{min} represent the largest and lowest values, and X_{norm} signifies the normalized data instance.

5.5.2 Adaptive Neuro-fuzzy Inference System

The ANFIS framework is established from the Takagi–Sugeno fuzzy system, whereby it is combined with artificial neural networks and fuzzy logic principles to design a nonlinear system. Initially, model estimation is done by quantifying the membership functions. For training the ANFIS, a set of data is then utilized to assess the overview of the anticipated framework. The if/then rules for a first-order Takagi–Sugen with one output (z) and two input parameters (x_1, x_2) are defined as specified in Eqs. (5.34 and 5.35):

$$\text{Rule1: If } x_1 \text{ is } \mathrm{A}_1 \text{ and } x_2 \text{ is } B_1 \text{ then } z_1 = p_1x_1 + q_1\,x_2 + r_1, \tag{5.34}$$

$$\text{Rule2: If } x_1 \text{ is } \mathrm{A}_2 \text{ and } x_2 \; is \; B_2 \text{ then } z_2 = p_2x_1 + q_2\,x_2 + r_2, \tag{5.35}$$

where A_1 and B_1 are membership functions, and p_i, q_i, and r_i represent subsequent variables.

As shown in Figure 5.1, the ANFIS technique consists of five layers and five functions that yield a target value as an amalgamation of input variables. The node functions are estimated during the network training stage.

Within Figure 5.1, A and B represent linguistic values of fuzzy sets, x is input, y is output, w is firing strength, $\bar{w}$ is normalized firing strength, z is fuzzy model output, and N reflects the number of observations in the training set data.

5.5.2.1 Layer 1: Input Membership Function Layer

This first layer of ANFIS is designed to be an adaptive layer that defines the membership degree of every input provided to the network based on the type of membership function, e.g. trapezoidal, generalized bell function,

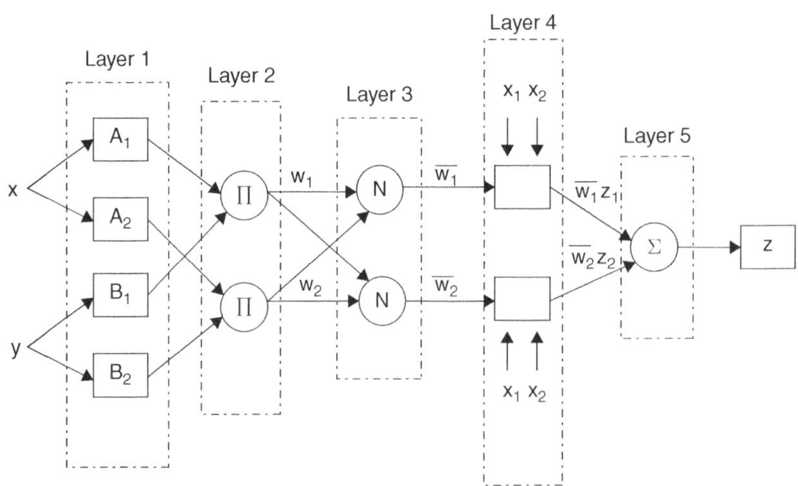

Figure 5.1 ANFIS architecture.

Gaussian or triangular membership functions. For the input variable of x, the output generated by this layer for each node can be expressed by Eq. (5.36):

$$O_{1,i} = \mu_{A_i}(x), \tag{5.36}$$

where $O_{1,i}$ represents the output of the ith node in the form of a membership degree between 0 and 1, and μ_{A_i} denotes the membership function.

For example, the Gaussian membership function is described in Eq. (5.37):

$$gaussian(x;c,\sigma) = e^{-\frac{1}{2}\left(\frac{x-c}{\sigma}\right)^2}, \tag{5.37}$$

where c represents *center* and σ denotes the *width*. These are known as premise parameters.

5.5.2.2 Layer 2: Product Layer

The nodes in this layer are called fixed nodes, which perform product Π operations to calculate the firing strength of a fuzzy rule w_i. In this layer, the node output $O_{2,i}$ represents combinations of membership degrees for input variables using the product operator as follows in Eq. (5.38), where x and y are the two inputs that have been given to the network, and μ_{A_i} and μ_{B_i} represents membership functions, i.e. trapezoidal, generalized bell function, Gaussian, or triangular membership functions.

$$O_{2,i} = w_i = \mu_{A_i}(x).\ \mu_{B_i}(y),\ \ i = 1,2. \tag{5.38}$$

5.5.2.3 Layer 3: Normalization Layer

All nodes in this layer are also fixed nodes that normalize the N firing strength of an incoming rule from a former layer against all rules existing in the knowledge base of this layer as in Eq. (5.39), where normalized firing strength of the ith rule is denoted as $\overline{w}_i$.

$$O_{3,i} = \overline{w}_i = \frac{w_i}{w_1 + w_2},\ \ i = 1,2. \tag{5.39}$$

5.5.2.4 Layer 4: Output Membership Function Layer

This fourth layer holds the consequent part (then) of the fuzzy rule with adaptive nodes and node function. The linear polynomial expression is used to determine the output $O_{4,i}$ of each rule in this layer as in Eq. (5.40), where normalized firing strength of each ith is denoted as $\overline{w}$, and p_i, q_i, r_i are the consequent parameters of that rule.

$$O_{4,i} = \bar{w}_i z_i \frac{w_i}{w_1 + w_2}(p_i x_1 + q_i x_2 + r_i), \quad i = 1,2. \tag{5.40}$$

5.5.2.5 Layer 5: Overall Output Layer

This layer generates the output of the entire ANFIS network in a single node that performs a summation of the rules' output in the previous layer, through Eq. (5.41).

$$O_{5,i} = \sum\nolimits_{i=1}^{2} \bar{w}_i f_i = \frac{\sum\nolimits_{i=1}^{2} \bar{w}_i f_i}{w_1 + w_2} \tag{5.41}$$

There are a variety of different training algorithms available for optimizing the parameters of the network to suit the state space that is analyzed. For the learning process, ANFIS typically uses a two-pass (forward and backward) learning algorithm to achieve minimum error value with the help of gradient descent (GD) and Least Square Estimator (LSE) methods. During the forward pass, the consequent parameters (p_i,q_i,r_i) of the network are optimized using LSE, while in the backward pass, membership function parameters are optimized and tuned using the GD approach [102].

5.5.3 SVM Approach

Support vector machines are functionally based on statistical theories and principles [104] but outfitted for regression analysis. Fundamentally, the approach of SVM is to fit a suitably approximated function to map the data points from the input parameter to a higher measurement. For a data set D = {(x_1, y_1), (x_2, y_2),...., (x_{Nt},y_{Nt})}, where $x_i \in R^m$ and $y_i \in R$, i = {1, 2, ..., N_t}, and m signifies the data observation points and the input variables. In such a data set, the SVM is described by Eq. 5.42:

$$\hat{y} = f(x) = \langle W, \psi(x) \rangle + B_s, \tag{5.42}$$

where, $\langle .,. \rangle$ W implies dot product, while B_s represents the variable weight factors, and $\varphi(x)$ signifies the mapping relation that shifts input data (x) from R^m into advanced regions to evaluate the target variable ($\hat{y}$) near the actual parameter (y). The optimum values of W and B_s are obtained by minimizing the cost function, as generalized in Eq. 5.43:

$$\text{Cost function} = \frac{1}{2}\|W\| + \frac{C}{N_t}\sum\nolimits_{i=1}^{Nt} L(f(x_i), y_i), \tag{5.43}$$

where L denotes the loss function predicted error, C designates the decision after coming across training errors.

Taking into account statistical risk, the cost function minimization serves to train the data through a convex optimization approach [103]. After applying Lagrange multipliers to the above issue, the ultimate regression function is given as Eq. 5.44:

$$f(x) = \sum_{i=1}^{N_t} \alpha_i - \alpha_i^* k(x_i, x) + B_s, \tag{5.44}$$

where α_i and α_i^* are Lagrange multipliers, and $k(x_i, x)$ signifies the kernel function. The kernel functions involved are typically (i) polynomial, (ii) the Gaussian radial basis function, and (iii) the sigmoid. These functions are represented by Eqs. (5.45, 5.46, and 5.47).

$$\text{Polynomial function}: k(x_i, x) = (\langle x_i, x\rangle + 1)^{d_p} \tag{5.45}$$

$$\text{Gaussian/RB function}: k(x_i, x) = \exp\left(-\frac{(\|x_i - x\|)^2}{2\Upsilon^2}\right) \tag{5.46}$$

$$\text{Sigmoid function}: k(x_i, x) = \tanh(\beta x_i, x\rangle + \nu) \tag{5.47}$$

Kernel method parameters are represented as d_p, Υ, β, and ν, which dictate how the learning process occurs (by affecting C and ε) through individually unique procedures, such as annealing, GA, harmony search, etc.

5.5.4 Genetic Algorithm

Developed by Goldberg through inspiration from Darwin's theory of evolution by natural selection [104], GA is a heuristic technique capable of delivering high-quality solutions. It is a stepwise procedure initiated by a randomized creation of nominated solutions in the variable section. After the conception of primary population data, the individual variables are estimated, and subsequent members are nominated using a fitness-based method. The nominated members are identified as parents and are brought together in a hybridized step to generate child members. In order to preserve GA diversity, child members are randomly selected to be slightly altered by a mutation process [104]. A new population is generated at each stage until the value converges.

In a study by Sadi et al. [100], ANFIS and SVM were used to model the hydrate-based desalination process and estimate its efficiency. The initial salinity and hydrate-gas equilibrium pressure input parameters were designated to determine the target value of desalination efficiency (η_{DeSal}), which is defined in Eq. 5.48 as:

$$\eta_{\text{DeSal}}(\%) = \frac{EC_i - EC_f}{EC_i} * 100, \tag{5.48}$$

where EC_i and EC_f are the electrical conductivity of the initial and final brine solution, respectively.
The consistency of the proposed models was assessed using graphical representations and quantitative analysis. Scatter plots and relative error figures were quantitatively determined with the coefficient of determination (R^2) and average absolute relative error. This analysis found that the triangular membership functions with five levels for initial salinity of produced water and three levels for gas hydrate equilibrium pressure provided the best ANFIS framework. The performance of SVM relies heavily on the variable specifications and type of kernel function. Sadi et al. implemented the Gaussian function [105], and the optimum values were figured using GA.

With optimized statistical parameters in place, the developed ANFIS and SVM models demonstrated high consistency and accuracy in determining the efficiency of desalination. However, SVM exhibited better performance than the ANFIS. The model prediction performance was further confirmed by values taken from experimental data [106]. Hence, a leverage mechanism was successfully implemented by [100] to detect any outliers, and it was determined that both ANFIS and SVM models are statistically adequate for estimating the desalination efficiency.

5.5.5 Conclusion

This section investigated the characteristics of ML models in hydrate-based desalination. The use of current and future experimentation results is invaluable for validating the models developed using these tools. Hybridization of ML techniques is suggested for finding the optimum process parameters, such as pollutant removal efficiency and higher water recovery.

While a promising technique, the capability of ML models should still be put to further scrutiny to guarantee their ability to meet the high-quality standards, particularly regarding discharge conditions. It is suggested that future investigations deliver a greater amount of experimental or online data to support ML techniques, to make the models more user-friendly, and enable the development of faster and more precise models for real-world applications of desalination.

Nomenclature

μ_w^H – chemical potential in the hydrate phase

μ_w^β – empty lattice hydrate phase

υ_m – Number of type m cage

C – Langmuir constant

Δh_w – change in enthalpy of hydrate water

Δv_w – difference in volume of vacant framework

I – ion concentration

T_f – Dissociation temperature in °K

ΔH_{fm} – Dissociation molar enthalpy in J/mol,

ΔC_{fm} – molar specific heat among the sub-cooled liquid and the hydrates in J/mol/K

a_w – water activity

T_0 – equilibrium temperature of hydrate in fresh water

T – equilibrium temperature of hydrate in electrolyte solution

ΔH_{diss} – Dissociation enthalpy of hydrate

n – hydration number

C_1, C_2, C_3 – coefficients

X – effective mole fraction

$\Delta\mu_w^{\beta-L}$ – chemical activity change among the reference and liquid state

$\Delta\mu_w^{\beta-H}$ – change among reference and hydrate phase

R – gas constant,

T – absolute temperature

θ_j^i – occupancy factor

D^* – Diffusion coefficient m^2/s

S_c – Schmidt number

C – concentration, mol/m^3

$\left(C_F\right)_i^{av}$ – average hydrate former concentration in the gas phase at the outlet of the I th cell mol/m^3

D – column diameter, m

D^* – diffusion coefficient, m^2/s

d_b – diameter of the air bubble, m

F_r – total cross-sectional area of the reactor, m^2

g – acceleration, m/s^2

h_i – height of the liquid phase in the ith cell, m

M_l – average molecular mass of liquid, kg/kmol

N – number of cells

P – pressure, atm

Q – throughput of the reactor with respect to gas hydrate, $kg/(m^3\ h)$

T – reaction temperature, K

t – temperature, °C

U_g – is linear flow rate of the gas phase, m/s

V_A – molar volume of the dissolved substance,m^3/mol

V_1^0 – total liquid phase volume in reactor, m^3

W_g – volumetric gas flow rate, m^3/s

W_l – volumetric flow rate of liquid, m^3/s

w – flow rate, m/s

β is mass transfer coefficient, m/s

ε – fraction of the cross section of the reactor, occupied by the gas phase

μ_l is dynamic viscosity coefficient, kg/(m s)

ν is kinematic viscosity coefficient, m^2/s

ρ_l, ρ_g – are densities of the liquid and gas, kg/m^3

σ – specific surface area of the interface, m^2/m^3

σ'– surface tension, kg/s^2

τ – time, min

C_G – concentration of hydrate former in $mole/m^3$

u – velocity in m/s

V_{cw} is the volume of water consumed in m^3

$\boldsymbol{V}$ is the volume of water in m^3

$\boldsymbol{S_{1,con}}$ – concentration difference of salt

$S_{G,\,out}$ – concentration difference of guest molecule

$R(t)$ – global reaction rate mol/s

k_r – local reaction rate constant m/s

m_H – mass of hydrate in kg,

M_H – molar mass of hydrate in g/mol

$g_{k,\alpha}$ – kth solute concentration distribution

$g_{k,\alpha}^{eq}$ – equilibrium distribution

e_α – discrete velocity

$J_{k,\alpha}$ – rest fraction

i – hydrate former

m – cage occupancy

α – liquid water/ice

f – fugacity

0 – reference condition

e_1 – e_4 – universal constant

z – ionic charge

x – mole fraction

β – reference state

References

1 Byk, S.S., Makogon, Y.F., and Fomina, V.I. (1980). *Gas hydrates* (*Gazovye gidraty*).

2 Yusifov, R.Y. (2008). Treatment of kinetic model of hydrate formation processes in the R-142B–Water– NaCl System (1.5 wt.%). Tez. dokl. nauchnoi konferen- tsii, posvyashchennoi 100-letnemu yubileyu

akademika Nagieva (Proc. Conf. Dedicated to Acad. Nagiev's 100th Birth Anniversary), Baku, 89.

3 Mohamadi-Baghmolaei, M., Mahmoudy, M., Jafari, D., Mohamadi-Baghmolaei, R., and Tabkhi, F. (2014). Assessing and optimization of pipeline system performance using intelligent systems. *Journal of Natural Gas Science and Engineering* 18, 64–76.

4 Mohamadi-Baghmolaei, M., Azin, R., Osfouri, S., Mohamadi-Baghmolaei, R., and Zarei, Z. (2015). Prediction of gas compressibility factor using intelligent models. *Natural Gas Industry B,* 2(4), 283–294.

5 Wilcox, W.I., Carson, D., and Katz, D. (1941). Natural gas hydrates. *Industrial and Engineering Chemistry*, 33(5), 662–665.

6 van der Waals, J. and Platteeuw, J. (2007). Clathrate solutions. *Advances in Chemical Physics*, 2, 1–57.

7 Garapati, N. and Anderson, B.J. (2014). Statistical thermodynamics model and empirical correlations for predicting mixed hydrate phase equilibria. *Fluid Phase Equilibria*, 373, 20–28.

8 Mohamadi-Baghmolaei, M., Azin, R., Sakhaei, Z., Mohamadi-Baghmolaei, R., and Osfouri, S. (2016). Novel method for estimation of gas/oil relative permeabilities. *Journal of Molecular Liquids*, 223, 1185–1191.

9 Sloan, E.D. and Koh, C.A. (2008). *Clathrate Hydrates of Natural Gases*. Boca Raton: CRC Press.

10 Rice, R.G. and Do, D.D. (2012). *Applied Mathematics and Modelling for Chemical Engineers*. John Wiley & Sons.

11 Anderson, J.D. and Wendt, J. (1995). *Computational Fluid Dynamics* (Vol. 206, p. 332). New York: McGraw-Hill.

12 Bishnoi, P.R. and Natarajan, V. (1996). Formation and decomposition of gas hydrates. *Fluid Phase Equilibria*, 117(1–2): 168–177.

13 Sun, C., Chen, G., Guo, T., Lin, W., and Chen, J. (2002). Kinetics of methane hydrate decomposition. *Journal of Chemical Industry and Engineering (China)*, 53: 899–903.

14 Vysniauskas, A. and Bishnoi, P.R. (1985). Kinetics of ethane hydrate formation. *Chemical Engineering Science*, 40: 299–303.

15 Lekvam, K. and Ruoff, P. (1993). A reaction kinetic mechanism for methane hydrate formation in liquid water. *Journal of the American Chemical Society*, 115: 8565–8569.

16 Boxall, J., Davies, S., Koh, C., and Sloan, E.D. (2009). Predicting when and where hydrate plugs form in oil-dominated flowlines. *SPE Projects, Facilities & Construction*, 4: 80–86.

17 Zerpa, L.E., Sloan, E.D., Sum, A.K., and Koh, C.A. (2012). Overview of CSMHyK: A transient hydrate formation model. *Journal of Petroleum Science & Engineering*, 98: 122–129.

18 Yang, D., Le, L.A., Martinez, R.J., Currier, R.P., and Spencer, D.F. (2011). Kinetics of CO2 hydrate formation in a continuous flow reactor. *Chemical Engineering Journal*, 172: 144–157.

19 Englezos, P., Kalogerakis, N., Dholabhai, P.D., and Bishnoi, P.R. (1987a). Kinetics of gas hydrate formation from mixtures of methane and ethane. *Chemical Engineering Science*, 42: 2659–2666.

20 Englezos, P., Kalogerakis, N., Dholabhai, P.D., and Bishnoi, P.R. (1987b). Kinetics of formation of methane and ethane gas hydrates. *Chemical Engineering Science*, 42: 2647–2658.

21 Skovborg, P. and Rasmussen, P. (1994). A mass transport limited model for the growth of methane and ethane gas hydrates. *Chemical Engineering Science*, 49: 1131–1143.

22 Herri, J.M., Pic, J.S., Gruy, F., and Cournil, M. (1999). Methane hydrate crystallization mechanism from in-situ particle sizing. *AIChE Journal*, 45: 590–602.

23 Clarke, M.A. and Bishnoi, P.R. (2005). Determination of the intrinsic kinetics of CO2 gas hydrate formation using in situ particle size analysis. *Chemical Engineering Science*, 60(3): 695–709.

24 Turner, D.J., Miller, K.T., and Sloan, E.D. (2009). Methane hydrate formation and an inward growing shell model in water-in-oil dispersions. *Chemical Engineering Science*, 64: 3996–4004.

25 Uchida, T., Ebinuma, T., Kawabata, J.I., and Narita, H. (1999). Microscopic observations of formation processes of clathrate-hydrate films at an interface between water and carbon dioxide. *Journal of Crystal Growth*, 204: 348–356.

26 Mori, Y.H. (2001). Estimating the thickness of hydrate films from their lateral growth rates: Application of a simplified heat transfer model. *Journal of Crystal Growth*, 223 (1–2): 206–212.

27 Peng, B.Z., Dandekar, A., Sun, C.Y., Luo, H., Ma, Q.L., Pang, W.X., and Chen, G.J. (2007). Hydrate film growth on the surface of a gas bubble suspended in water. *The Journal of Physical Chemistry B*, 111: 12485–12493.

28 Mochizuki, T. and Mori, Y.H. (2006). Clathrate-hydrate film growth along water/hydrate-former phase boundaries—numerical heat-transfer study. *Journal of Crystal Growth*, 290: 642–652.

29 Hashemi, S., Macchi, A., and Servio, P. (2007). Gas hydrate growth model in a semibatch stirred tank reactor. *Industrial & Engineering Chemistry Research*, 46: 5907–5912.

30 Bergeron, S. and Servio, P. (2008). Reaction rate constant of propane hydrate formation. *Fluid Phase Equilibria*, 265: 30–36.

31 Salamatin, A.N., Hondoh, T., Uchida, T., and Lipenkov, V.Y. (1998). Post-nucleation conversion of an air bubble to clathrate air–hydrate crystal in ice. *Journal of Crystal Growth*, 193: 197–218.

32 Wang, X., Schultz, A.J., and Halpern, Y. (2002). Kinetics of methane hydrate formation from polycrystalline deuterated ice. *The Journal of Physical Chemistry A*, 106: 7304–7309.

33 Staykova, D.K., Kuhs, W.F., Salamatin, A.N., and Hansen, T. (2003). Formation of porous gas hydrates from ice powders: Diffraction experiments and multistage model. *The Journal of Physical Chemistry B*, 107: 10299–10311.

34 Shindo, Y., Lund, P.C., Fujioka, Y., and Komiyama, H. (1993a). Kinetics of formation of CO2 hydrate. *Energy Conversion and Management*, 34: 1073–1079.

35 Shindo, Y. et al. (1993b). Kinetics and mechanism of the formation of CO2 hydrate. *International Journal of Chemical Kinetics*, 25 (9): 777–782.

36 Shindo, Y., Sakaki, K., Fujioka, Y., and Komiyama, H. (1996). Kinetics of the formation of CO2 hydrate on the surface of liquid CO2 droplet in water. *Energy Conversion and Management*, 37: 485–489.

37 Lund, P.C., Shindo, Y., Fujioka, Y., and Komiyama, H. (1994). Study of the pseudo-steady-state kinetics of CO2 hydrate formation and stability. *International Journal of Chemical Kinetics*, 26: 289–297.

38 Dalmazzone, D., Hamed, N., and Dalmazzone, C. (2009). DSC measurements and modelling of the kinetics of methane hydrate formation in water-in-oil emulsion. *Chemical Engineering Science*, 64(9): 2020–2026.

39 Teng, H., Yamasaki, A., and Shindo, Y. (1996). Stability of the hydrate layer formed on the surface of a CO2 droplet in high-pressure, low-temperature water. *Chemical Engineering Science*, 51: 4979–4986.

40 Freer, E.M., Selim, M.S., and Sloan, E.D., Jr. (2001). Methane hydrate film growth kinetics. *Fluid Phase Equilibria*, 185: 65–75.

41 Mu, L., Li, S., Ma, Q.L., Zhang, K., Sun, C.Y., Chen, G.J., and Yang, L.Y. (2014). Experimental and modeling investigation of kinetics of methane gas hydrate formation in water-in-oil emulsion. *Fluid Phase Equilibria*, 362: 28–34.

42 Rempel, A.W. and Buffett, B.A. (1997). Formation and accumulation of gas hydrate in porous media. *Journal of Geophysical Research: Solid Earth*, 102: 10151–10164.

43 Yin, Z., Chong, Z.R., Tan, H.K., and Linga, P. (2016). Review of gas hydrate dissociation kinetic models for energy recovery. *Journal of Natural Gas Science and Engineering*, 35: 1362–1387.

44 Liu, X. and Flemings, P.B. (2007). Dynamic multiphase flow model of hydrate formation in marine sediments. *Journal of Geophysical Research: Solid Earth* 112(B3).

45 Uddin, M., Coombe, D., Law, D., and Gunter, B. (2008). Numerical studies of gas hydrate formation and decomposition in a geological reservoir. *Journal of Energy Resources Technology*, 130(3).

46 Zerpa, L.E., Rao, I., Aman, Z.M., Danielson, T.J., Koh, C.A., Sloan, E.D., and Sum, A.K. (2013). Multiphase flow modeling of gas hydrates with a simple hydrodynamic slug flow model. *Chemical Engineering Science*, 99: 298–304.

47 Ribeiro, C.P., Jr and Lage, P.L. (2008). Modelling of hydrate formation kinetics: State-of-the-art and future directions. *Chemical Engineering Science*, 63: 2007–2034.

48 Yin, Z., Khurana, M., Tan, H.K., and Linga, P. (2018). A review of gas hydrate growth kinetic models. *Chemical Engineering Journal*, 342: 9–29.

49 Barrer, R.M. and Stuart, W.I. (1957). Non-stoichiometric clathrate compounds of water. *Proceedings of the Royal Society of London. Series A, Mathematical and Physical Sciences*, 243(1233): 172–189.

50 Holder, G.D., Corbin, G., and Papadopoulos, K.D. (1980). Thermodynamic and molecular properties of gas hydrates from mixtures containing methane, argon and krypton. *Industrial & Engineering Chemistry Fundamentals*, 19: 282–286.

51 John, V.T., Papadopoulos, K.D., and Holder, G.D. (1985). A Generalized model for predicting equilibrium conditions for gas hydrates. *AIChE Journal*, 31: 252–259.

52 Peng, D.Y. and Robinson, D.B. (1976). A new two-constant equation of state. *Industrial and Engineering Chemistry Fundamentals*, 15: 59–64.

53 Soave, G. (1972). Equilibrium constants from a modified Redlich-Kwong equation of state. *Chemical Engineering Science*, 27: 1197–1203.

54 Valderrama, J.O. (1990). A generalized Patel-Teja equation of state for polar and nonpolar fluids and their mixtures. *Journal of Chemical Engineering of Japan*, 23: 87–91.

55 Nasrifar, K. and Bolland, O. (2006). Simplified hard-sphere and hard-sphere chain equations of state for engineering applications. *Chemical Engineering Communications*, 193: 1277–1293.

56 Kontogeorgis, G.M., Voutsas, E.C., Yakoumis, I.V., and Tassios, D.P. (1996). An equation of state for associating fluids. *Industrial & Engineering Chemistry Research*, 35: 4310–4318.

57 Chapman, W.G., Gubbins, K.E., Jackson, G., and Radosz, M. (1989). SAFT: Equation-of-state solution model for associating fluids. *Fluid Phase Equilibria*, 52: 31–38.

58 Michelsen, M.L. (1990). A modified Huron-Vidal mixing rule for cubic equations of state. *Fluid Phase Equilibria*, 60: 213–219.

59 Dahl, S. and Michelsen, M.L. (1990). High-pressure vapor-liquid equilibrium with a UNIFAC-based equation of state. *AIChE Journal*, 36: 1829–1836.

60 Parrish, W.R. and Prausnitz, J.M. (1972). Dissociation pressures of gas hydrates formed by gas mixtures. *Industrial & Engineering Chemistry Process Design and Development*, 11: 26–35.

61 Javanmardi, J., Moshfeghian, M., and Maddox, R.N. (1998). Simple method for predicting gas-hydrate-forming conditions in aqueous mixed-electrolyte solutions. *Energy & Fuels*, 12: 219–222.

62 Javanmardi, J., Moshfeghian, M., and Maddox, R.N. (1998). Simple method for predicting gas-hydrate-forming conditions in aqueous mixed-electrolyte solutions. *Energy & Fuels*, 12 (2): 219–222.

63 Nasrifar, K., Moshfeghian, M., and Maddox, R.N. (1998). Prediction of equilibrium conditions for gas hydrate formation in the mixtures of both electrolytes and alcohol. *Fluid Phase Equilibria*, 146: 1–13.

64 Babu, P., Nambiar, A., He, T., Karimi, I.A., Lee, J.D., Englezos, P., and Linga, P. (2018). A review of clathrate hydrate-based desalination to strengthen energy–water nexus. *ACS Sustainable Chemistry & Engineering*, 6(7): 8093–8107.

65 Nallakukkala, S. and Lal, B. (2021). Seawater and produced water treatment via gas hydrate: Review. *Journal of Environmental Chemical Engineering*, 9(2). doi: 10.1016/j.jece.2021.105053.

66 He, T., Nair, S.K., Babu, P., Linga, P., and Karimi, I.A. (2018). A novel conceptual design of hydrate based desalination (HyDesal) process by utilizing LNG cold energy. *Applied Energy*, 222: 13–24. doi: 10.1016/j.apenergy.2018.04.006.

67 Ho-Van, S., Bouillot, B., Douzet, J., Babakhani, S.M., and Herri, J.M. (2018). Experimental measurement and thermodynamic modeling of cyclopentane hydrates with NaCl, KCl, $CaCl_2$, or NaCl-KCl present. *AIChE Journal*, 64(6): 2207–2218.

68 Corak, D., Barth, T., Hoiland, S., Skodvin, T., Larsen, R., and Skjetne, T. (2011). Effect of subcooling and amount of hydrate former on formation of cyclopentane hydrates in brine. *Desalination*, 278 (1–3): 268–274. doi: 10.1016/j.desal.2011.05.035.

69 Han, S., Rhee, Y.W., and Kang, S.P. (2017). Investigation of salt removal using cyclopentane hydrate formation and washing treatment for seawater desalination. *Desalination*, 404: 132–137. doi: 10.1016/j.desal.2016.11.016.

70 Ho-Van, S., Bouillot, B., Douzet, J., Babakhani, S.M., and Herri, J.M. (2018). Implementing cyclopentane hydrates phase equilibrium data and simulations in brine solutions. *Industrial & Engineering Chemistry Research*, 57(43): 14774–14783.

71 Miyawaki, O., Saito, A., Matsuo, T., and Nakamura, K. (1997). Activity and activity coefficient of water in aqueous solutions and their relationships with solution structure parameters. *Bioscience, Biotechnology, and Biochemistry*, 61(3): 466–469. doi:10.1271/bbb.61.466.

72 Hu, Y., Lee, B.R., and Sum, A.K. (2017). Universal correlation for gas hydrates suppression temperature of inhibited systems: I. Single salts. *AIChE Journal*, 63(11): 5111–5124.

73 Hu, Y., Lee, B.R., and Sum, A.K. (2018). Universal correlation for gas hydrates suppression temperature of inhibited systems: II. Mixed salts and structure type. *AIChE Journal*, 64(6): 2240–2250. doi:10.1002/aic.16116.

74 He, T., Chong, Z.R., Babu, P., and Linga, P. (2020). Techno-economic evaluation of cyclopentane hydrate-based desalination with liquefied natural gas cold energy utilization. *Energy Technology*, 8(8): 1900212. doi:10.1002/ente.201900212.

75 Najibi, H., Momeni, K., and Sadeghi, M.T. (2015). Theoretical and experimental study of phase equilibrium of semi-clathrate hydrates of methane+ tetra-n-butyl-ammonium bromide aqueous solution. *Journal of Natural Gas Science and Engineering*, 27: 1771–1779. doi:10.1016/j.jngse.2015.11.002.

76 Sun, S., Peng, X., Zhang, Y., Zhao, J., and Kong, Y. (2017). Stochastic nature of nucleation and growth kinetics of THF hydrate. *The Journal of Chemical Thermodynamics*, 107: 141–152. doi:10.1016/j.jct.2016.12.026.

77 Cai, L. (2016). Desalination via formation of binary clathrate hydrates. PhD Dissertation, Princeton University.

78 Ho-Van, S., Bouillot, B., Douzet, J., Babakhani, S.M., and Herri, J.M. (2019). Cyclopentane hydrates – A candidate for desalination. *Journal of Environmental Chemical Engineering*, 7(5): 103359. doi:10.1016/j.jece.2019.103359.

79 Li, F., Chen, Z., Dong, H., Shi, C., Wang, B., Yang, L., and Ling, Z. (2018). Promotion effect of graphite on cyclopentane hydrate based desalination. *Desalination*, 445: 197–203. doi:10.1016/j.desal.2018.08.011.

80 Ling, Z., Shi, C., Li, F., Fu, Y., Zhao, J., Dong, H., ... Song, Y. (2020). Desalination and Li+enrichment via formation of cyclopentane hydrate. *Separation and Purification Technology*, 231: 115921. doi: 10.1016/j.seppur.2019.115921.

81 Cha, J.H. and Seol, Y. (2013). Increasing gas hydrate formation temperature for desalination of high salinity produced water with secondary guests. *ACS Sustainable Chemistry & Engineering*, 1(10): 1218–1224.

82 Zhang, Y., Sheng, S.M., Shen, X.D., Zhou, X.B., Wu, W.Z., Wu, X.P., and Liang, D.Q. (2017). Phase equilibrium of cyclopentane + carbon dioxide binary hydrates in aqueous sodium chloride solutions. *Journal of Chemical and Engineering Data*, 62(8): 2461–2465. doi:10.1021/acs.jced.7b00404.

83 Babakhani, S.M., Ho-Van, S., Bouillot, B., Douzet, J., and Herri, J.M. (2020). Phase equilibrium measurements and modelling of mixed cyclopentane and carbon dioxide hydrates in presence of salts. *Chemical Engineering Science*, 214: 115442.

84 Herri, J.-M., Bouchemoua, A., Kwaterski, M., Fezoua, A., Ouabbas, Y., and Cameirao, A. (2011). Gas hydrate equilibria for CO_2–N_2 and CO_2–CH_4 gas mixtures—Experimental studies and thermodynamic modelling. *Fluid Phase Equilibria*, 301(2): 171–190. doi:10.1016/j.fluid.2010.09.041.

85 Jamaluddin, A.K.M., Kalogerakis, N., and Bishnoi, P.R. (1991). Hydrate plugging problems in undersea natural gas pipelines under shutdown conditions. *Journal of Petroleum Science and Engineering*, 5(4): 323–335.

86 Zhao, J.K. (2005). Study on flow properties of hydrate slurry in multiphase pipeline. PhD dissertation. China University of Petroleum, Beijing.

87 Shi, B.H., Gong, J., Sun, C.Y., Zhao, J.K., Ding, Y., and Chen, G.J. (2011). An inward and outward natural gas hydrates growth shell model considering intrinsic kinetics, mass and heat transfer. *Chemical Engineering Journal*, 171(3): 1308–1316.

88 Aliev, A.M., Yusifov, R.Y., Tairov, A.Z., Sarydzhanov, A.A., Mirzoeva, R.Y., and Yusifov, Y.G. (2011). Mathematical modeling of seawater desalination by the gas hydrate method. *Theoretical Foundations of Chemical Engineering*, 45(2): 185–189.

89 Yang, Y., Liu, Z., Yao, J., Zhang, L., Ma, J., Hejazi, S.H., and Ngarta, T.D. (2018). Flow simulation of artificially induced microfractures using digital rock and lattice Boltzmann methods. *Energies*, 11(8): 2145.

90 Liu, H., Valocchi, A.J., and Kang, Q. (2012). Three-dimensional lattice Boltzmann model for immiscible two-phase flow simulations. *Physical Review E*, 85(4): 046309.

91 Wang, Q., Han, D., Wang, Z., Ma, Q., and Wang, D. (2019). Lattice Boltzmann modeling for hydrate formation in brine. *Chemical Engineering Journal*, 366: 133–140.

92 Kang, Q., Zhang, D., Lichtner, P.C., and Tsimpanogiannis, I.N. (2004). Lattice Boltzmann model for crystal growth from supersaturated solution. *Geophysical Research Letters*, 31 (21).

93 Nina, N., El-Sayed, M.M., Sanghvi, T., and Yalkowsky, S.H. (2000). Estimation of the effect of NaCl on the solubility of organic compounds in aqueous solutions. *Journal of Pharmaceutical Sciences*, 89(12): 1620–1625.

94 Lu, G., DePaolo, D.J., Kang, Q., and Zhang, D. (2009). Lattice Boltzmann simulation of snow crystal growth in clouds. *Journal of Geophysical Research: Atmospheres*, 114 (D7).

95 Sullivan, S.P., Sani, F.M., Johns, M.L., and Gladden, L.F. (2005). Simulation of packed bed reactors using lattice Boltzmann methods. *Chemical Engineering Science*, 60(12): 3405–3418.

96 Heddam, S., Bermad, A., and Dechemi, N. (2012). ANFIS-based modeling for coagulant dosage in drinking water treatment plant: A case study. *Environmental Monitoring and Assessment*, 184(4): 1953–1971.

97 Wei, X., (2013). Modeling and Optimization of Wastewater Treatment Process with a Data-Driven Approach. PhD Thesis. University of Iowa, Iowa City, IA, USA.

98 Nadiri, A.A., Shokri, S., Tsai, F.T.C., and Moghaddam, A.A. (2018). Prediction of effluent quality parameters of a wastewater treatment plant using a supervised committee fuzzy logic model. *Journal of Cleaner Production*, 180: 539–549.

99 Sadi, M. and Shahrabadi, A. (2018). Evolving robust intelligent model based on group method of data handling technique optimized by genetic algorithm to predict asphaltene precipitation. *Journal of Petroleum Science and Engineering*, 171: 1211–1222.

100 Sadi, M., Fakharian, H., Ganji, H., and Kakavand, M. (2019). Evolving artificial intelligence techniques to model the hydrate-based desalination process of produced water. *Journal of Water Reuse and Desalination*, 9(4): 372–384.

101 Araromi, D.O., Majekodunmi, O.T., Adeniran, J.A., and Salawudeen, T.O. (2018). Modeling of an activated sludge process for effluent prediction—a comparative study using ANFIS and GLM regression. *Environmental Monitoring and Assessment*, 190(9): 1–17.

102 Talpur, N., Abdulkadir, S.J., and Hasan, M.H. (2020). A deep learning based neuro-fuzzy approach for solving classification problems. In 2020 International Conference on Computational Intelligence (ICCI), 167–172. IEEE.

103 Vapnik, V.N. (1995). *The Nature of Statistical Learning Theory*. New York, NY, USA: Springer.

104 Goldberg, D.E. (1989). *Genetic Algorithms in Search, Optimization and Machine Learning*. Reading, MA, USA: Addison-Wesley Professional.

105 Ghorbani, M., Zargar, G., and Jazayeri-Rad, H. (2016). Prediction of asphaltene precipitation using support vector regression tuned with genetic algorithms. *Petroleum*, 2(3): 301–306.

106 Rousseeuw, P.J. and Leroy, A.M. (1987). *Robust Regression and Outlier Detection*. New York, NY, USA: John Wiley and Sons.

6

Gas Hydrates in Wastewater Treatment

Adeel Ur Rehman, Dzulkarnain B Zaini, and Bhajan Lal

6.1 Ecosystem Approach to Pollution Control

Ecology is the systematic study of the abundance, distributions, and interactions of organisms within the environment. In general understanding, it is a study based on the populations and communities within the ecosystem and their intra- and interrelationships [1]. Ecology is not analogous to the environment, environmental science, or environmentalism [2], though physiology, evolution, genetics, and behavioral sciences are closely linked to it [3].

Similar to the other natural sciences, a conceptual understanding of ecology can be discovered in its further details, including:

- adaptations of life processes
- abundance and distribution of organisms
- energy and materials trailing of living communities
- the successful progress of ecosystems
- biodiversity abundance and distribution in environmental contexts [4].

Ecology has developed into a biological subdiscipline concerned with the broader study of life but is distinct from the study of natural history, which primarily focuses on the description and study of species. The fields of conservation biology, wetland management, management of natural resources (agriculture, forestation, and fishing), environmental management (urban ecology), community health, economics, and many other unmentioned facets of fundamental and applied sciences have seen many practical applications from the findings of ecology at large [5–8].

The environment comprises two major components: (i) living and (ii) non-living. Ecology is the study of the complex interrelationships between these two elements and the influence that disruptions in either have on the natural

Gas Hydrate in Water Treatment: Technological, Economic, and Industrial Aspects,
First Edition. Bhajan Lal and Sirisha Nallakukkala.

equilibrium. If the balance is to be asserted, an ecological system (ecosystem) approach to controlling external influences is, therefore, required.

The following three classes make up the types of different contaminants that can reach the environment through wastewater:

- biologically accumulative (persistent)
- degradable
- non-degradable (conservative).

Complex organics (e.g. sewage or industrial waste) and dead organisms, which can gradually decompose microbially, are examples of *degradable* pollutants. Degradable pollutants also include substances that undergo physical degradation or decay, such as radioactive isotopes, though some have such a long half-life that they may as well be considered non-degradable. Degradation is a time–rate process, with the rate generally following first-order kinetics.

Some organic matter can quickly become degradable, while others may be harder to degrade. Microbial degradation of organic matter results in oxygen consumption, which can provide an indirect method of estimating the amount of organic matter present through the biochemical oxygen demand (BOD) test.

Non-degradable (conservative) substances are inert to biological action and do not degrade with time. These include inorganic chemicals (e.g. chlorides), heavy metals, and certain refractory organics such as polychloride biphenyls (PCB), dichlorodiphenyltrichloroethane (DDT), etc. Biologically accumulative (persistent) substances are those that tend to accumulate in the food chain. They include mercury, cadmium, arsenic, lead, manganese, pesticides, radioactive isotopes, and others [2].

6.2 Interaction of Wastewater with the Ecosystem

Alongside the non-living components of an ecosystem, such as water, gases, minerals, soil, chemicals, etc., the living component of an ecosystem consists of three principal groups: (i) primary producers (plants, vegetation, algae), (ii) consumers (fish, mammals), and (iii.) decomposers (worms, micro-organisms like bacteria).

The planet, as well as all of its land and aquatic habitats, are examples of ecosystems and eco-subsystems. A healthy ecosystem is one in which producers, consumers, and decomposers exist in equilibrium. The decomposers break down the degradable organics into stable inorganic end-products, such as nitrates and phosphates, which are converted to organic material by the primary producers in the presence of solar energy and water. The organic material supplied by the primary producers is then consumed by the primary consumers (cattle, fish, insects), which, in turn, constitute the food supply for various carnivores and finally for human beings. The unconsumed fraction of producer organic material, organic wastes generated by the succession of consumers, and the dead consumers themselves, end up as organic matter, which feeds aerobic and anaerobic decomposers who convert the matter to stable inorganic end-products (the nutrients). These nutrients then cascade again through the producers, the consumers, and back to decomposers.

The human activities that consume Earth's natural resources generate liquid, solid, and gaseous wastes, which must ultimately also enter the food chains. Among anthropogenic wastes, some are degradable while others are non-degradable, and both may lead to pollution. Non-degradable substances accumulate in the food chain over time and impact the health of higher level consumers, such as humans. However, degradable wastes, discharged in volumes greater than the ecosystem can process, can be just as damaging. When the natural rate of recycling and the ecological balance in a system are adversely affected by pollution, adverse effects may cascade throughout the whole of the system because of the interdependence of the system's individual components. For example, in a river where pollutant loads exceed the river's self-purification capacity, the loss of primary producers and consumers may destroy fisheries. Similarly, a lake or coastal region may become too enriched in nutrients, leading to excessive algal growth, loss of other producer species due to shading from the algal mass, and low oxygen conditions due to decomposition of the excess biological material. In terrestrial systems, polluted soil may eventually lose its porosity and moisture rentention capacity, thus becoming useless for agriculture. The effect of a radioactive isotope from the atmosphere may enter a water body and eventually accumulate in algae, fish, and finally in humans. Though all organisms can eliminate most waste through natural processes (e.g. in the case of humans, elimination occurs through sweat, urine, feces, nails, hair, etc.), during the time that a substance resides in an organism (or organ), physical decay or chemical breakdown may also occur.

There are no boundaries between the soil, water, and gaseous phases of the environment, meaning that pollution occurring in one phase can be transferred to another. Pollutants that enter the atmosphere may either degrade over time or remain inert and thus not decrease over time. This condition will result in adverse economic and health effects. Many of these effects only become apparent in the long term; hence, they tend to be neglected until it is too late [3].

6.3 Sources of Wastewater

The common sources of wastewater [4] are shown in Figure 6.1.

6.3.1 Agricultural Wastewater

The primary source of agricultural wastewater is the surplus of water that flows off the fields during surface irrigation at the lower extremity of the furrows, boundaries, reservoirs, and flooded areas. This is also known as tailwater irrigation.

Effluent from harvested crops is also a source of polluted agricultural water. These plants produce significant quantities of industrial/agricultural wastewater, usually containing higher organic matter concentrations. Wastewater is also transported from such plants to a nearby municipal wastewater treatment plant. Due to the lack of sanitary wastes in agricultural plants, it is best to treat their waste separately and make it suitable for reuse. A list of common agricultural pollutant classes and the relative contributions to the total from crops, livestock, and aquaculture is presented in Table 6.1.

Both wastewater sources can be recovered into valuable use – usually in farmhouses near the wastewater site. Runoff from the base of the furrows can be used without treatment or pumping for irrigating areas at lower elevations. In certain situations, a pump will absorb and store the runoff in pools for later reuse. Owing to its chemical content, this water does not penetrate the aquifer, though nearby lagoons should still be fastened with waterproof clay or a membrane line, and the water should be reused as much as possible [5].

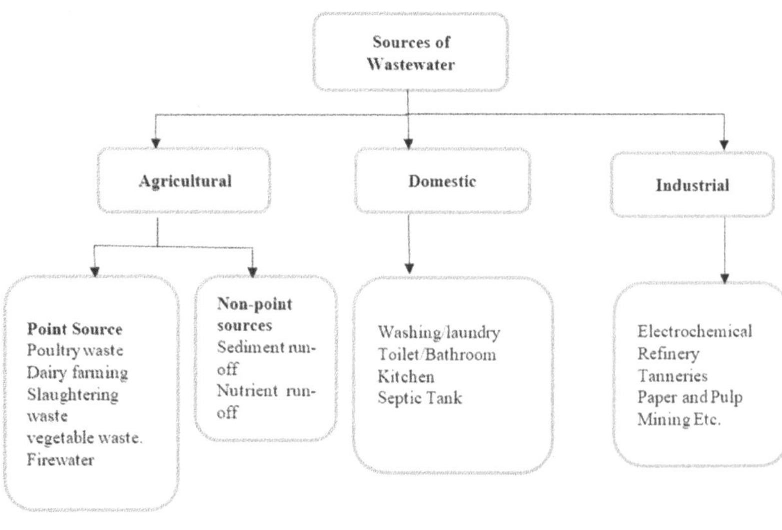

Figure 6.1 Source of wastewater.

Table 6.1 Classification of agricultural water pollutants and their contributions in the three main production agricultural systems.

Pollutant category	Indicators/examples	Relative contribution by:		
		Crops	Livestock	Aquaculture
Nutrients	In particular, the nitrogens and phosphorus present as nitrate, ammonia, or phosphates in chemical and organic fertilizers and in animal excretions.	***	***	*
Pesticides	Insecticides, herbicides, fungicides, and bactericides, including carbamates, organophosphates, organochlorine pesticides, pyrethroids, and others – many, such as DDT, are banned in most countries, but continue to be illegally and persistently used.	***	N/A	N/A
Salts	Ions of sodium, chloride, potassium, magnesium, sulfate, calcium, and bicarbonate, and total dissolved solids detected in water, either directly or indirectly, as electric conductance.	***	*	*
Sediment	Measured in water as a total suspended solid or nephelometric turbidity unit – particularly during harvesting by pond drainage.	***	***	*
Organic matter	Substances that require chemical or biochemical oxygen (e.g. organic materials such as plant material and livestock excretion).	*	***	**
Pathogens	Indicators of bacteria and pathogen. (e.g. *Escherichia coli*, gross coliforms, and enterococcal fecal coliforms).	*	***	*
Metals	Selenium, lead, copper, mercury, arsenic, and manganese.	*	*	*
Emerging pollutants	Drug residues, hormones, and feed additives.	N/A	***	**

6.3.2 Municipal Wastewater

Domestic waste includes suspended, colloidal, and dissolved organic and inorganic substances. The dramatic change in industrialization and the climate have a pronounced impact on the characteristics of wastewater deposited by consumers and ecosystems at large. Contents of industrial waste alone have dramatically changed the nature of wastewater in public sewers. The volume of water used per person is often influenced by the accumulation of wastewater as there are relatively narrow limits to the amount of waste contributed per person in certain cultures. As a result, wastewater characteristics vary not only between towns, but between seasons and hours depending upon the water source within a specific area [6]. The waste is primarily organic, containing a variety of nutrients for micro-organisms, including carbon, nitrogen, and phosphorus. It is readily biodegradable as it continues as waste through sewages, thereby changing some of its characteristics with time.

A higher BOD indicates that the water is generally of lower quality. These values generally average around 54 g per person per day within a reasonably efficient system. In some developing areas, the BOD values may be only 30–40 g per person per day as all the sewage produced in the area may not be entering the sewer system. In places where combined sewers are used, the BOD values may be about 40% higher, or 77 g per person per day [7].

In the case of offices, factories, schools, etc., where there is a part-time occupancy, the BOD values are generally taken as half of 54 g per person day or even less. For restaurants and cafeterias, each meal served may be taken to contribute a quarter of the per capita BOD, and theaters and cinemas contribute another sixth of that. Hotels and hospitals, on the other hand, may contribute as much as 1.5–2.5 times more than the usual 54 g of BOD per person per day. Domestic sewage is the primary source of nutrients such as nitrogen and phosphorous; most industrial wastes (except those from the food and fertilizer industries) contain relatively minor sources of these nutrients [8].

6.3.3 Industrial Wastewater

The characteristics of industrial wastewater depend on the raw materials, processes, and other factors, rendering it difficult to generalize. As such, wastewater characteristics differ widely from industry to industry; and even within the same industry. Some industries and their possible pollutant outcomes are listed in Table 6.2.

Industrial wastewater containing high concentrations of heavy metals, i.e. densities greater than 5 g/cm^3, are among the most concerning for environmental researchers. Most elements in this class are known to be highly

Table 6.2 Industrial sources of pollutants.

Industry	Pollutants
Iron and steel	Acids, oil, metals, phenols, and cyanides
Textiles and leather	Chromium, solid sludges, and sulfates
Pulp and paper	Solids, chlorinated organic compounds
Petrochemical refineries	Chromium, phenols, and Mineral oils
Chemicals	Heavy metals, organic chemicals, and cyanide
Microelectronics	Organic chemicals
Mining	Metals, acids, and salts

water-soluble, carcinogenic, and toxic. They can be ingested by absorbing through skin, swallowed and transmit by another medium and retained in the human body and can cause significant adverse health effects such as cancer, organ damage, damage to the nervous system, and in serious cases, death. It also decreases development and growth. Heavy metal present in industrial effluent include copper, silver, zinc, cadmium, gold, mercury, lead, chromium, iron, nickel, tin, arsenic, selenium, molybdenum, cobalt, manganese, and aluminum. These pose significant threats to people as well as the fauna and flora of the waterbodies receiving them [9].

Heavy metals like cadmium, zinc, lead, chromium, nickel, copper, vanadium, platinum, silver, and titanium are produced in electroplating, electrolysis depositions, conversion-coating, anodizing-cleaning, milling, and etching industries. Heavy metal wastes like tin, lead, and nickel is heavily tied to printed circuit board (PCB) manufacturing. Wastewater containing arsenic is generated in large amounts from the wood processing industries which use copper–arsenate wood treatment. Inorganic paints and pigments require the production of dyes that contain cadmium sulfide and chromium compounds. In petroleum refining, conversion catalysts are composed largely of nickel, vanadium, and chromium, which carry a large risk if improperly disposed of. Photographic processing produces effluent with high concentrations of silver and ferrocyanide.

In the industries mentioned above, a large amount of hazardous materials, sludges, and waste residue discharged requires effective treatment prior to being disposed into large bodies of water [2]. Since these metals are high in toxicity, recent stringent legislation has put great focus on the removal of heavy metals from wastewater. These comprise restrictions on heavy metal forms and amounts that may be found in wastewater discharged. Table 6.3

Table 6.3 The maximum contaminant level standards for the most hazardous heavy metals.

Metal	Source	Toxicity Effect	Permission Level (mg/L)
Arsenic	Pesticides, fungicides, metal smelters	skin manifestations, liver and kidney damage, loss of appetite	0.02
Cadmium	Welding, electroplating, pesticides, fertilizers, Cd-Ni batteries	kidney damage, renal disorder, human carcinogen	0.06
Chromium	Paints, electro plating, and metallurgy	lung damage and limitation of the respiratory system	0.05
Mercury	Pesticides, batteries, paper industry	rheumatoid arthritis, diseases of the kidneys, circulatory system, and nervous system	.01(vapor)
Lead	Paint, pesticides, smoking, automobile emission, mining	fetal brain damage, kidneys disease	0.15
Nickel	Electrochemical industries	damage to kidneys, liver, and lungs	0.1

summarizes the US Environmental Protection Agency's maximum contaminant level requirements for certain heavy metals [10]. Therefore, metal-contaminated wastewater must be treated before discharge into the environment. Conventional treatment methods such as coagulation, chemical precipitation, membrane operations, ion exchange, electrodeposition, solvent extraction, foam flotation, cementation, and activated carbon adsorption can contribute to the removal of heavy metals from inorganic effluents.

6.4 Impact of Wastewater on Ecology

Wastewater disposal to coastal waters is complex, as the composition of discharges vary greatly by source and, for some sources, seasonally. The ecological impacts of individual discharges, as well as cumulative effects of multiple discharge sources, are complicated by both biological interactions within the systems and the physical characteristics, such as salinity, tides, waves, and currents, that define those systems. Moreover, human interactions with these system must be considered.

Fish, both from estuaries as well as the open sea, constitute an important source of livelihood and nourishment for the relatively large coastal populations of India. When exposed to all the conservative and toxic pollutants contained in wastewaters, fish act like a boomerang by bringing back those pollutants to the people's dining tables. Such is the power of the food chains. The degradable organics, however, undergo biodegradation in the sea in the same manner as they do in a river, with the only difference being that in the case of a river, all the problems occur downstream for someone else whereas, in the case of the sea, the same community suffers.

In the coastal areas of many countries, each river that empties into the sea carries all the polluting wastes, nutrients, and debris from the upstream land which it drains. Similarly, the drains that carry storm runoff from a coastal city generally discharge their wastes into the coastal waters. These discharges also carry nutrients and other wastes that wash into the drains from the streets, parking lots, parks, and yards. On the other hand, sanitary sewers first discharge into a treatment plant and later to the sea via outfalls. In the case of an overpopulated city, overloaded sanitary sewers are relieved through a bypass that flows into the storm drains. Thus, even some raw sewage directly enters the coastal waters. Swimming in such waters is a positive hazard. The BOD values of sea waters normally range from 1 to 3 mg/L, but in the case of polluted shorelines, they may reach 30–40 mg/L on occasion. When traveling by air, one often clearly sees the relatively bright greenish color of the algae-laden coastal waters just adjoining the land mass and the bluish color of the waters of the open sea [11].

Global warming is also raising the sea levels all around India and all low-lying coastal areas are likely to exhibit its adverse health and other ill effects in due course. The water quality objectives that need to be met in the case of sea disposal are essentially meant to protect people's health and promote the beneficial uses of coastal waters by minimizing their impact on the micro-environment. For instance, odors, and the presence of floatables (oil, solid wastes) may cause aesthetic problems such as the discoloration of water, thereby affecting recreational and navigational uses and land values. Other ecological effects arise from oxygen depletion, excessive primary productivity, reduction in water transparency, species diversity, and important health effects resulting from the bioaccumulation of pollutants in the marine food chain and affects on the taste of the fish [12].

In many coastal cities, the municipal wastewater is carried out into the sea by outfalls of adequate length and discharged into the sea either at the end of the pipe or through a set of diffusers. Dissolved oxygen requirements must be met, especially in the vicinity of the discharge point where the plume rises vertically and then flows away horizontally. Diffusers tend to mix the low-density wastewater well with the higher density sea water such that the resulting density of the plume makes it rise only up to the pycnocline level, thus leaving it submerged below the sea surface, unseen by the general public. From the viewpoint of health, apart from the bioaccumulation in fish, coliform requirements also must be met at bathing beaches and shellfish grounds [13].

The control of pollutants is not just a matter concerning waste treatment. It involves an entire set of approaches based on the philosophy that the overall cost to a nation should be minimized. It is not important as to who spends money for waste treatment (be it the government or private industry) since eventually, the money comes from the people in the form of either taxes or higher costs of the manufactured products. It is simply important to optimize the total money spent by a nation on pollution control. It is, of course, sensible to adopt what has come to be called the PPP, or the polluter pays principle. But the discussions here will be limited to technical rather than fiscal and organizational aspects [14].

Among the different approaches available for pollution control, modern trends favor the following method [15]:

- Select, modify, and develop a manufacturing technology which is, for all intents and purposes, a cleaner technology from the pollution viewpoint, as it gives minimum mass and volume of wastes (important in most industries); this is also called low-waste technology.
- Produce wastes that can be readily treated and handled (e.g. avoid the use of the arsenic processes in fertilizer manufacturing, mercury compounds in agriculture and paper manufacturing, and industries based on hard detergents).
- Conserve water, raw products, and other resources (through the reuse of water, by following good housekeeping practices). Control at the source is one of the best methods of pollution control and eliminates the cost of treating unnecessary wastes.
- Conserve and minimize the energy requirements of various processes to save on fuel costs.
- Locate the new industries carefully so that treatment costs are minimized. For example, facilities near the coast or along large rivers may benefit from the large volumes of water readily available for dilution. Also, facilties should avois valleys with air inversion conditions, practice land disposal of wastes where feasible, and discharge wastes into public sewers in cities.

6.5 Current Technologies for Addressing Wastewater Issues

Inorganic, heavy metal ion removal practices such as chemical precipitation, electrochemical, and gas hydrate-based extraction can be accomplished with traditional treatment processes, as shown in Figure 6.2. These processes have considerable limitations, such as insufficient removal, high energy requirements, and toxic sludge production.

Still, adequate and diligent consideration should be paid to the removal technologies for heavy metals to protect human health, plants, animals, soil, and the environment. Most physical and chemical removal methods necessitate dealing with large amounts of toxic sludge, ruin circumferent ecosystems, and are costly [16].

6.5.1 Chemical Precipitation

Chemical precipitation is widely employed for the removal of heavy metals from inorganic effluent. Removal by chemical precipitation is shown in Figure 6.3 and can be expressed conceptually by Eq. (6.1):

$$M^{2+} + 2(OH)^{-} \rightarrow M(OH)_2 \downarrow, \tag{6.1}$$

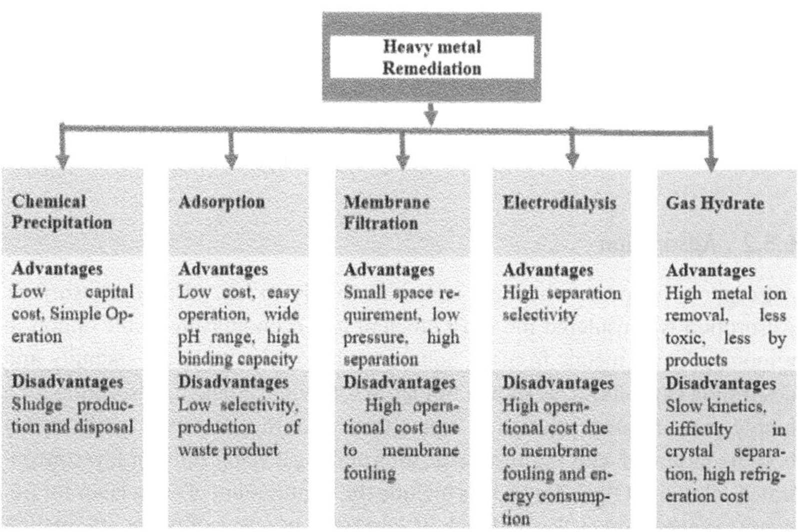

Figure 6.2 Heavy metal removal techniques.

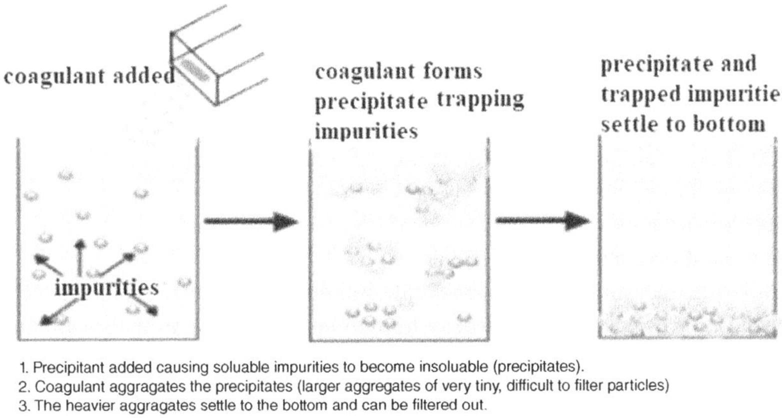

Figure 6.3 Removal principle of chemical precipitation.

where M^{2+} and OH^- represent the dissolved metal ions and the precipitant, respectively, while $M(OH)_2$ is the insoluble metal hydroxide.

The main parameter which significantly improves the removal of heavy metals by chemical precipitation is pH correction to basic conditions (pH 9–11). Lime and limestone are the most commonly employed precipitant agents due to their availability and low cost in most countries. Lime precipitation can be used to effectively treat inorganic effluent with a metal concentration higher than 1000 mg/L. Other benefits include the ease of the procedure, low cost of equipment, and relatively safe operations. In order to decrease the metal content to an appropriate level for discharge, however, a large amount of chemical precipitation is needed. Other disadvantages are sluggish metal deposition, inadequate settlement, metal precipitate accumulation, excessive sludge production, additional treatment procedures for the sludge produced, and long-term environmental effects of sludge disposal [17, 18].

6.5.2 Adsorption

Adsorption happens when an adsorbate binds to an adsorbent's surface. Adsorption is considered the most efficient and economically viable option in removing heavy metals from an aqueous solution due to its reversibility and desorption capabilities, as shown in Figure 6.4. While efficient, adsorption for a highly concentrated solution is not effective as the adsorbent will quickly become saturated with the adsorbate, thus making it viable only for low concentrated solutions. Other drawbacks include the requirement of extra labor for frequent regeneration cycles and lack of selectivity in terms of metal attenuation [19]. Adsorption is therefore not practical in large-scale heavy metal rejection.

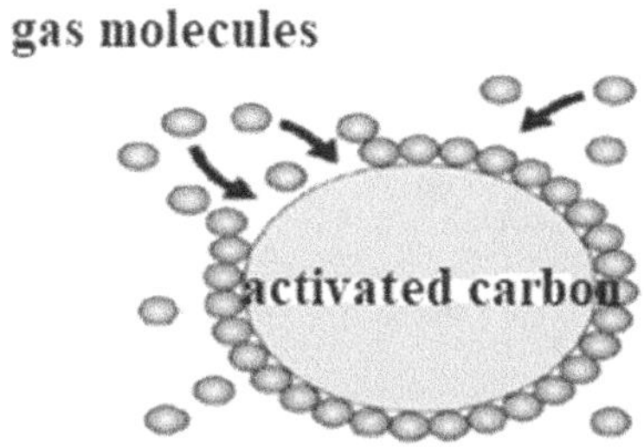

Figure 6.4 Removal principle of adsorption.

6.5.3 Membrane Technologies

For water with high pollutant concentrations, the use of membrane technologies for heavy metal extraction is another proven option. This process utilizes the phenomenon of concentration gradients, as shown in Figure 6.5. Microfiltration, nanofiltration, and reverse osmosis are membrane separation processes used to treat mine waters. Higher pressure drop across the membrane and frequent membrane replacements are the considerable limitations of this technique [20].

6.5.4 Electrodialysis

Electrolytic or electrowinning recovery is one of the many methods used in removing metals from process or wastewater sources. As shown in Figure 6.6, The method uses electricity to pass ions through streams of the heavy metal-bearing anode and a cathode platform. Metallic, positively charged ions adhere to the negatively charged cathodes leaving behind a recoverable metal deposit. A significant disadvantage to this method is corrosion buildup, requiring electrode replacement [21]. A hydrate technology for the removal of heavy metal

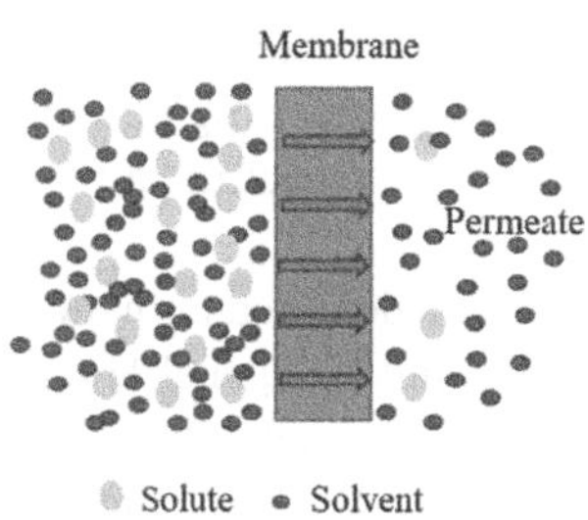

Figure 6.5 Removal principle of membrane separation.

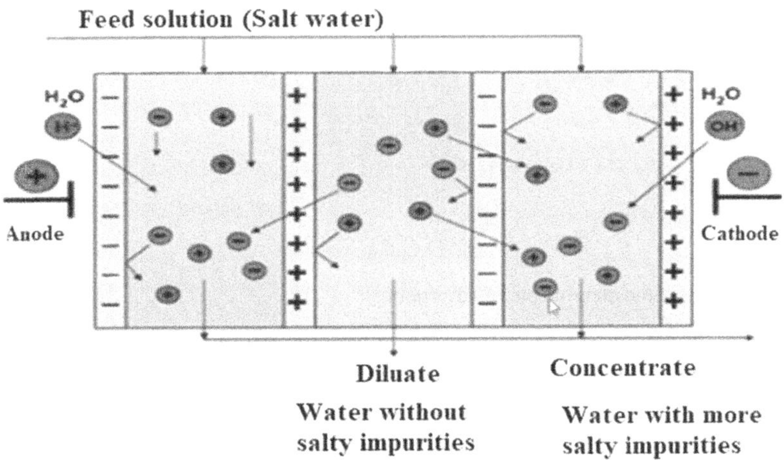

Figure 6.6 Removal principle of electrodialysis.

ions from the aqueous solution is possible since clathrates exclude almost all the inorganic salt and heavy metal ions from the hydrate crystal produced. However, disposal of the heavy metal concentrate presents environmental and human health issues [22].

6.6 Gas Hydrates

Gas hydrates (known as flammable ice) are nonstochiometric, ice-like structures created by trapping gas (guest) molecules in hydrogen-bonded water molecules (host) under low-temperature and high-pressure conditions, as shown in Figure 6.7. Hydrate-based separation allows for the removal of almost all the inorganic salt and heavy metal ions from an aqueous solution due to the many significant physical properties of clathrates.

The steps included in the hydrate-based wastewater treatment process are: (i) formation of gas hydrates, (ii) separation of hydrates, (iii) post-treatment to enhance the quality of water (e.g. washing, centrifuging), and (iv) hydrate dissociation to form purified water and the recycling of freed gas.

In general, there are three common types of structures identified for natural gas hydrates. They are the cubic I (sI), cubic II (sII), and hexagonal H (sH) structure. The size and shape of the hydrate cavities in the cages determine structure, while the type and shape of the guest gas molecules define the type of hydrate formed [23].

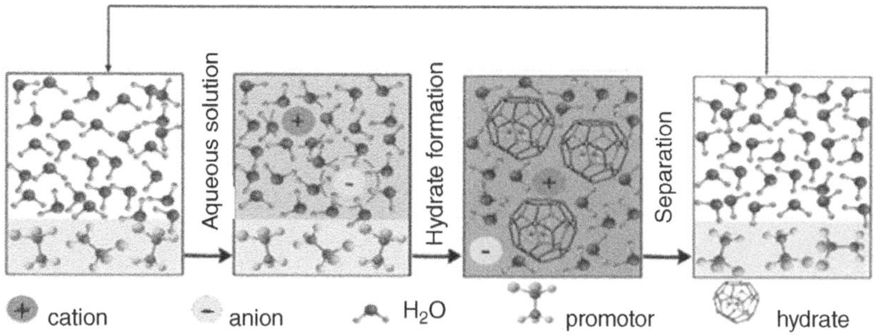

Figure 6.7 Principles of separation of heavy metal ions via gas hydrate technique.

6.6.1 Formation Process of Gas Hydrates

Gas hydrate formation is a crystallization process consisting of nucleation and crystal growth [24–26]. As such, hydrate formation is a physical, not chemical, process in which the guest molecule moves freely around the cavity created by the host water molecules. The driving force, also called supersaturation, is one of the effective parameters in gas hydrate formation [27, 28]. The difference between the Gibbs free energies of the solution and crystal phases expresses the driving force. A variety of driving forces have been introduced into the literature for the nucleation and hydrate crystallization growth process.

6.6.2 Gas Hydrate Growth Process

Hydrate growth occurs following nucleation, indicating a change from stable nuclei to solid hydrate. Conditions for the growth of stable hydrate nuclei to solid hydrate, i.e. the temperature, pressure, agitation, and interfacial region, vary by guest species. This process is shown in Figure 6.8. The pressure sharp decrease in pressure indicates hydrate formation and growth. When the pressure reaches a static value, the hydrate formation process is complete. Note that the hydrate formation step is represented by a concomitant and sudden increase in the system temperature because it is an exothermic phase. The completion of the formation process is depicted by a constant temperature and some degree of subcooling [29].

6.6.3 Kinetics of Hydrate Formation

Numerous studies have been performed to evaluate the feasibility and efficiency of different gas hydrate processes. In order to further research, the kinetics of hydrate formation, the impact of wastewater on steam, stirrer, flow

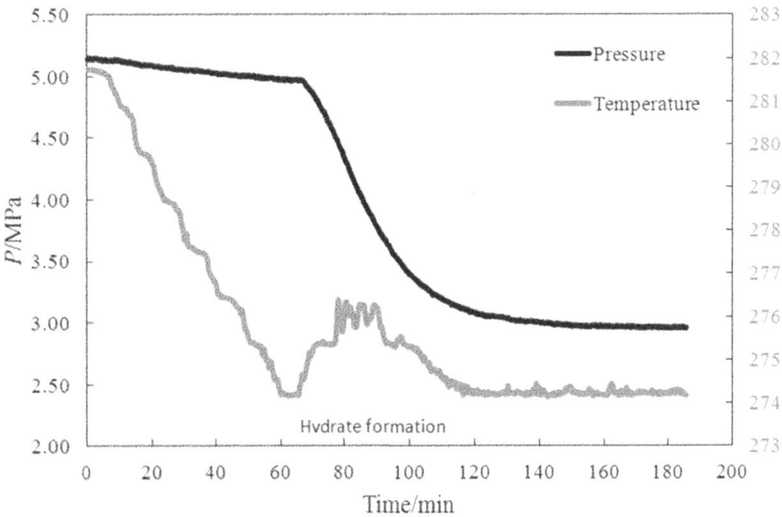

Figure 6.8 Hydrate growth formation.

rate, and pressure is examined. These experiments will lead to optimizing the process in terms of water recovery, enrichment factor, salinity, and water quality.

6.6.3.1 Effects of Salt During Hydrate Formation

It is understood that the presence of salts effects the temperature and pressure parameters [30–34]. A hydrate phase equilibrium diagram of carbon dioxide (CO_2) in pure water and with different sodium chloride (NaCl) concentration solutions [35–38] is exhibited in Figure 6.9. In the plot, the salt presence shifts the phase equilibrium to lower pressure and temperature ranges. Increased salt concentrations increase the pressure required for hydrate formation at a given temperature, thus reducing the driving force for the formation of hydrates and slowing hydrate growth [38–41].

Salts reduce the efficiency of hydrate formation by increasing the balance pressures at a certain temperature or by reducing the balance temperatures at certain pressures [42]. The degree of supercooling is simply calculated as $T - T_{eq}$, where T_{eq} depends on the concentration and type of salt in solution [43]. Salt suppresses the formation of hydrate, which decreases the nucleation rate and growth due to the decrease in the cavity in the aqueous phase [44]. The inhibition effect of salts in the hydrate phase is mainly determined by ion size and electrical charge, based on the investigated system conditions [45]. The intensity of inhibition is directly proportional to the number of electrical charges and inversely related to ionic size. In order to enhance the operating

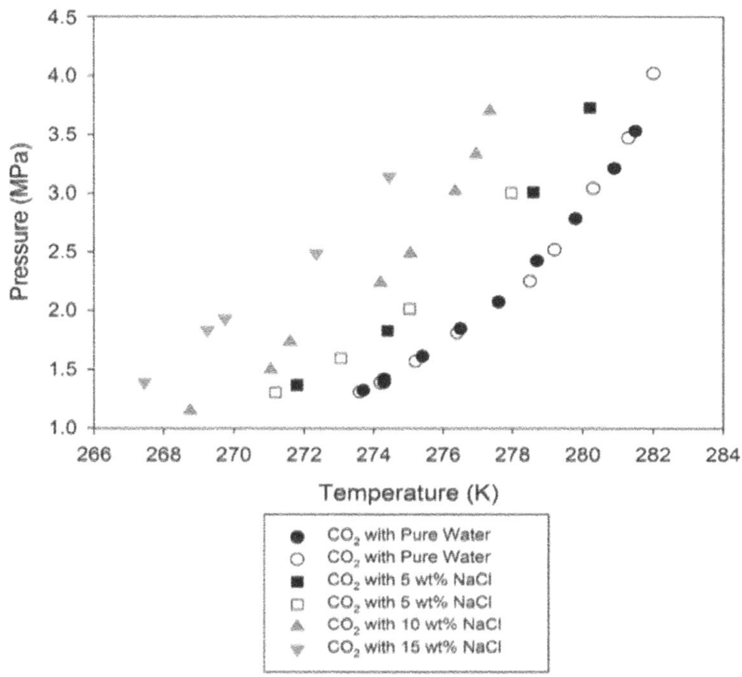

Figure 6.9 Hydrate phase equilibrium of CO_2 in the presence of pure water & various strengths of brine solutions [35–38].

temperature, immiscible hydrate formers like cyclopentane and cyclohexane were used in conjunction with CO_2. Hydrate formation rates were 16 and 22 times higher, respectively, than that associated with using the CO_2 hydrate former alone [46].

6.6.3.2 Effect of Water to Gas Ratio

Dong et al. [47] analyzed the effect of various R141b to wastewater effluent ratios on heavy metal ion removal. The removal efficiency reached 90.82% for Cu^{2+} at a 6 : 1 ratio. The removal effciency decreased at lower R141b concentrations due to excessive slurry formation. Song et al. [48] report removal efficiencies of 88.01–90.82% for Zn^{2+}, Cr^{3+}, Ni^{2+}, and Cu^{2+} with a R141b volume ratio of six and a washing operation (range of 67.82–71.87% without washing). During hydrate growth, the concentration of ions excluded from the hydrates increases. As concentration increases some molecules will necessarily become trapped in the porous crystalline structure and make their way into the dissociated water, thus decreasing removal efficiency. Increasing the former (R141b) concentration retains more liquid water, thus reducing ionic concentration and entrapment. In both studies, increased volume ratio showed a tendency to improve the efficiency of removal, but lower the dissociated water yield and residual enrichment factor.

6.6.3.3 Effect of Pressure During Hydrate Formation

Kang et al. [49] investigated the removal efficiencies of cations and anions at various hydraulic pressures using CO_2 hydrate pellets in seawater desalination. In their tests, methane (CH_4) hydrate pellets compressed at the low hydraulic pressure (< 8 MPa) were exposed to atmospheric pressure and sometimes broken. The comparative pressure for CH_4 hydrate pellets was lower than that for CO_2 hydrate pellets because the formation pressure of CH_4 hydrate (4.6 MPa) is higher than that of CO_2 hydrate (2.9 MPa). For the same reason, when CH_4 hydrate slurries in the connection pipe were compressed and squeezed by the strokes of the dual cylinder operation, the hydraulic pressure was not enough to make pellets, unlike the CO_2 hydrate pellets. It also appears that the efficiency of ion removal depends not only on the gas-forming hydrate but also physical property (i.e. crystal system, compressibility, and critical pressure for pelletizing) of the gas hydrate.

6.6.3.4 Effect of Stirrer during Hydrate Formation

In a study by Wijayanti [50], a significant pressure drop occurred at the stirrer speed of 400 rpm, though the lowest pressure drop occurs at a stirrer speed of 200 rpm. Because of this rotation speed, the molecules of water agitation in which the gas and water molecules interact at high speeds become heavier, so that the gas can be absorbed more easily into the water molecule cavities.

6.6.4 Hydrate Dissociation

6.6.4.1 Water Recovery

The volumetric removal efficacy of the process is demonstrated by water recovery. The amount of water in the feed solution involved in this process is transformed into hydrate, but the volume normally recovered from the hydrate crystal is less than 100%. Water recovery reflects the process volumetric efficiency and can be calculated by using Eq. (6.2) below, as given in the literature [51]:

$$\text{Water recovery } \% = \frac{\left(\textit{Volume of water converted to hydrate}\right) * (F_h)}{\textit{Volume of feed solution}} * 100, \quad (6.2)$$

where F_h is the fraction of hydrate formed that is recoverable at the hydrate crystal separation step from the brine. In the studied cases, hydrate crystals were not separated from the brine, therefore it is assumed that $F_h = 1$. Nambiar et al. [47] reported that very poor water recovery (~2%) was achieved by 10% propane in N_2 mixtures, in the presence of NaCl 3 wt% in the solution, and 10% propane in argon gas mixtures. With the 10% propane in the CO_2 gas mixture, the water recovery due to the presence of salt was found to be less than 10%. Hence, the 10% propane in the CO_2 gas mixture is figured to be somewhat suitable for the process when using porous media.

6.6.4.2 Removal Efficiency, Enrichment Factor, and Yield

Salt rejection represents the effectiveness of the process in removing contaminants. The salt rejection or removal efficiency is calculated as in Eq. (6.3) [45, 52].

$$\text{Removal Efficiency} = \frac{C_{A0} - C_A}{C_{A0}} * 100 \tag{6.3}$$

where C_{A0} is the initial concentration of effluent, and C_A is the effluent concentration in the dissociated water. The ion concentration in the feed seawater and the dissociation water measures the percent rejection of each ion. Increased device performance translates to greater rejection. To characterize water output, the yield of dissociated water is calculated as in Eq. (6.4):

$$\text{Water Yield} = \frac{V_1}{V_0} * 100 \tag{6.4}$$

where V_0 is the initial volume of effluent, and V_1 is the volume of dissociated water from hydrate dissociation. Kang et al. [49] reported that 71–94% of each cation and 73–83% of each anion were removed in a single-stage process without pretreatment using CO_2.

6.6.5 Kinetic Models of Gas Hydrate Growth

The production of gas hydrates occurs immediately following nucleation and is considered to be a complex interfacial phenomenon, which requires multicomponent (i.e. gas, water, and hydrate) distribution in multiple phases (i.e. gas, aqueous, and hydrate) at multiscale levels (i.e. macroscopic-scale and molecular-scale) [53]. The analysis of this phenomenon includes the knowledge of mass and heat transfer, fluid flow, and intrinsic kinetics. At a larger scale, the rate of growth in gas hydrates was usually measured by the rate of gas consumption computed by pressure (P) and temperature (T) measurements [37, 54] as well as other direct visualization techniques on the hydrate film thickness and morphology [55–58]. On a microscopic level, hydrate growth can be considered as a combination of three different factors [59]:

- Mass transfer from liquid water and gas molecules to the growing hydrate surface
- The intrinsic kinetics of hydrate production at the surface of hydrate
- Transfer of heat away from the crystal growing surface

Accordingly, a kinetic model can be structured based on the controlling mechanism postulated: (i) mass transfer, (ii) intrinsic kinetics, (iii) heat transfer, or (iv) a combination of all three. Since the growth of hydrates is usually seen as an interfacial phenomenon [60], the motion of the guest gas molecule should be taken into account in the rate equation. Thus, the common rate expression is typically composed of mass transfer rate terms in addition to the commonly known intrinsic kinetics [25, 26, 61, 62]. Furthermore, heat transfer analysis from classical nucleation theory also provides insight into the thickness and growth rate of the initial gas hydrate film [63, 64]. These mechanisms are often interconnected and can dominate hydrate growth, depending on the hydrate formation condition. To date, most models have not been based on first principles. Moreover, model formulation is based primarily on the hydrated multiphase reactor configuration (gas–liquid–hydrate, liquid–liquid–hydrate) fluid interaction patterns, and the techniques of gas hydrate formatiom. No uniform model accounts for all the kinetic hydrate steps that have been found so far concerning water recovery.

References

1 Odume, O.N. (2017). Ecosystem approach to managing water quality. In *Water Quality* (Hlanganani Tutu ed.) Rijeka, Croatia: Intech.

2 Bunn, S.E. (2016). Grand challenge for the future of freshwater ecosystems. *Frontiers in Environmental Science* 4: 21.

3 Capps, K.A. (2019). Wastewater infrastructure and the ecology and management of freshwater systems. *Acta Limnologica Brasiliensis* 31.

4 Robinson, K.G., Robinson, C.H., and Hawkins, S.A. (2005). Assessment of public perception regarding wastewater reuse. *Water Science and Technology: Water Supply* 5 (1): 59–65.

5 Raschid-Sally, L. and Jayakody, P. (2009). *Drivers and characteristics of wastewater agriculture in developing countries: Results from a global assessment*, 127. IWMI.

6 Jetten, M.S.M., Horn, S.J., and Van Loosdrecht, M.C.M. (1997). Towards a more sustainable municipal wastewater treatment system. *Water Science & Technology* 35 (9): 171–180.

7 Arceivala, S.J. and Asolekar, S.R. (2006). *Wastewater Treatment for Pollution Control and Reuse*. New Delhi: Tata McGraw-Hill Professional.

8 Chan, Y.J., Chong, M.F., Law, C.L., and Hassell, D.G. (2009). A review on anaerobic–aerobic treatment of industrial and municipal wastewater. *Chemical Engineering Journal* 155 (1–2): 1–18.

9 Sörme, L. and Lagerkvist, R. (2002). Sources of heavy metals in urban wastewater in Stockholm. *The Science of the Total Environment* 298 (1–3): 131–145.

10 Gunatilake, S.K. (2015). Methods of removing heavy metals from industrial wastewater. *Methods* 1 (1): 14.

11 Siebe, C. and Cifuentes, E. (1995). Environmental impact of wastewater irrigation in central Mexico: An overview. *International Journal of Environmental Health Research* 5 (2): 161–173.

12 Grönlund, S.E. (2019). Indicators and methods to assess sustainability of wastewater sludge management in the perspective of two systems ecology models. *Ecological Indicators* 100: 45–54.

13 Kamika, I., Coetzee, M., Mamba, B.B., Msagati, T., and Momba, M.N.B. (2014). The impact of microbial ecology and chemical profile on the enhanced biological phosphorus removal (EBPR) process: A case study of Northern Wastewater Treatment Works, Johannesburg. *International Journal of Environmental Research and Public Health* 11 (3): 2876–2898.

14 Foulon, J., Lanoie, P., and Laplante, B. (2002). Incentives for pollution control: Regulation or information? *Journal of Environmental Economics and Management* 44 (1): 169–187.

15 Alzamora Rumazo, C., Pryor, A., Ocampo Mendoza, F., Campos Villareal, J., Robledo, J.M., and Rodriguez Mercado, E. (2000). Cleaner production in the chemical industry. *Water Science & Technology* 42 (5–6): 1–7.

16 Ur Rehman, A., Zaini, D.B., and Lal, B. (2020). Application of gas hydrate based technique in wastewater treatment-a mini review. *Third International Conference on Separation Technology 2020 (Icost 2020)*, Advances in Engineering Research, volume 200. Atlantis Press. 249–254.

17 López-Maldonado, E.A., Oropeza-Guzman, M.T., Jurado-Baizaval, J.L., and Ochoa-Terán, A. (2014). Coagulation–flocculation mechanisms in wastewater treatment plants through zeta potential measurements. *Journal of Hazardous Materials* 279: 1–10.

18 Kurniawan, T.A., Chan, G.Y.S., Lo, W.-H., and Babel, S. (2006). Physico–chemical treatment techniques for wastewater laden with heavy metals. *Chemical Engineering Journal* 118 (1–2): 83–98.

19 Reddad, Z., Gerente, C., Andres, Y., and Le Cloirec, P. (2002). Adsorption of several metal ions onto a low-cost biosorbent: Kinetic and equilibrium studies. *Environmental Science & Technology* 36 (9): 2067–2073.

20 Chen, L. and Chen, Q. (2003). Industrial application of UF membrane in the pretreatment for RO system [J]. *Journal of Membrane Science & Technology* 4: 9.

21 Huang, C., Xu, T., Zhang, Y., Xue, Y., and Chen, G. (2007). Application of electrodialysis to the production of organic acids: State-of-the-art and recent developments. *Journal of Membrane Science* 288 (1–2): 1–12.

22 Gaikwad, N., Nakka, R., Khavala, V., Bhadani, A., Mamane, H., and Kumar, R. (2020). Gas hydrate-based process for desalination of heavy metal ions from an aqueous solution: Kinetics and rate of recovery. *ACS ES&T Water* 1 (1). doi: 10.1021/acsestwater.0c00025.

23 Sloan, E.D., Jr and Koh, C.A. (2007). *Clathrate Hydrates of Natural Gases*, 3rd ed. Boca Raton FL: CRC press.

24 Englezos, P., Kalogerakis, N., Dholabhai, P.D., and Bishnoi, P.R. (1987). Kinetics of formation of methane and ethane gas hydrates. *Chemical Engineering Science* 42 (11): 2647–2658.

25 Bishnoi, P.R. and Natarajan, V. (1996). Formation and decomposition of gas hydrates. *Fluid Phase Equilibria* 117 (1–2): 168–177.

26 Skovborg, P. and Rasmussen, P. (1994). A mass transport limited model for the growth of methane and ethane gas hydrates. *Chemical Engineering Science* 49 (8): 1131–1143.

27 Mullin, J.W. (2001). *Crystallization*. Elsevier.

28 Vysniauskas, A. and Bishnoi, P.R. (1983). A kinetic study of methane hydrate formation. *Chemical Engineering Science* 38 (7): 1061–1072.

29 Chacin, M.C.G., Hughes, R.G., Civan, F., and Taylor, C.E. (2004). Phenomenological modeling of hydrate formation and dissociation. In: *Advances in the Study of Gas Hydrates*, (eds. C.E. Taylor and J.T. Kwan) 27–41. Springer.

30 Nambiar, A., Babu, P., and Linga, P. (2019). Improved kinetics and water recovery with propane as co-guest gas on the hydrate-based desalination (hydesal) process. *ChemEngineering* 3 (1): 31.

31 Ohgaki, K., Makihara, Y., and Takano, K. (1993). Formation of CO2 hydrate in pure and sea waters. *Journal of Chemical Engineering of Japan* 26 (5): 558–564.

32 Qi, Y., Wu, W., Liu, Y., Xie, Y., and Chen, X. (2012). The influence of NaCl ions on hydrate structure and thermodynamic equilibrium conditions of gas hydrates. *Fluid Phase Equilibria* 325: 6–10.

33 Ngema, P.T., Petticrew, C., Naidoo, P., Mohammadi, A.H., and Ramjugernath, D. (2014). Experimental measurements and thermodynamic modeling of the dissociation conditions of clathrate hydrates for (refrigerant+ NaCl+ water) systems. *Journal of Chemical & Engineering Data* 59 (2): 466–475.

34 Clarke, M.A., Majumdar, A., and Bishnoi, P.R. (2004). Experimental investigation of carbon dioxide hydrate formation conditions in the presence of KNO3, MgSO4, and CuSO4. *Journal of Chemical & Engineering Data* 49 (5): 1436–1439.

35 Fan, -S.-S. and Guo, T.-M. (1999). Hydrate formation of CO2-rich binary and quaternary gas mixtures in aqueous sodium chloride solutions. *Journal of Chemical & Engineering Data* 44 (4): 829–832.

36 Mohammadi, A.H., Afzal, W., and Richon, D. (2008). Gas hydrates of methane, ethane, propane, and carbon dioxide in the presence of single NaCl, KCl, and CaCl2 aqueous solutions: Experimental measurements and predictions of dissociation conditions. *Journal of Chemical Thermodynamics* 40 (12): 1693–1697.

37 Dholabhai, P.D., Kalogerakis, N., and Bishnoi, P.R. (1993). Equilibrium conditions for carbon dioxide hydrate formation in aqueous electrolyte solutions. *Journal of Chemical & Engineering Data* 38 (4): 650–654.

38 Englezos, P. and Hall, S. (1994). Phase equilibrium data on carbon dioxide hydrate in the presence of electrolytes, water soluble polymers and montmorillonite. *The Canadian Journal of Chemical Engineering* 72 (5): 887–893.

39 Englezos, P. (1993). Clathrate hydrates. *Industrial & Engineering Chemistry Research* 32 (7): 1251–1274.

40 Uchida, T., Ikeda, I.Y., Takeya, S., Ebinuma, T., Nagao, J., and Narita, H. (2002). CO2 hydrate film formation at the boundary between CO2 and water: Effects of temperature, pressure and additives on the formation rate. *Journal of Crystal Growth* 237: 383–387.

41 Chong, Z.R., Chan, A.H.M., Babu, P., Yang, M., and Linga, P. (2015). Effect of NaCl on methane hydrate formation and dissociation in porous media. *Journal of Natural Gas Science and Engineering* 27: 178–189.

42 Mekala, P., Babu, P., Sangwai, J.S., and Linga, P. (2014). Formation and dissociation kinetics of methane hydrates in seawater and silica sand. *Energy & Fuels* 28 (4): 2708–2716.

43 Simmons, B.A., Bradshaw, R.W., Dedrick, D.E., Cygan, R.T., Greathouse, J.A., and Majzoub, E.H. (2008). "Desalination utilizing clathrate hydrates (LDRD final report).," Citeseer.

44 Lederhos, J.P., Long, J.P., Sum, A., Christiansen, R.L., and Sloan, E.D., Jr (1996). Effective kinetic inhibitors for natural gas hydrates. *Chemical Engineering Science* 51 (8): 1221–1229.

45 Park, K. *et al.* (2011). A new apparatus for seawater desalination by gas hydrate process and removal characteristics of dissolved minerals (Na+, Mg2+, Ca2+, K+, B3+). *Desalination* 274 (1–3): 91–96.

46 Cha, J.-H. and Seol, Y. (2013). Increasing gas hydrate formation temperature for desalination of high salinity produced water with secondary guests. *ACS Sustainable Chemistry & Engineering* 1 (10): 1218–1224.

47 Dong, H., Fan, Z., Wang, B., Xue, S., Zhao, J., and Song, Y. (2017). Hydrate-based reduction of heavy metal ion from aqueous solution. *Energy Procedia* 105: 4706–4712.

48 Song, Y. *et al.* (2016). Hydrate-based heavy metal separation from aqueous solution. *Scientific Reports* 6 (1): 1–8.

49 Kang, K.C., Linga, P., Park, K., Choi, S.-J., and Lee, J.D. (2014). Seawater desalination by gas hydrate process and removal characteristics of dissolved ions (Na+, K+, Mg2+, Ca2+, B3+, Cl−, SO42−). *Desalination* 353: 84–90.

50 Wijayanti, W. (2018). The stirrer rotation effects of gas hydrate performance in the hydrate crystallizer. *AIP Conference Proceedings* 1983 (1): 20012.

51 Veluswamy, H.P., Kumar, S., Kumar, R., Rangsunvigit, P., and Linga, P. (2016). Enhanced clathrate hydrate formation kinetics at near ambient temperatures and moderate pressures: application to natural gas storage. *Fuel* 182: 907–919. doi: 10.1016/j.fuel.2016.05.068.

52 Babu, P. Nambiar, A., He, T., Karimi, I.A., Lee, J.D., Englezos, P., & Linga, P. (2018). A review of clathrate hydrate based desalination to strengthen energy–water nexus. *ACS Sustainable Chemistry & Engineering* 6 (7): 8093–8107.

53 Fakharian, H., Ganji, H., and Naderifar, A. (2017). Saline produced water treatment using gas hydrates. *Journal of Environmental Chemical Engineering* 5 (5): 4269–4273.

54 Fakharian, H., Ganji, H., Naderifar, A., Mofrad, H.R., and Kakavand, M. (2019). Effect of gas type and salinity on performance of produced water desalination using gas hydrates. *Journal of Water Reuse and Desalination* 9 (4): 396–404.

55 Li, X.S., Yang, B., Zhang, Y., Li, G., Duan, L.P., *et al.*, (2012). Experimental investigation into gas production from methane hydrate in sediment by depressurization in a novel pilot-scale hydrate simulator. *Applied Energy* 93: 722–732.

56 Li, B., Li, X.-S., and Li, G. (2014). Kinetic studies of methane hydrate formation in porous media based on experiments in a pilot-scale hydrate simulator and a new model. *Chemical Engineering Science* 105: 220–230.

57 Peng, B.Z., Dandekar, A., Sun, C.Y. Luo, H., Ma, Q.L., Pang, W.X., and Chen, G.J. (2007). Hydrate film growth on the surface of a gas bubble suspended in water. *The Journal of Physical Chemistry B 111* (43): 12485–12493.

58 Yin, Z., Khurana, M., Tan, H.K., and Linga, P. (2018). A review of gas hydrate growth kinetic models. *Chemical Engineering Journal* 342: 9–29.

59 Li, S.L., Sun, C.Y., Liu, B., Feng, X.J., Li, F.G., Chen, L.T., and Chen, G.J. (2013). Initial thickness measurements and insights into crystal growth of methane hydrate film. *AIChE Journal 59* (6): 2145–2154.

60 Li, S.-L., Sun, C.-Y., Liu, B., Li, Z.-Y., Chen, G.-J., Sum, A.K. (2014). New observations and insights into the morphology and growth kinetics of hydrate films. *Scientific Reports* 4 (1): 1–6.

61 Xu, C.-G. and Li, X.-S. (2015). Research progress on methane production from natural gas hydrates. *RSC Advances* 5 (67): 54672–54699.

62 Kim, H.C., Bishnoi, P.R., Heidemann, R.A., and Rizvi, S.S.H. (1987). Kinetics of methane hydrate decomposition. *Chemical Engineering Science* 42 (7): 1645–1653.

63 Mochizuki, T. and Mori, Y.H. (2006). Clathrate-hydrate film growth along water/hydrate-former phase boundaries—numerical heat-transfer study. *Journal of Crystal Growth* 290 (2): 642–652.

64 Yang, S.H.B., Babu, P., Chua, S.F.S., and Linga, P. (2016). Carbon dioxide hydrate kinetics in porous media with and without salts. *Applied Energy* 162: 1131–1140.

7

Artificial Intelligence in Water Treatment Process Optimization

Jai Krishna Sahith and Bhajan Lal

7.1 Introduction

Intelligent process optimization for drinking and wastewater treatment facilities allows for the automation of the treatment process and the provision of real-time decision assistance to treatment plant operators. Historically, treatment plants have been run in accordance with mechanisms from established first principles in the form of state–space models. Problematically, the quality of the source water (i.e. influent) at each treatment plant can vary substantially, and thus different treatment procedures, chemical dosages, and so on are needed to achieve the desired results.

For consumer satisfaction and to be deemed fit for consumption, treated water must fulfill criteria related to aesthetics (taste, odor, and color). More importantly, endproduct water quality must meet government health and environmental standards. The material (i.e. chemicals), energy, and labor costs associated with meeting these water quality criteria are high. Businesses, water utilities, and municipalities spend roughly US$76 billion per annum globally to run treatment facilities. Plant operators can get valuable knowledge that helps them to fine-tune plant operations to satisfy water quality criteria at a lesser cost; however, this takes time and domain expertise to achieve. The operational efficiency of a treatment plant relies on the expertise and judgment of its operators. However, carrying over the experience of previous operators to future personnel can be difficult, particularly when documentation is limited.

Implementation of artificial intelligence (AI) technology on a digital copy of the treatment processes allow for considerable cost savings in treatment plant procedures. Self-learning AI systems can analyze trends from historical and

Gas Hydrate in Water Treatment: Technological, Economic, and Industrial Aspects,
First Edition. Bhajan Lal and Sirisha Nallakukkala.

present sensor data in real-time to manage the processes and give performance forecasts and action recommendations. Overall, AI technology can improve the consistency of water quality and the efficiency of water treatment plants, i.e. meet the water quality standards for a greater percentage of the operating lifespan. This offers water utilities and municipalities an opportunity to increase consumer happiness, attain requisite business demands, and reduce environmental consequences.

The application of AI in water treatment, however, is still in its early stages. One significant issue is the development of a model capable of replicating the chemistry of water treatment operations within a treatment facility. Without a digital recreation based on sound theory, the AI may give poor decision assistance when confronted with input data that differs considerably from past data. Another issue is the prerequisite for historical data, which is dependent on having an adequate number of working sensors during the treatment procedure. Without the right amount or quality of historical data, the AI can struggle to achieve the right solutions for optimizing the treatment processes.

7.2 Background Information

Wastewater treatment facilities (WWTPs) have historically been constructed in accordance with conservative design principles and regulations based on traditional design practices. Procedures are frequently transferred from operator to operator without regard for innovative optimization that might enhance performance or save costs. Experience has generally shown that WWTPs frequently have far more capacity than the rated capacity provided during the design stage.

When creating a predictive model from a water treatment system data set, the typical first step is to preprocess the data to address any missing data, redundant observations, and records containing incorrect data. Traditional data analysis techniques have several limitations, including the inability to discover singularities inside too large of a data jumble. Technological advancements, on the other hand, may provide new ways of producing and storing knowledge.

Part of the traditional AI methods, namely knowledge discovery in databases (KDDB) and its tools and approaches for problem-solving, has been improved since its adoption in 1989 [1]. The KDDB method involves automated detection and recognition of patterns in a database, i.e. collecting relevant, unknown information that might be used in a decision-making process without the need for prior formulation of a hypothesis [2].

Within the field of ecological mining, various data mining methods, including decision trees (DTs), artificial neural networks (ANNs), and genetic algorithms (GA), have been employed to discover patterns in water quality information [3]. Although DTs have not been widely utilized in ecological modeling, they do offer the benefit of directly stating regularities and making it easier to check for ecological validity [4–6].

Currently, treated water quality is assessed using analytical methods after the treatment process is complete, which is a rather limited approach. Using data mining-based models in combination with a decision support system is preferable for quality control. In practice, these models are valuable in determining the conditions that cause high levels of manganese and turbidity in treated water, allowing us to predict issues and create contingency plans to avoid them [7].

7.3 Optimization of Water Treatment Plants

The operation of a WWTP is frequently influenced by several physical, chemical, and biological variables. Effluent measurements are not adequate for tracking these during operation. Rather, predicting these characteristics based on the quality of the influent water is more helpful, as it allows the operator to control the system and take necessary safeguards before a problem arises [8]. The use of modern control strategies is a must due to increasingly stringent regulation of effluent quality.

The primary focus of WWTPs has often been meeting water quality requirements to maintain public confidence and avoid legal fines, with less emphasis on energy conservation. As a result, few WWTPs are constructed with energy efficiency goals in mind [9, 10]. However, this attitude has shifted in recent years toward achieving the EU 20-20-20 climate and energy targets outlined in Directive 2009/28/EU. Several technologies have been used by WWTPs in recent years to enhance treatment efficiency and comply with the increasingly stringent discharge regulations. Proper mathematical models in wastewater treatment engineering can significantly assist engineers as they work toward optimizing WWTPs. Mathematical models can describe energy consumption and production, as well as biological activities and hydraulic and settling phenomena, in the WWTP's various treatment units. Simulated ideal operational settings can then be suggested to the management firm [11]. Furthermore, scenarios of various operational modes and weather circumstances can be analyzed and utilized as a decision-making tool to enhance the management company's preparation for more demanding situations.

Traditional wastewater treatment creates enormous amounts of primary and secondary sludge, which are referred to collectively as sewage sludge. According to the literature, the specific production of secondary, or waste-activated (WAS), sludge ranges between 22 and 30 g VS (volatile solids) per inhabitant per day. Both primary sludge and WAS should be carefully treated before final disposal since they include decomposable organic waste, pathogens, and dangerous insect eggs. Furthermore, numerous studies have shown that sludge management expenses might account for up to 50–60% of overall WWTP operating costs [12, 13].

In a WWTP's sludge line, sewage sludge is typically thickened to average total solid (TS) levels in the range of 3–6% before being treated anaerobically to reduce volumes and make the sludge more biologically stable [14]. Anaerobic digestion processes are typically carried out in a mesophilic temperature range (35–40°C), implying that heat input is required to raise the sludge temperature and compensate for the digester's heat loss to the outside environment [15]. The procedure for sludge stabilization should aim to achieve a neutral equilibrium between positive and negative components in the energy analysis. Primary sludge is composed of readily biodegradable materials and emits an average of 0.280 Nm^3 methane (CH_4) per kilogram VS added [16]. WAS, on the other hand, produces very little CH_4 within its hydraulic retention time in conventional mesophilic processes (18–20 d). Specific CH_4 production (SMP) in WAS seldom exceeds 0.100 Nm^3/kg VS added. The structure of WAS, in which the biodegradable material is either contained in the microbial cell wall or entangled in an extracellular polymeric matrix, contributes to the restricted CH_4 generation [17]. Several treatments (dubbed pretreatments since they are administered prior to digestion) can be employed to make the biodegradable material of WAS more available to the anaerobic process and, as a result, boost the CH_4 generation [18].

Borzooei et al. carried out a study on optimizations ranging from energy saving to environmental effect reduction. They proposed a multi-objective, integrated method based on modeling and experimental research for enhancing the energy efficiency of Castiglione Torinese, Italy's largest WWTP. In this method, wastewater and sludge treatment units are extensively studied and detailed steps are discussed elsewhere to identify potential solutions to improve the system's energy efficiency and the ability to adapt to any WWTP [19].

Historically, the management of wastewater treatment facilities (WWTPs) has been centered on lowering operating costs while meeting effluent discharge limitations. Running parameters, such as solids retention time, aeration rate, and internal recycling flow rate, have the greatest impact on effluent quality and plant operating costs. Optimal operation is a critical component of modern wastewater treatment management performance [20, 21]. Because of the increased emphasis on sustainable infrastructure operations, decreasing greenhouse gas (GHG) emissions must necessarily become part of optimization goals. As a result, the operation of WWTPs becomes more complicated, and the potential for trade-offs between the intrinsic jobs of wastewater treatment (i.e. aquatic pollutant removal at low cost) and the sustainable plant tasks (e.g. reduce global warming impact) increases. Because all concerns may not be addressed fully, compromises will have to be made to achieve good overall performance.

So far, system analysis of activated sludge process management techniques has been primarily concerned with quantifying environmental and economic performance [22, 23]. The major difficulty in assessing these performance criteria is that the quantities generally have different units and magnitudes; thus, their values may not be comparable. It is considerably easier for decision-makers to examine outcomes and discover the most successful aspects of an operational strategy when the system objectives can be measured using a simple quantitative term. Therefore, an efficient method permits quantitative comparisons of management and control techniques using various choice criteria.

Using an optimization method with an integrated index is one way to determine the best set of options for managing the complicated nonlinear system characteristic of WWTPs. Optimization challenges related to the analysis and operational design of WWTPs have spurred interest from several scholars. Most recent research has concentrated on either model calibration (estimation of model parameters) [24, 25] or optimization of process design and control in various configurations [26–30]. However, only a few have attempted to address wastewater treatment difficulties by executing optimum operations with various objectives using optimization algorithms. Furthermore, no attempt has been made to assess plant-wide performance using optimization approaches. Finally, while much emphasis has been given to reducing GHG emissions in wastewater treatment facilities, no evaluation of system cost and water quality performance, as well as GHG emissions, has been completed using optimization algorithms.

Kim et al. used optimization approaches to discover optimal operating conditions for lowering both operating costs and GHG emissions while meeting effluent discharge restrictions. This was achieved by using a performance index that integrated numerous criteria into a single index. The environmental effects produced by pollutant discharges and the rate of GHG emissions were quantified and then translated into normalized scores, along with operational expenses. The integrated index may make it easier to evaluate system performance and compare the effects of various operating circumstances while also providing the ability to forecast both direct and indirect carbon dioxide (CO_2), nitrous oxide (N_2O), and CH_4 emissions [31].

7.4 Application of Artificial Neural Networks for Freshwater Treatment

The primary goals of a drinking water treatment plant are to: (i) produce safe drinking water for consumption (free of pathogenic microorganisms and toxic compounds), (ii) produce appealing drinking water (free of color, taste, and odor), and (iii) to avoid solids accumulation, corrosion, and bacterial growth in distribution and transport pipelines. These must be accomplished at the lowest possible cost and with the least possible environmental effect [32]. Water treatment often includes clarification and disinfection. The traditional purification system (coagulation/sedimentation and filtration) is mainly employed to remove turbidity. Sedimentation is the process of separating solid liquids using gravitational settling to remove suspended particles.

Filtration is the process of passing water through a porous material, leaving the filtrate to be of higher quality than the influent. To increase particle agglomeration, the coagulation process employs chemical coagulation (aluminum or iron salt). Chemical dosing, fast mixing, and gradual mixing are all part of the chemical coagulation method. Any anomalous or excessive turbidity in raw water quality will require a corresponding adjustment in the quantity of chemical coagulant used and pH. As a result, the coagulant may be under or overdosed, resulting in additional chemical expenses. Because of the complexity of drinking water treatment processes, ANN has gained attraction as a plant modeling option.

The history, capabilities, types, structure, and learning algorithm of neural networks have been discussed at length [33–35]. ANNs have been successfully utilized in many fields of water technology as data-driven empirical models, such as:

- Bioleaching of metals in municipal sludge [36].
- Rate of wastewater influx forecast [37].
- Cryptosporidium and Giardia peak concentrations [38].
- Membrane technology modeling [39].
- Breakthrough time prediction in a fixed-bed adsorption system [40].
- Modeling coagulant [41].
- Inflow after backwashing and during ultrafiltration [42].
- To locate nonpoint sources of fecal pollution [43].

According to Haykin [44], a neural network is "a massively parallel distributed processor with a natural proclivity for storing experience information and making it available for application." Neural networks are only one branch of AI, though it is often synonymously referred to as machine learning. The ANN is based on a model of the human neurological system, which is made up of informational nodes (called neurons) that are linked. A conventional three-layer back propagation neural network with N input nodes, L hidden nodes, and one output node is typical. The weight matrix, bias vector, and transfer function of a layer of neurons influence its behavior, and while the optimal number of hidden neurons can be determined by trial and error, the tuning of other hyperparameters is potentially more beneficial. The output layer's number of neurons equals the number of intended outputs, and the hidden layer transmits information from the input layer to the output layer.

Ogwueleka et al. created an ANN model for the process and cost optimization of drinking water treatment operations [45]. Cost reduction, process safety improvement, and high stability were observed. The results of the ANN model demonstrated that it is a viable technique for improving overall process performance and cost-effectiveness in drinking water treatment.

7.5 Application of Artificial Neural Networks for Wastewater Treatment

ANNs are commonly utilized for evaluating and modeling nonlinear data in the form of input–output relationships. Multivariate statistical approaches are typically the most common approach; however, they are only useful when there are linear connections between variables [46]. ANNs, on the other hand, can evaluate partial data sets, imprecise or missing information, and very complicated and imprecise issues in which people often make choices based on intuition [47, 48].

Wastewater treatment influent characteristics such as biochemical oxygen demand (BOD), suspended solids (SS), ammonia nitrogen (NH_3-N), and chemical oxygen demand (COD) are key variables in determining the volume and intensity of municipal wastewater pollution. As a result, their concentrations in wastewater have an impact on effluent quality [49, 50]. The linkage of these parameters, influent and effluent, is always a concern for wastewater treatment plant (WWTP) process engineers. However, the input to the WWTP varies, as do the parameter concentrations, and these variations impede the development of a connection between input and output parameters needed for forecasting the effluent strength [51]. Furthermore, all WWTPs are unique since the features of the influents are time-dependent and depend on the nature of the local lifestyle. The presence of bioorganic components alters a plant's performance, introducing more variability. These variables add to the difficulty of monitoring and managing the WWTP process. Such issues might cause damage to the bioreactors or even the ecosystem if untreated wastewater makes its way into the environment [52].

The primary goal of each WWTP is to remove contaminants that are present in wastewater. If contaminants are released into waterways before treatment, they can have severe consequences for humans and aquatic ecosystems. Due to the complicated structure of the WWTP, complying with new limits of wastewater effluent quality established by law enforcement agencies proves to be difficult. Maintaining the effluent quality may be accomplished by building a model for forecasting plant performance based on historical observations of key plant parameters [53]. Monitoring and controlling influent parameters requires a plant-specific strategy, necessitating an understanding of the individual plant's performance as well as the factors that influence the influent characteristics such as time, season, and the nature of the local people's lifestyle. As in freshwater operations, the option of using ANN was adopted for analyzing the dynamic connections between the data in a WWTP [54].

Jami et al. used the ANN toolbox within MATLAB to forecast the performance of a WWTP by developing prediction models using two years' worth of measured historical data. The information was collected from a sewage treatment plant in Malaysia. The influent BOD, SS, and COD were used as input parameters, while the output parameters were a mixture of the effluent properties. Networks with a single input and a single output were compared to those with a single input and many outputs. The ANNs were created for both raw and filtered data, and the outcomes for both networks were compared. It was discovered that data filtering is required to develop a better ANN model. According to the regression study, networks with one hidden layer and 20 neurons are the best for the single input–single output method. The proposed multiple inputs–single output ANN models may be used to investigate how wastewater characteristics such as BOD, COD, and SS interact with one another [55].

Tümer and Edebali examined the modeling of a wastewater treatment facility in Konya, Turkey using ANNs with different topologies. The plant's treatment efficiency was calculated by comparing input values of pH, temperature, COD, total dissolved solids (TSS), and BOD to output values of TSS. The model's performance was examined using mean squared error (MSE) and correlation coefficient (R^2). It was found that the ANN can forecast plant performance with an R^2 of up to 0.96 between the observed and projected output variables [56].

Messaoud et al. examined the Ain Beida wastewater treatment facility using an ANN with several designs in MATLAB. The plant's treatment efficiency was calculated by taking into consideration the input values of pH, temperature, conductivity, COD, TSS, five-day BOD (BOD_5), total nitrogenous (TN) content, and total phosphorus (TP) content, as well as the output values of COD, TSS, BOD_5, TN, and TP. Similarly, the model's performance was examined using MSE and R^2. The resulting networks could predict plant performance with R^2 correlations of 0.9934 for BOD_5, 0.9654 for COD, 0.9639 for TSS, 0.9453 for TN, and 0.9428 for P-PO_4 [57].

Luis et al. suggested an intelligent system that analyzes process data to anticipate its behavior and help in decision-making. A multilayer perceptron neural network with two hidden layers and 22 neurons was used to build the system and process variable analysis, time-series decomposition, correlation, and autocorrelation algorithms as the preprocessing steps. Within a 24h window, the model estimated the bioreactor's COD projection by a mean absolute percentage error of 10.8%, placing this work between the acceptable ranges provided in the literature [58].

7.6 Other Artificial Intelligence Techniques for Wastewater Treatment

Based on various issues associated with the measurement of wastewater quality parameters such as BOD, COD, and TSS, Bayat-Varkeshi et al. set out to discover the optimal ANN model and predict complex wastewater quality data by accessing the ANN outcomes. The prediction of BOD, COD, and TSS removal and selection of optimal topology was performed by two models: (i) an artificial neural network–genetic algorithm (ANN–GA) and (ii) a co-active neuro-fuzzy logic inference system (CANFIS). The developed model improved efficiency in wastewater treatment since it predicted the output parameters based on the wastewater input characteristics. The results showed that the GA approach delivers more accurate results than the fuzzy logic technique. Despite the model's poorer performance, its reduced number of input parameters provides a cost advantage in sampling, as well as widens its use to instances where only a limited number of samples are available [59].

7.7 Application on Gas Hydrate Plants

The gas hydrate desalination process can be optimized on the following factors:

- Guest gas
- Subcooling conditions
- Type and amount of promoter
- Salt rejection rate.

These key factors can be optimized through experimentation and by developing a correlation that satisfies the optimized conditions for maximum water recovery. Experimental results can be further used for developing a unified AI model for the optimization process, similar to how it was performed for conventional desalination plants.

References

1 Fayyad, U.M., Piatetsky-Shapiro, G., Smyth, P., and Uthurusamy, R., eds. (1996). *Advances in Knowledge Discovery and Data Mining*, American. Cambridge MA USA: AAAI Press.

2 Thuraisingham, B. (1998). *Data Mining: Technologies, Techniques, Tools, and Trends*. Boca Raton, FL: CRC Press.

3 Berk, R.A. (2005). Data Mining within a Regression Framework. In *Data Mining and Knowledge Discovery Handbook* (pp. 231-255). Boston, MA USA: Springer.

4 Chau, K.W. (2006). A review on integration of artificial intelligence into water quality modelling. *Marine Pollution Bulletin.* 52: 726–733. doi: 10.1016/j.marpolbul.2006.04.003.

5 Kuo, J.T., Hsieh, M.H., Lung, W.S., and She, N. (2007). Using artificial neural network for reservoir eutrophication prediction. *Ecological Modelling.* 200: 171–177. doi: 10.1016/j.ecolmodel.2006.06.018.

6 Koroteev, D. and Tekic, Z. (2021). Artificial intelligence in oil and gas upstream: Trends, challenges, and scenarios for the future. *Energy AI* 3: 100041. doi: 10.1016/j.egyai.2020.100041.

7 Turban, E., Aronson, J.E., and Liang, T.-P. (2004). *Decision Support Systems and Intelligent Systems*, 7e. USA: Prentice-Hall, Inc.

8 Lee, D.S. and Park, J.M. (1999). Neural network modeling for on-line estimation of nutrient dynamics in a sequentially-operated batch reactor. *Journal of Biotechnology.* 75: 229–239. doi: 10.1016/S0168-1656(99)00171-6.

9 Ahmetović, E., Ibrić, N., and Kravanja, Z. (2014). Optimal design for heat-integrated water-using and wastewater treatment networks. *Applied Energy.* 135: 791–808. doi: 10.1016/j.apenergy.2014.04.063.

10 Panepinto, D., Fiore, S., Zappone, M., Genon, G., and Meucci, L. (2016). Evaluation of the energy efficiency of a large wastewater treatment plant in Italy. *Applied Energy.* 161: 404–411. doi: 10.1016/j.apenergy.2015.10.027.

11 Borzooei, S., Zanetti, M.C., Lorenzi, E., and Scibilia, G. (2017). Performance investigation of the primary clarifier- Case study of Castiglione Torinese. *Lecture Notes in Civil Engineering.* 4: 138–145. doi: 10.1007/978-3-319-58421-8_21.

12 Coma, M., Rovira, S., Canals, J., and Colprim, J. (2013). Minimization of sludge production by a side-stream reactor under anoxic conditions in a pilot plant. *Bioresource Technology.* 129: 229–235. doi: 10.1016/j.biortech.2012.11.055.

13 Pilli, S., More, T., Yan, S., Tyagi, R.D., and Surampalli, R.Y. (2015). Anaerobic digestion of thermal pre-treated sludge at different solids concentrations – Computation of mass-energy balance and greenhouse gas emissions. *Journal of Environmental Management.* 157: 250–261. doi: 10.1016/j.jenvman.2015.04.023.

14 Ruffino, B., Campo, G., Genon, G., Lorenzi, E., Novarino, D., Scibilia, G., and Zanetti, M. (2015). Improvement of anaerobic digestion of sewage sludge in a wastewater treatment plant by means of mechanical and thermal pre-treatments: Performance, energy and economical assessment. *Bioresource Technology*. 175: 298–308. doi: 10.1016/j.biortech.2014.10.071.

15 Hreiz, R., Adouani, N., Jannot, Y., and Pons, M.N. (2017). Modeling and simulation of heat transfer phenomena in a semi-buried anaerobic digester. *Chemical Engineering Research & Design*. 119: 101–116. doi: 10.1016/j.cherd.2017.01.007.

16 Ruffino, B., Cerutti, A., Campo, G., Scibilia, G., Lorenzi, E., and Zanetti, M. (2019). Improvement of energy recovery from the digestion of waste activated sludge (WAS) through intermediate treatments: The effect of the hydraulic retention time (HRT) of the first-stage digestion. *Applied Energy*. 240: 191–204. doi: 10.1016/j.apenergy.2019.02.061.

17 Grübel, K. and Suschka, J. (2015). Hybrid alkali-hydrodynamic disintegration of waste-activated sludge before two-stage anaerobic digestion process. *Environmental Science and Pollution Research*. 22: 7258–7270. doi: 10.1007/s11356-014-3705-y.

18 Cano, R., Pérez-Elvira, S.I., and Fdz-Polanco, F. (2015). Energy feasibility study of sludge pretreatments: A review. *Applied Energy*. 149: 176–185. doi: 10.1016/j.apenergy.2015.03.132.

19 Borzooei, S., Campo, G., Cerutti, A., Meucci, L., Panepinto, D., Ravina, M., Riggio, V., Ruffino, B., Scibilia, G., and Zanetti, M. (2019). Optimization of the wastewater treatment plant: From energy saving to environmental impact mitigation. *The Science of the Total Environment*. 691: 1182–1189. doi: 10.1016/j.scitotenv.2019.07.241.

20 Ostace, G.S., Cristea, V.M., and Agachi, P.Ş. (2011). Cost reduction of the wastewater treatment plant operation by MPC based on modified ASM1 with two-step nitrification/denitrification model. *Computers & Chemical Engineering*. 35: 2469–2479. doi: 10.1016/j.compchemeng.2011.03.031.

21 Yoon, S.H., Kim, H.S., and Yeom, I.T. (2004). The optimum operational condition of membrane bioreactor (MBR): Cost estimation of aeration and sludge treatment. *Water Research*. 38: 37–46. doi: 10.1016/j.watres.2003.09.001.

22 Benedetti, L., Dirckx, G., Bixio, D., Thoeye, C., and Vanrolleghem, P.A. (2008). Environmental and economic performance assessment of the integrated urban wastewater system. *Journal of Environmental Management*. 88: 1262–1272. doi: 10.1016/j.jenvman.2007.06.020.

23 Quadros, S., Rosa, M.J., Alegre, H., and Silva, C. (2010). A performance indicators system for urban wastewater treatment plants. *Water Science and Technology*. 62: 2398–2407. doi: 10.2166/wst.2010.526.

24 Fang, F., Ni, B.J., and Yu, H.Q. (2009). Estimating the kinetic parameters of activated sludge storage using weighted non-linear least-squares and accelerating genetic algorithm. *Water Research.* 43: 2595–2604. doi: 10.1016/j.watres.2009.01.002.

25 Wu, Y. and Liu, S. (2012). Automating calibration, sensitivity and uncertainty analysis of complex models using the R package Flexible Modeling Environment (FME): SWAT as an example. *Environmental Modelling and Software.* 31: 99–109. doi: 10.1016/j.envsoft.2011.11.013.

26 Balku, S. and Berber, R. (2006). Dynamics of an activated sludge process with nitrification and denitrification: Start-up simulation and optimization using evolutionary algorithm. *Computers & Chemical Engineering.* 30: 490–499. doi: 10.1016/j.compchemeng.2005.10.014.

27 Bournazou, M.N.C., Hooshiar, K., Arellano-Garcia, H., Wozny, G., and Lyberatos, G. (2013). Model based optimization of the intermittent aeration profile for SBRs under partial nitrification. *Water Research.* 47: 3399–3410. doi: 10.1016/j.watres.2013.03.044.

28 Holenda, B., Domokos, E., Rédey, Á., and Fazakas, J. (2008). Dissolved oxygen control of the activated sludge wastewater treatment process using model predictive control. *Computers & Chemical Engineering.* 32: 1270–1278. doi: 10.1016/j.compchemeng.2007.06.008.

29 Maere, T., Verrecht, B., Moerenhout, S., Judd, S., and Nopens, I. (2011). BSM-MBR: A benchmark simulation model to compare control and operational strategies for membrane bioreactors. *Water Research.* 45: 2181–2190. doi: 10.1016/j.watres.2011.01.006.

30 Rivas, A., Irizar, I., and Ayesa, E. (2008). Model-based optimisation of wastewater treatment plants design. *Environmental Modelling and Software.* 23: 435–450. doi: 10.1016/j.envsoft.2007.06.009.

31 Kim, D., Bowen, J.D., and Ozelkan, E.C. (2015). Optimization of wastewater treatment plant operation for greenhouse gas mitigation. *Journal of Environmental Management.* 163: 39–48. doi: 10.1016/j.jenvman.2015.07.005.

32 Van Dijk, J.C. and Van Der Kooij, D. (2004). Water Quality 21 research programme for water supplies in The Netherlands. *Water Science and Technology: Water Supply.* 4: 181–188. doi: 10.2166/ws.2004.0107.

33 Göb, S., Oliveros, E., Bossmann, S.H., Braun, A.M., Nascimento, C.A.O., and Guardani, R. (2001). Optimal experimental design and artificial neural networks applied to the photochemically enhanced Fenton reaction. *Water Science and Technology.* 44: 339–345. doi: 10.2166/wst.2001.0321.

34 Loke, E., Warnaars, E.A., Jacobsen, P., Nelen, F., and Do Céu, M. (1997). Almeida, Artificial neural networks as a tool in urban storm drainage. *Water Science and Technology.* 36: 101–109. doi: 10.1016/S0273-1223(97)00612-4.

35 Lek, S., Guiresse, M., and Giraudel, J.L. (1999). Predicting stream nitrogen concentration from watershed features using neural networks. *Water Research.* 33: 3469–3478. doi: 10.1016/S0043-1354(99)00061-5.

36 Laberge, C., Cluis, D., and Mercier, G. (2000). Metal bioleaching prediction in continuous processing of municipal sewage with Thiobacillus ferrooxidans using neural networks. *Water Research.* 34: 1145–1156. doi: 10.1016/S0043-1354(99)00246-8.

37 El-Din, A.G. and Smith, D.W. (2002). A neural network model to predict the wastewater inflow incorporating rainfall events. *Water Research.* 36: 1115–1126. doi: 10.1016/S0043-1354(01)00287-1.

38 Neelakantan, T.R., Brion, G.M., and Lingireddy, S. (2001). Neural network modelling of Cryptosporidium and Giardia concentrations in the Delaware River, USA. *Water Science and Technology.* 43: 125–132. doi: 10.2166/wst.2001.0723.

39 Strugholtz, S., Panglisch, S., Gebhardt, J., and Gimbel, R. (2008). Neural networks and genetic algorithms in membrane technology modelling. *Journal of Water Supply: Research and Technology-Aqua.* 57: 23–34. doi: 10.2166/aqua.2008.008.

40 Basheer, I.A. and Najjar, Y.M. (2005). Designing and analyzing fixed-bed adsorption systems with artificial neural networks. *Journal of Environmental Sciences.* 23: 291–312. doi: 10.2190/6auv-9eh6-wky8-g5je.

41 Maier, H.R., Morgan, N., and Chow, C.W.K. (2004). Use of artificial neural networks for predicting optimal alum doses and treated water quality parameters. *Environmental Modelling and Software.* 19: 485–494. doi: 10.1016/S1364-8152(03)00163-4.

42 Teodosiu, C., Pastravanu, O., and Macoveanu, M. (2000). Neural network models for ultrafiltration and backwashing. *Water Research.* 34: 4371–4380. doi: 10.1016/S0043-1354(00)00217-7.

43 Brion, G.M. and Lingireddy, S. (1999). A neural network approach to identifying non-point sources of microbial contamination. *Water Research.* 33: 3099–3106. doi: 10.1016/S0043-1354(99)00025-1.

44 Haykin, S. (2002). *Adaptive filter theory*. Publishing House of Electronics Industry. Pearson 4 (96): 469–490.

45 Science, C. (2009). Optimization of drinking water treatment processes using artificial neural network. *Nigerian Journal of Technology.* 28: 16–25.

46 Oliveira-Esquerre, K.P., Seborg, D.E., Bruns, R.E., and Mori, M. (2004). Application of steady-state and dynamic modeling for the prediction of the BOD of an aerated lagoon at a pulp and paper mill Part I. Linear approaches. *Chemical Engineering Journal.* 104: 73–81. doi: 10.1016/j.cej.2004.05.011.

47 Şencan, A. and Kalogirou, S.A. (2005). A new approach using artificial neural networks for determination of the thermodynamic properties of fluid couples. *Energy Conversion and Management.* 46: 2405–2418. doi: 10.1016/j.enconman.2004.11.007.

48 Ráduly, B., Gernaey, K.V., Capodaglio, A.G., Mikkelsen, P.S., and Henze, M. (2007). Artificial neural networks for rapid WWTP performance evaluation: Methodology and case study. *Environmental Modelling and Software.* 22: 1208–1216. doi: 10.1016/j.envsoft.2006.07.003.

49 Jasim, N.A. (2020). The design for wastewater treatment plant (WWTP) with GPS X modelling. *Cogent Engineering.* 7. doi: 10.1080/23311916.2020.1723782.

50 Hamed, M.M., Khalafallah, M.G., and Hassanien, E.A. (2004). Prediction of wastewater treatment plant performance using artificial neural networks. *Environmental Modelling and Software.* 19: 919–928. doi: 10.1016/j.envsoft.2003.10.005.

51 Hong, Y.S.T., Rosen, M.R., and Bhamidimarri, R. (2003). Analysis of a municipal wastewater treatment plant using a neural network-based pattern analysis. *Water Research.* 37: 1608–1618. doi: 10.1016/S0043-1354(02)00494-3.

52 Mjalli, F.S., Al-Asheh, S., and Alfadala, H.E. (2007). Use of artificial neural network black-box modeling for the prediction of wastewater treatment plants performance. *Journal of Environmental Management.* 83: 329–338. doi: 10.1016/j.jenvman.2006.03.004.

53 Shetty, G.R. and Chellam, S. (2003). Predicting membrane fouling during municipal drinking water nanofiltration using artificial neural networks. *Journal of Membrane Science.* 217: 69–86. doi: 10.1016/S0376-7388(03)00075-9.

54 Hanbay, D., Turkoglu, I., and Demir, Y. (2008). Prediction of wastewater treatment plant performance based on wavelet packet decomposition and neural networks. *Expert Systems with Applications.* 34: 1038–1043. doi: 10.1016/j.eswa.2006.10.030.

55 Jami, M.S., Husain, I.A.F., Kabashi, N.A., and Abdullah, N. (2012). Multiple inputs artificial neural network model for the prediction of wastewater treatment plant performance. *Australian Journal of Basic and Applied Sciences.* 6: 62–69.

56 Tümer, A.E. and Edebali, S. (2015). An artificial neural network model for wastewater treatment plant of Konya. *International Journal of Intelligent Systems and Applications in Engineering* 3 (4): 131–135.

57 Djeddou, M., Aouatef, H., and Loukam, M. (2019). Wastewater treatment plant performances modelling using artificial neural networks. ecoSTP18 Conference Ecotechnologies for Wastewater Treatment At: London, Ontario, Canada, 1–10.

58 Arismendy, L., Cárdenas, C., Gómez, D., Maturana, A., Mejía, R., Christian, G., and Quintero, M. (2020). Intelligent system for the predictive analysis of an industrial wastewater treatment process. *Sustainability* 12. doi: 10.3390/SU12166348.

59 Varkeshi, M.B., Godini, K., ParsiMehr, M., and Vafaee, M. (2019). Predicting the performance of gorgan wastewater treatment plant using ANN-GA, CANFIS, and ANN models. *Avicenna Journal of Environmental Health Engineering* 6: 92–99. doi: 10.34172/ajehe.2019.12.

8

Standard Analytical Techniques for Analysis of Wastewater

Sirisha Nallakukkala and Bhajan Lal

8.1 Methods, Scope, and Their Applications

The standard measurement approaches discussed here are used to analyze a wide range of waters like domestic, saline, groundwater, and industrial wastewater. Due to the specificity of wastewater treatment processes, analytical techniques are classified primarily based on the water type. Though attempts have been made to present universally applicable strategies, when different approaches are unavoidably essential for specimens of varying composition, the rationale for choosing the best approach is explained in detail. With unusual concentrations, the method should be modified by analysts.

For sludges and sediments, specially purposed methods are also proposed, with procedures and principles made to fit a number of scenarios. Nevertheless, these techniques might require alterations or be unsuitable for chemical sludge.

Most techniques, and the underlying principle of operations and applications, are discussed as well, though systems for analyzing bulk water treatment chemicals are not incorporated. The discussions made are intended as a foundation for the education and training of users. The deliberations presented are meant to assist, supplement, and smooth the development of more targeted test procedures.

8.2 Physical Properties of Water

The most common water sources are seawater, brackish groundwater, industrial wastewater, surface water containing specified dissolved solids, and formation

Gas Hydrate in Water Treatment: Technological, Economic, and Industrial Aspects,
First Edition. Bhajan Lal and Sirisha Nallakukkala.

water. The physical properties of water are associated with the appearance of water, namely, the color, temperature, turbidity, taste, electrical conductivity, and odor. Water must be free from all impurities that are offensive to the sense of sight, taste, or smell to be suitable for use. The physical characteristic of turbidity is particularly vital in these aspects. Turbidity is a measure of the clarity of water and is measured in turbidity units.

8.2.1 Color

Surface and ground waters have different colors, primarily resulting from the presence of organic matter, suspended particles, and aquatic humic substances. The presence of lignin, tannin, dyes, and other organic or inorganic substances causes the color of industrial wastewater. Visual comparison and spectrophotometric single-wavelength methods are used to measure color caused chiefly by natural organic matter. The spectrophotometric multiwavelength, tristimulus spectrophotometric, weighted-ordinate spectrophotometric techniques can be used to measure color resulting from any dissolved chemical substances that give color in the visible spectrum. These methods are specifically employed for colored waters [1].

8.2.1.1 Visual Comparison Method

In this method, color is determined with identifiable concentrations of the sample. The (platinum-cobalt) Pt-Co method is appropriate in treating natural, potable, and wastewaters. The color of the water is strongly pH-dependent and rises as the pH of water rises. Calculation of color units is given as in Eq. (8.1):

$$\text{Colour} = \frac{A * 50}{B} \tag{8.1}$$

where A = dilute sample color, and B = dilution sample, in mL.

8.2.1.2 Spectroscopic Single-wavelength Method

In the spectroscopic single-wavelength method, the color is estimated at a wavelength ranging from 450-465 nm with Pt-Co solutions [1–3]. It is applicable for natural, potable, and wastewaters. In this method, the color is measured without particulate matter removal. During measurement, the wavelength is set to develop standard curves, ensuring that the length of the cell path is the same as that utilized in the standard curve analysis. The sample color is estimated using absorbance, standard curve linking absorbance, and color unit. The unknown concentration of the sample is determined through the absorbance value obtained based on the calibration curve generated between absorbance and concentrations.

8.2.1.3 Spectrophotometric Multiwavelength Method

This process uses multiwavelength spectrophotometry for multiple chromophores to obtain the color of a filtered sample based on the light transmission characteristics. The technology has wide application in clinical chemistry [4] but is also applicable to both domestic and industrial wastewater samples. This approach involves at least one wavelength per component with an equal number of equations, in the form of Eq. (8.2):

$$A_\lambda = \sum_i \varepsilon_{\lambda i} b c_i \tag{8.2}$$

where A_λ is the observed absorbance at wavelength λ, $\varepsilon_{\lambda i}$ relates to the extinction coefficient of the ith component at a wavelength λ, b is the path length, and c_i is the concentration of the ith component.

Dual-wavelength spectrophotometry has seen commercial applications and proven advantageous for the following reasons [5].

- It is able to estimate the small change in absorbance in the existence of large absorbance background.
- Colorimetry without reference is possible if total absorbance is less than 0.01.
- With both wavelengths scanned, derivative spectra are produced.
- Special relative spectra can be generated by scanning one wavelength and keeping the second constant.
- In one vessel, two reactions can take place independently.

8.2.1.4 Tristimulus Spectrophotometric Method

The tristimulus method is the most well-established process according to the Commission Internationale de l'Elcairage (CIE). This method delivers three color values represented as X, Y, and Z coordinates. Conventional CIE color system approaches are based on reflectance factors and are meant to measure the surface color appearance of solid/opaque substances. They are not directly appropriate to analyze transparent or semi-transparent liquids. However, the tristimulus method helps to quantitatively approximate physiological color mixing.

The tristimulus method calculates luminance and three tristimulus values directly from the percentage transmittance value at three wavelengths: 590, 540, and 438 nm. The ChemScan Process Analyzer detects these measurements and analyzes the points on a chromaticity diagram. Outcomes are given in terms of dominant wavelength (hue), percentage luminance, and percent purity. Alternatively, the values can be converted to an American Dye Manufacturers Institute (ADMI) single number color difference value using the Adams-Nickerson color difference calculation [6]. Differences between sample measurements and standards can also be made using American Public Health Association (APHA) color values.

8.2.1.5 ADMI Weighted-ordinate Spectrophotometric Method

The three qualities of color are hue (colors such as red, blue, green, or yellow), chroma (intensity of color brightness/dullness), and value (brightness/darkness). The ADMI weighted-ordinate method measures the value of color, independent of the hue and the chroma, as a transmittance value. This system is applicable to colored water and wastewater. If two colors are visually determined to vary from the colorless state to a similar degree, their ADMI color standards will be alike [7]. It has an effective operating range of 400-700 nm.

8.2.2 Turbidity

The clearness of water is an important parameter intended for human consumption and is a vital determinant based on the color condition and the productivity of the aquatic system. Turbidity is caused by suspended colloidal debris, organic and inorganic material, and microorganisms in the water. Essentially, it is a measure of a liquid's relative clarity. More specifically, it is a measure of the amount of light dispersed by suspended material in the water. The higher the turbidity, the greater the intensity of scattered light. Several factors influence the cloudiness of water, such as phytoplankton and algal growth, sediment inputs from erosion, waste disposal, and runoff from urban areas.

Turbidity is measured in nephelometric turbidity units (NTU) using a nephelometer or turbidimeter. These tools estimate the light intensity dispersed at a 90° angle as a light beam passes through a water sample. Jackson turbidity units were commonly used in the past but are now considered archaic. The TSS (total suspended solids) concentration, otherwise a tedious and difficult metric to quantify, can be estimated using turbidity measurements. Nephelometers are especially preferred for measuring low turbidities because they are essentially unaffected by slight variations in design parameters.

The maximum permissible turbidity in drinking water, as specified by WHO, is below 5 NTU and preferably less than 1 NTU. One reason to control turbidity is purely for aesthetics; water that looks clean is likely to be more enjoyable to consume. However, removing turbidity is also necessary to successfully sterilize water for drinking purposes. This raises the cost of treating surface waters. The suspended particles in turbid water also aid in the attachment of heavy metals and a variety of other toxic organic compounds and pesticides.

8.2.2.1 Nephelometric Method

In this approach, the strength of the light distributed by the sample under specific circumstances is compared to the light intensity dispersed by a standard reference. The turbidity increases as the strength of dispersed light rises. The major standard reference suspension is made up of formazin polymer. The turbidity of a specific concentration of formazin suspension is 4000 NTU and can be measured in any kind of water that is devoid of debris and coarse silt.

A laboratory nephelometer consists of a light source for illuminating the sample and one or more photoelectric detectors to indicate the light intensity scattered at a 90° angle to the incident light path. The apparatus sensitivity should allow it to detect turbidity differences as low as 0.02 NTU in the water, within turbidities less than 1 NTU. For low turbidities, multiple ranges may be required to get acceptable coverage and sensitivity. Even when the same suspension is used for calibration, changes in equipment design will induce discrepancies in measured turbidity values.

Measurement techniques that minimize the effects of variations in the instrument are air bubbles and stray light. The sample must be measured immediately after extraction to avoid temperature variations, particle flocculation, and sedimentation. If aggregates appear to be flocculating, agitation should be used to break them up. When the sample is diluted or the temperature changes, the suspended particles within might disintegrate or alter their features. Before measuring the sample, remove any air or other entrained gases [8].

8.2.3 Odor

Both odor and taste result from a stimulating material coming into contact with the human receptor cell. The stimuli are chemical in nature; therefore, odor and taste are frequently referred to as chemical senses. Water, as a neutral medium, can easily trigger our receptors both through taste and olfactory senses. Water is odorless in its purest form, and any unpleasant tastes and odors are primarily imparted by organic and inorganic substances. Such substances may come from industrial or municipal waste, decomposed vegetable matter, and the breakdown of disinfectants or their derivatives.

Odor examinations are used to provide a subjective explanation of odor strength and approximate quantitative measures. One such examination is the threshold odor examination, which is built on a concept of bounds and measures intensity [9]. While this method is not generally favored [10], it has advantages [11]. Sensory tests can be used to assess the quality of both processed and unprocessed water and regulate odor during the treatment process.

They can also be used to evaluate the efficacy of various treatments and track down the source of contamination.

8.2.3.1 Threshold Odor Test

The odor threshold is estimated by diluting the sample with odor-free water until the least perceptible odor is reached. The highest dilution of the specimen that produces a discernible odor is called the threshold odor number (TON) and is estimated as in Eq. (8.3):

$$\text{TON} = \frac{A+B}{A} \tag{8.3}$$

where A is the tested sample, and B is odorless water in mL.

This method finds application for samples varying from odorless natural water to industrial wastes with exceedingly high thresholds. High thresholds are not problematic because concentrations are reduced proportionately before being tested. Chlorination is used for most tap and wastewaters and is a common source of odor complaints. It is advised to test the odor of the chlorinated sample and then retest after dechlorination.

Temperature affects the threshold odor readings, with discernability of odor increasing with temperature. A sample temperature of 60°C allows identification of scents not typically detected in most tap and unprocessed water sources; thus, the standard temperature for the hot threshold odor test is 60°C. However, the hot odor examination might not be appropriate for some applications, such as when the odor is too transient or in instances of extreme heat transfer. In these cases, the test should be carried out at a conventional temperature of 40°C.

$$\text{TON} = \frac{A+B}{A}$$

8.2.4 Taste

Flavor is a group of gustatory, olfactory, and trigeminal sensations triggered by chemical activation of sensory nerve endings in the tongue, nasal cavity, and mouth. Bitter, salty, sour, and sweet are the major identifiable gustatory sensations that occur by chemical activation of sensory nerve terminals in the tongue's papillae and soft palate. The taste, odor, or mouth-sensation perceived depends on what chemical compounds exist in the water sample placed into the mouth for sensory examination. There are three procedures for the estimation of taste in water samples: the flavor threshold test (FTT), flavor rating assessment (FRA), and flavor profile analysis (FPA) [10]. The FTT is the first of its kind and is widely utilized, especially for detecting whether the overall flavor of a processed sample water differs significantly from a standard sample. The FRA is notably useful for establishing if a sample of processed water is safe for daily consumption, while the FPA is best used to identify and characterize specific flavors in water samples.

8.2.4.1 Flavor Threshold Test

The FTT is used to quantify detectable taste. This method is used to compare the sample flavor to that of a diluent which serves as the reference water sample. The flavor threshold number (FTN) is the highest sample dilution that yields a clear distinction. The FTN is calculated in the following way using Eq. (8.4):

$$\mathrm{FTT} = \frac{A+B}{A} \tag{8.4}$$

where A is the volume of sample, and B is the volume of reference water in mL.

8.2.4.2 Flavor Rating Assessment

The FRA test method is used to estimate acceptability for everyday consumption. It is essentially a blind taste study of various water samples, with their responses taken in the form of a survey. This approach has been used to establish standards for appropriate levels of mineral content in drinking water. Up to ten samples can be evaluated in a single rating session. Each customer or assessor is given a list of nine statements regarding water that range from favorable to unfavorable conditions and is tasked with choosing the statement that most closely expresses their acceptance of the sample's taste [12]. The statements are:

i) I would be very glad to take this water as my everyday drinking water.
ii) I would be glad to take this water as my everyday drinking water.
iii) I am certain that I could take this water as my everyday drinking water.
iv) I might accept this water as my everyday drinking water.
v) Perhaps I might take this water as my everyday drinking water.
vi) I don't think I could take this water as my everyday drinking water.
vii) I could not take this water as my everyday drinking water.
viii) I could at no time drink this water.
ix) I cannot take this water in my mouth, and I could never drink it.

8.2.4.3 Flavor Profile Analysis

FPA is one of the numerous procedures to examine the taste and odor of water samples. The FPA is used to assess the key tastes and smells in drinking water. This analysis is not utilized to assess the sample, estimate the sample preferences or estimate whether a water sample is fit for community drinking. FPA is implemented to identify the taste and odor properties of a sample, which can subsequently be linked to or correlated with consumer impressions. Without dilution, FPA establishes the degree or experience of flavor or odor.

There are two types of FPA measures, namely recognition and detection thresholds. The least strength at which a stimulus may be recognized is known as the recognition threshold. The detection threshold is defined as a concentration range below which the odor or taste of a substance will not be detectable under any practical conditions and above which individuals with a normal sense of smell or taste will notice the odor or flavor of a substance. The concentrations of odor detection thresholds are usually lower than the concentrations of odor recognition thresholds [13].

Sensory detection thresholds are conceptually akin to chemical technique detection levels. The sensory threshold, like a detection level, is imprecise and difficult to evaluate, as minute variations in test settings, contamination, or background interference can significantly impact results. However, sensory thresholds are typically far more variable than method detection levels [14]. Together, sensory and analytical tests are numerical explanations of measured information. This could explain why drinking water utilities have failed to adopt a feasible way of calculating odor thresholds in drinking water.

8.2.5 Acidity

Acidity is a measure of a water's total quality that can only be understood in terms of specific chemicals if the sample's chemical composition is known. According to the method of determination, strong, weak acids like carbonic, acetic acids, and hydrolyzing salts such as Fe or aluminum sulfates might assist in acidity. Acids are corrosive and influence the rate of chemical reactions, chemical species, and biological processes. This measurement also indicates a variation in the water source quality.

8.2.5.1 Titration Method (Acidity Measurement)

Titration is based on the principle that the H^+ ions present in a sample will react by either hydrolysis or solute dissociation when an alkali, or titrant, is added. The endpoint is the point at which the volume of titrant is sufficient to react completely with the H^+ ions present, as denoted by a color change in a secondary acid-base system added called an indicator. The endpoint pH, or acidity, is determined as the negative logarithm of the indicator concentration. Calibration data developed by measuring the pH of the sample after measuring the titrant allows for identification of the inflection point and buffering capacity as well.

The most accurate endpoint for a single acidic species, as in the standardization of reagents, is attained from the inflection point of a calibration curve. The pH at which the curvature changes from convex to concave, or inversely, is known as the inflection point.

Wastes from industries like acid mine drainage or additional solutions comprising small quantities of hydrolyzable inorganic metal ions like Fe, Al, or Mn are treated with hydrogen peroxide (H_2O_2) to confirm oxidation of any reduced form of polyvalent cations and later heated to accelerate hydrolysis. Ideally, the acidity titration endpoint should correlate to the point of stoichiometric equivalence for the neutralization of the acids. The equivalence pH point will differ depending on the sample, the number of inflection points considered, and the usage of data.

8.2.6 Alkalinity

The acid-neutralizing capacity reflects the total of all titratable bases in a water sample and is measured by the sample's alkalinity. The calculated value can differ notably depending on the endpoint. Similar to acidity, alkalinity is a measure of water's collective property that can only be explained in terms of specific substances if the composition of the sample is known.

Alkalinity is vital in several natural and wastewater applications and treatments. As the alkalinity of surface water is primarily determined by the concentrations of carbonates, bicarbonates, and hydroxides, it is used to estimate the concentrations of these compounds. The calculated values may also comprise borates, phosphates, silicates, and other bases that also contribute to the measured values if they are present. The alkalinity of Be, Mg, Ca, Sr, Ba, and Ra is important in evaluating the water's appropriateness for irrigation purposes. Measurements of alkalinity are used to interpret and control the treatment of wastewater. The alkalinity of unprocessed domestic wastewater samples is typically less than the water supply, though it can be higher in certain cases. Supernatant alkalinities in anaerobic digesters exist in the range of 2000–4000 mg calcium carbonate ($CaCO_3$)/L.

8.2.6.1 Titration Method (Alkalinity Measurement)

The dissociation of solute or hydrolysis of hydroxyl ions [OH^-] in a sample due to reactions with a standard acid guides the principle behind this relatively simple method. With low alkalinity samples (e.g. lower than 20 mg $CaCO_3$/L), an extrapolation method is established on the proportionality of [H^+] to excess titrant after the point of equivalence is reached. The quantity of acid used to achieve a 0.30 pH reduction is precisely calculated. This change in pH corresponds to a doubling of the hydrogen ion concentration; thus, the equivalence point can be extrapolated backward based on the linear relationship between the excess titrant and pH [15, 16]. Particularly when alkalinity is caused by the presence of carbonates or bicarbonates, the pH at the point of equivalence is determined by the carbon dioxide (CO_2) concentration. The concentration of CO_2 is determined by the overall carbonate species initially existing along with any loss that occurred in the course of the titration process. It is calculated using Eq. (8.5):

$$\text{Alkalinity, mg CaCO}_3/\text{L} = \frac{A * N * 50000}{ml\ sample} \tag{8.5}$$

where A is the standard acid volume (mL), and N is the normality of acid. The results from phenolphthalein indicators and total alkalinity provide a method for a stoichiometric arrangement of three alkaline bases found in water, namely carbonate, bicarbonate, and hydroxide, without the presence of organic or inorganic acids. It also assumes that bicarbonate and hydroxide alkalinities are incompatible. As they are measured on a stoichiometric basis, concentrations of ions vary notably from actual concentrations, particularly at pH >10. The following are used as rules of thumb:

- The alkalinity of carbonate (CO_3^-) exists once the phenolphthalein alkalinity is greater than zero but lower than total alkalinity.
- The alkalinity of OH^- exists if phenolphthalein alkalinity exceeds half of the total alkalinity.
- The alkalinity of bicarbonate (HCO_3^-) exists if the phenolphthalein alkalinity is below half the total alkalinity.

Equations (8.6 and 8.7) can be used to calculate these relationships, where P and T denote phenolphthalein alkalinity and total alkalinity:

$$CO_3^{2-} = 2P - 2\left[OH^-\right] \tag{8.6}$$

$$HCO_3^- = T - 2P + \left[OH^-\right] \tag{8.7}$$

If the phenolphthalein endpoint is difficult to achieve, it needs to be checked to determine phenolphthalein alkalinity as $CaCO_3$ from the nomographic estimations of $[CO_3^{2-}]$ and $[OH^-]$ as shown in Eq. (8.8).

$$P = \frac{1}{2}\left[CO_3^{2-}\right] + \left[OH^-\right] \tag{8.8}$$

8.2.7 Calcium Carbonate Saturation

To avoid scaling in pipelines and equipment, $CaCO_3$ saturation indices (SI) are often utilized to analyze scale-forming and scale-dissolving trends in heat exchangers or water heaters. SI can be used to minimize corrosion in iron, steel, and cement piping but is not applicable for controlling issues related to Pb and Cu metals. $CaCO_3$ precipitates in water that is oversaturated with it and dissolves in water that is undersaturated [17]. Saturated waters (in equilibrium with $CaCO_3$) tend to neither dissolve nor precipitate $CaCO_3$. The saturation line determines whether the solution is likely or unlikely to precipitate. Total dissolved solids (TDS) must also be used to determine ionic strength. $CaCO_3$ SI is divided into two categories:

- Indicators for determining whether water is capable of dissolving/precipitating $CaCO_3$,
- Indicators for estimating the amount of $CaCO_3$ precipitated/dissolved.

$CaCO_3$ precipitation or dissolution behaviors are used to determine whether water is saturated, undersaturated, or oversaturated. The most extensively utilized indices are SI, driving force index, relative saturation (RS), saturation ratio, and the Ryznar stability index.

8.2.7.1 Saturation Index Basis

SI, shown in Eq. (8.9), is the most widely implemented index [18]. The RS and SI indices are related; however, while the semi-empirical RS provides better outcomes, it is less consistent than the SI.

$$\mathrm{SI} = pH - pHs \tag{8.9}$$

The calculated pH is represented as *pH*, and *pHs* is the pH of water in equilibrium with $CaCO_3$ at the existing $[Ca^{2+}]$ and bicarbonate ion $[HCO_3^-]$.

If the water is oversaturated with $CaCO_3$, then SI becomes positive; if it is negative, then the water is unsaturated, and if SI is 0, water and $CaCO_3$ are in equilibrium.

8.2.7.2 Saturation Index by Experimental Determination

SI can be determined experimentally by saturometers, which measure the degree of $CaCO_3$ saturation in saltwater. In a closed container with a pH electrode, analysts equilibrate $CaCO_3$ with water in a temperature regulating bath. When $CaCO_3$ precipitates, pH drops. When it dissolves, pH increases. Equilibrium is attained when there is no further change in pH. The carbonate concentration measures RS by measuring the starting pH, concentration of calcium, and end pH value. This method has the advantage of being able to approach equilibrium by measuring pH, thereby reducing the ambiguity regarding approaching equilibrium. This approach is sensitive in the pH range of 7.5 to 9.5 and has been utilized both in situ and in laboratories.

SI can also be measured by equilibrating water with $CaCO_3$ in a closed isothermal system with known pH, calcium content, and alkalinity. Initial calcium, pH, and alkalinity data are used to calculate the activity of $CaCO_3$ prior to equilibrium. After equilibrium attainment, the $CaCO_3$ activity is used to find a $CaCO_3$ solubility product constant (K_s) and is calculated using the change in alkalinity that happened in the course of equilibrium. RS is obtained by dividing the original activity product by K_s. This technique has the advantage of not making any assumptions about the $CaCO_3$ phase's identity. This method, however, is difficult in determining when equilibrium is attained and is more challenging than the saturometry method [19].

8.2.7.3 Calcium Carbonate Precipitation Potential for Alkalinity Measurement

The $CaCO_3$ precipitation potential (CCPP) is the amount of $CaCO_3$ that could be precipitated theoretically from oversaturated or dissolved waters by unsaturated waters during equilibrium. It is used to measure the water's proclivity to precipitate/dissolve $CaCO_3$, as well as the amount of $CaCO_3$ dissolved or precipitated.

The quantity that precipitates or dissolves might be higher or lower for a variety of reasons, including temperature and approximations of ionic strength, the presence of threshold inhibitors, inability to account for complex species, erroneous assumptions about the solid-phase formation, variations in concentrations between the bulk and local surroundings, erroneous solubility constants, and dissociation constants and their measurements. The possibility that equilibrium cannot even be established may also cause variation [20]. Undersaturated waters have a negative CCPP, while saturated waters have zero CCPP, and oversaturated waters have a positive CCPP.

8.2.8 Hardness

Water hardness is considered a measurement of a liquid's ability to precipitate soap. Soap is precipitated mainly by the presence of Ca^{2+} and Mg^{2+}. Additional polyvalent cations can precipitate soap as well but are usually in complex forms with organic ingredients, so their impact on water hardness is difficult to quantify. Total hardness is defined as the summation of Ca^{2+} and Mg^{2+} concentrations indicated as $CaCO_3$ in mg/L. Based on the treatment and source, the hardness can range from zero to hundreds of mg/L.

8.2.8.1 Calcium Carbonate Precipitation Potential for Hardness Measurement

The CaCO3 precipitation method can be applied to determine the hardness of water experimentally by ethylenediaminetetraacetic acid (EDTA) titration that estimates Ca2+ and Mg2+ions. Other indicators of CaCO3 saturation, such as Saturation Index and Ryznar index, can only predict a thermodynamic driving force for precipitation or dissolution of CaCO3. When compared to these indicators, there exists a chemical equilibrium in CCPP, which gives a more accurate representation of CaCO3 over- or undersaturation. The total hardness is a sum of the calcium and magnesium content under certain coefficients, as shown in Eq. (8.10):

$$\text{Total Hardness} = 2.50{*}X_{Ca} + 4.12{*}X_{Mg} \tag{8.10}$$

where, the total hardness in mg/L $CaCO_3$ is determined from X_{Ca}, the calcium concentration in mg/L Ca^{2+}, and X_{Mg}, the magnesium concentration in mg/L Mg^{2+}.

8.2.8.2 EDTA Titrimetric Method

When ethylenediaminetetraacetic acid (EDTA) and sodium salts are introduced to a solution containing inorganic cations, they form a chelated soluble complex. The solution turns a wine-red color when a small quantity of dye, such as calmagite or Eriochrome black T, is mixed into an aqueous solution comprising Ca^{2+} and Mg^{2+} ions at a pH of 10.0 ± 0.1. When EDTA is used as a titrant, the Ca^{2+}, Mg^{2+} turn into complexes, and the color of the solution changes from red to blue, indicating the titration's endpoint. From there, the hardness is estimated according to Eq. (8.11). A satisfying endpoint requires the presence of magnesium ions. A small quantity of complexometrically neutral Mg salt of EDTA is mixed with the buffer to ensure the automatic introduction of enough Mg, thereby eliminating the requirement for a blank correction. As the pH rises, the sharpness of the endpoint rises as well. However, the pH cannot be elevated indefinitely without risk of precipitating $CaCO_3$ or $Mg(OH)_2$ and changes in the color of dye at higher pH. The pH of 10.0 ± 0.1 is a good compromise. A titration interval of 5 min is suggested to minimize the precipitation of $CaCO_3$.

$$\text{Hardness} = \left(A{*}B{*}1000\right)/\left(\text{mL of sample}\right) \tag{8.11}$$

In this equation, the hardness is measured in mg/L $CaCO_3$, *A* is the titrated sample, and *B* is the equivalent $CaCO_3$ to titrant, in mg.

8.2.9 Conductivity

The conductivity (k) of an aqueous mixture is a measure of the mixture's ability to transport electric current. This capability is dependent on the overall concentration, valency, mobility, and measured temperature of ions. Most inorganic compounds are good conductors; however, organic mixtures that do not separate are poor conductors. Conductance (*G*) is the inverse of resistance (R), where R is measured in ohm. A solution's conductance is calculated between two chemically inert and spatially fixed electrodes. The measurement is made with an alternating current pulse to prevent polarization at the surface of the electrode [21]. A solution's conductance is proportionate to the surface area of electrode A and in reverse proportionality to the distance between the electrodes (L) in cm. The proportionality constant k is given in Eq. (8.12).

$$G = k\left(\frac{A}{L}\right) \tag{8.12}$$

Conductivity is a distinctive feature of the electrolytic fluid among the electrodes. Units for the constant k are 1/ohm-cm, or mho/cm. Micromhos per centimeter (μmhocm^{-1}) is the standard unit of conductivity measurement. The siemens (S) is the reciprocal of the ohm, allowing conductivity units of mS/m.

Ions have positive and negative charges, which carry electricity. Electrolytes break into positively charged (cation) and negatively charged (anion) particles when they dissolve in water. The concentrations of each positive and negative charge remain equal as the dissolved compounds split in water. It follows then that, although the increase in ions increases the conductivity of water, the water remains electrically neutral [22]. The conductivity of seawater is known to be extremely high.

8.2.10 Salinity

The total concentration of all dissolved salts in water is known as salinity. As electrolytes dissolve, they generate ionic particles, each with a positive and negative charge. As a result, salinity has a significant impact on conductivity. While a thorough chemical analysis can be used to determine salinity, this process is complicated, time-consuming, and unlikely to provide a complete measure of all salts present. Seawater cannot simply be evaporated to a dry salt mass measurement as chlorides are lost during the process [23]. Thus, indirect estimation of a physical parameter (density, conductivity, refractive index, or speed of sound) is more commonly used to determine salinity. In the past, hydrometric and argentometric methods were used to determine the salinity of saltwater. Due to their convenience and acceptably high precision, the conductivity and density approaches have since been adopted as standard methods. For precise field and laboratory measurements, these are the recommended approaches.

Salinity and TDS are used to calculate the electrical conductivity of water, which helps to indicate the water's purity [25]. Purer water has a lower conductivity. The measure of salinity is defined by the practical salinity scale. Practical salinity is a unitless quantity approximately equivalent to the mass fraction of dissolved solute in seawater but is not interchangeable with absolute salinity. Practical salinity is an analog for the conductivity of seawater, adjusted for temperature and pressure, while absolute salinity is defined as the mass fraction of dissolved salts in seawater in units of g/kg.

It is feasible to estimate the density of natural waters quickly using an accurate vibrating flow densimeter. The method entails introducing the sample over a vibrating tube wrapped in an isothermal jacket. The density of the solution (ρ) is directly related to the square of the vibration period (τ), as shown in Eq. (8.13):

$$(\rho) = A + B(\tau^2) \tag{8.13}$$

where A and B are estimated through calibration. The difference between the sample density and pure water is given in Eq. (8.14):

$$\rho - \rho_0 = B(\tau^2 - \tau_0^2) \tag{8.14}$$

where τ and τ_0 are the vibration periods of the sample and water, respectively.

8.2.11 Solids

Solids in this context refer to suspended or dissolved materials in domestic, industrial, surface, and saline waters. Some degree of dissolved material, specifically dissolved nutrients and essential minerals, is necessary in natural and drinking waters. At elevated levels, however, these compounds can have a variety of negative effects on water and effluent quality. Higher levels of solids are associated with turbidity, water hardness, odors, and taste issues. Moreover, waters with high concentrations of dissolved solids can cause adverse physiological reactions. Therefore, solid content analyses are critical for monitoring and managing water and wastewater treatment operations and determining whether regulatory requirements are met.

8.2.11.1 Total Dissolved Solids

TDS is described as the number of total solids in the water sample that pass across a filter media through a pore size of 2 microns or less and is calculated by Eq. (8.15):

$$\text{TDS, mg/L} = \frac{(A - B) * 100}{sample\ volume\ mL} \tag{8.15}$$

where A is the sum of dish weight and dried residue mass, and B is dish mass, in mg.

8.2.11.2 Total Suspended Solids

TSS is the overall solids that remain on the filter after an aqueous sample has been filtered and estimated using Eq. (8.16):

$$\text{Total Suspended Solids}(\text{mg/L}) = \frac{(A - B) * 100}{sample\ volume\ mL} \tag{8.16}$$

where A is the sum of final filtrate and dried residue mass, and B is the filter mass, in mg.

Total solids (dissolved and suspended) weight loss after ignition for a particular duration at a specific temperature is known as volatile solids. The organic content is analyzed through total organic carbon (TOC), biochemical oxygen demand (BOD), and chemical oxygen demand (COD). Volatile solids are calculated according to Eq. (8.17):

$$\text{Volatile solids}\left(\text{mg/L}\right) = \frac{\left(A - B\right) * 100}{sample\ volume\ mL} \tag{8.17}$$

where A is the sum of final residue weight and dish prior to ignition, and B is the mass of final residue and the dish after ignition, in mg.

8.2.12 Asbestos

Asbestos refers to naturally existing, inorganic, extremely fibrous silicate minerals that are effortlessly separated into thin, elongated, flexible strands as soon as they are processed or crushed. Asbestos has seen use for its insulating and special permeability properties, though it is now considered a health hazard and is no longer in conventional use. A transmission electron microscope (TEM) is used to calculate the asbestos structure concentration, given as the number of structures per liter of water. In TEM analysis, a monochromatic beam of electrons is accelerated through a potential of 40–100 kilovolts (kV) and passes through a strong magnetic field that functions as a lens in the TEM. A modern TEM has a resolution of roughly 0.2 nm. In a solid, this is the usual separation between two atoms. This resolution is 1000 times that of a light microscope and almost 500 000 times that of the human eye. With the help of TEM analysis, morphology, size, shape, arrangement of particles, crystallographic information based on diffracted electrons, order and arrangement of atoms, detection of atomic-scale defects, and compositional data such as chemical identity can be obtained. Asbestos structures containing fibers and concentrations of more than 10 μm are determined with Eq. (8.18):

$$\text{The concentration of asbestos, structures/L} = \frac{N * A_f * D}{G * A_G * V_s} \tag{8.18}$$

where N is the counted asbestos structures, A_f is the final effective sampling area of the filter (mm^2), D is the factor of dilution, G is the number of openings of the grid, A_G is the area of the openings of the grid (mm^2), and Vs is the sample volume (L).

8.2.13 Oxidation–Reduction Potential

Several chemical constituents in drinking water, wastewaters, and highly aquatic environments are facilitated by oxidation and reduction, or redox, reactions [26, 27]. Redox conditions significantly impact the mobilities and reactivities of essential elements, such as Fe, S, N, C, and several other metallic elements, in biological systems. Because the chemical reactions relating to both protons and electrons are dependent on pH and redox potential (Eh), chemical reactions in aqueous media are characterized by Eh and pH, along with the activity of the dissolved chemical species. Eh, similar to pH, is a measure of intensity. It does not describe the system's oxidation or reduction capacity, i.e. poise.

Although measuring Eh in water is reasonably straightforward, there are limitations in interpretation of the values. The limiting aspects comprise irreversible reactions, poisoning of the electrode, multiple redox species and reactions, very minute exchange of currents, and inactive redox compounds. Eh results obtained in the field have a poor correlation with Eh values computed from the existing redox couples. Nonetheless, when correctly done and interpreted, redox potential measurements are valuable in establishing a better understanding of water chemistry.

To perform the electrometric measurements via potentiometric determination of electron activity (intensity) requires an inert indicator electrode and an appropriate electrode at the reference section. In theory, the indicator electrode will behave as an electron acceptor in solutions containing chemical species that have been oxidized or reduced. When the system is in redox equilibrium, the potential difference between the ideal and the reference electrode indicates the system's redox potential. In natural waters, no one indicator electrode will perform perfectly. For Eh measurements, platinum electrodes are most usually employed. However, like other materials such as gold and graphite, each electrode has its own limitations [18].

For typical laboratory and field use, the conventional hydrogen reference electrode is fragile and impractical. As a result, reference electrodes made of silver, silver chloride, or calomel are widely utilized. The difference in potential between the reference electrode and the standard hydrogen electrode is compensated for redox potential measurement.

8.2.14 Tests and Methods on Sludges

This section contains a set of tests that are specific to sludges or slurries.

8.2.14.1 Oxygen Consumption Rate

Oxygen consumption rate test assesses the rate at which the biological sample solution, such as the activated sludge, consumes oxygen. It is primarily used in the laboratory and pilot-plant investigations along with full-scale treatment plant operations and is estimated using Eq. (8.19), where the oxygen consumption rate is in the unit of (mg/g)·h^{-1}.

$$\text{Specific rate of } O_2 \text{ consumption} = \frac{oxygen\ consumption\ rate\left(\frac{mg}{L}\right)/min}{volatile\ suspended\ solids,\ g/l} * \frac{60\ min}{hr} \tag{8.19}$$

8.2.14.2 Sludge Volume Index

The sludge volume index (SVI) is the volume in milliliters engaged by 1 g of suspension after settling for 30 min. SVI defined in Eq (8.20), is commonly used to track the settling properties of the activated sludge process and further biological suspensions.

$$\text{SVI} = \frac{settled\ sludge\ volume\left(\frac{ml}{L}\right) * 1000}{suspended\ solids,\ mg/l} \tag{8.20}$$

8.2.14.3 Specific Gravity

The mass ratio of sludge and distilled water in equal volumes is defined as the specific gravity of a sludge. It is estimated by comparing the mass of a known quantity of a sample of homogeneous sludge at a certain definite temperature to the mass of the distilled water of the same volume at 4°C.

8.2.14.4 Zone Settling Rate

With high concentrations of suspended solids, the suspensions settle collectively as a zone. This type of settling is characterized by the formation of a distinct interface between the supernatant liquid and the sludge zone and occurs under quiescent conditions. The height of the sludge interface is calculated and monitored with respect to time. The data on settling might be utilized in the operation, design, and estimation of the settling basins for zone settling suspensions [28]. The rate of zone settling is determined by the concentration of suspended particles, suspension properties, vessel dimensions, and laboratory setup. A plot of height interfaces vs. time will give the settling rate by determination of its slope.

8.2.14.5 Time for Capillary Suction

The time for capillary suction test determines the rate at which the sludge releases water. The numerical quantity of how quickly water is lost in sludge, normally in seconds, is useful to know in planning dewatering sludge procedures, estimating sludge conditioning and dosage, and estimating the coagulation effect on the rate at which water discharges from sludge. This test requires the sludge sample to be placed in a cylinder containing chromatography paper. The liquid is drawn from the sludge by capillary action. The time it takes for the liquid to move a certain distance is automatically measured by measuring the change in conductivity at two contact sites that are properly spread out and in association with the chromatography paper. The rate of water drainage is then determined by the elapsed time. This examination is implemented to describe the majority of sludge dewatering procedures.

8.2.15 Anaerobic Sludge Digester Gas Analysis

CO_2 and methane (CH_4) are the main gases formed during the anaerobic breakdown of wastes, along with small amounts of nitrogen (N_2), hydrogen (H_2), oxygen (O_2), and hydrogen sulfide (H_2S). The standard analytical procedure is to examine the gases formed, evaluate their fuel value, and monitor the process for treatment. Due to the fairly higher percentages of these gases, the relative amounts of N_2, CH_4, and CO_2, are usually of the most interest and simple to estimate.

For gas analysis, two approaches are implemented: gas chromatography and volumetric methods. CO_2, H_2, CH_4, and O_2 can all be determined using volumetric analysis. Though the equipment involved is simple, the procedure itself is time-consuming. This approach is to analyze digester gases as well as CH_4 in the water. The primary benefit of gas chromatography is its speed.

8.2.15.1 Volumetric Method

Methodologically, a measured gas volume is sequentially placed through a series of solutions to remove gases. The sample first passes through a potassium hydroxide (KOH) solution to eliminate CO_2, then through a solution of alkaline pyrogallol to eliminate O_2, and next, an overheated cupric oxide (CuO) to oxidize H_2. Finally, CH_4 is converted to CO_2 and water via catalytic oxidation. The produced CO_2 volume during combustion is used to calculate the amount of CH_4 that was initially present. The fraction of N_2 is estimated as the difference between 100% of the total gas value and the sum of all other analyzed gases. In most cases, the combustible gases found in the digester are CH_4 and H_2. Equations (8.21–8.25) are used to compute the percentage of each gas by volume.

$$\%CO_2 = \frac{(V_1 - V_2)*100}{V_1} \tag{8.21}$$

$$\%O_2 = \frac{(V_2 - V_3) * 100}{V_1} \tag{8.22}$$

$$\%H_2 = \frac{(V_3 - V_4) * 100}{V_1} \tag{8.23}$$

$$\%CH_4 = \frac{V_4 * (V_8 - V_9) * 100}{V_1 * V_5} \tag{8.24}$$

$$\%N_2 = 100 - (\%CO_2 + \%O_2 + \%H_2 + \%CH_4) \tag{8.25}$$

8.2.15.2 Gas Chromatographic Method

Gas chromatography (GC) partitions individual components of the sample between two phases: a mobile phase, which contains carrier gas, and a stationary phase, made up of either packed or capillary columns. N_2, argon-methane mixture (Ar-CH_4), helium (He), or H_2 are used as carrier gases [29]. The sample is volatilized and carried through the stationary columns by the carrier gas.

In the case of a packed column, the stationary phase consists of a liquid covered by an inert granular solid known as column packing and retained in glass tubing made of borosilicate. The column is placed in an oven, with a heated injector block at the entrance and a detector at the outlet. The injector block, oven, and detector all have precise and consistent temperature control. The controlled variables are concentration and stationary phase material, column diameter and length, carrier gas flow, oven temperature, and type of detector. When the sample is put inside the column, the vaporized organic compounds are transported by the carrier gas via the column. They move at various speeds through the column due to changes between the stationary and mobile phase partition coefficients. A calibration curve for each gas component needs to be developed for correct detection results, as some components of gas do not yield equal detector responses on the molar basis or weight basis. The calibration can be done with pure gases or prepared synthetic samples.

8.3 Analysis of Inorganic Metal Constituents

Metals present in wastewater have a wide variety of effects, from useful to problematic and potentially toxic. Some of the elements present are required for plant and animal life, while others have the potential to harm drinkers and induce complications for wastewater treatment systems. For some metals, concentration deems whether they are advantageous or toxic. Preliminary treatment of wastewater is frequently required to ensure that metals are in a form conducive for the analytical methods. In this section, alternative strategies for pretreatment are discussed, along with a variety of methods used to analyze metal constituents. Descriptions and details of the colorimetric and instrumental analysis techniques, including atomic absorption spectroscopy (AAS), flame photometry, anodic stripping voltammetry (ASV), inductively coupled plasma–mass spectrometry (ICP–MS), and inductively coupled plasma emission spectrometry (ICP-OES)) techniques are given [18].

AAS techniques include flame, electrothermal, hydride, and cold vapor methods. Flame methods are usually useful in complex-matrix and clean samples at moderate concentrations (0.1 to 10-mg/l). For numerous Group I and II elements, flame photometry yields good findings at higher concentrations. In relatively clean matrixes, anodic stripping provides great sensitivity for numerous elements. Electrothermal analyses can often boost sensitivity if matrix-related issues are not of any concern. ICP techniques can be used across a wide linear range and are particularly sensitive to refractory materials. Colorimetric methods are used to determine individual metal ions when interferences are not expected to affect method accuracy.

8.3.1 Conductivity

Metals and metallic compounds may be inorganic, organic, particulate, or dissolved. Pretreatment is usually required for samples containing particles or organic substances before the spectroscopic examination. Single-phased, colorless, clear samples with a turbidity of lower than 1 NTU, and no odor, can be examined using AAS (flame or electrothermal vaporization) or ICP–MS for metals lacking the digestion process.

8.3.2 Dissolved and Suspended Metals Filtration

To analyze suspended or dissolved metals, filter the sample during the collection with a plastic filtering system, either by vacuum or pressure-sealing with a plastic or fluorocarbon filter provision. If suspended metals are to be estimated, then filter the sample for the dissolved metals without centrifuging. Keep the filter and then digest it for the measurement of the metals suspended.

8.3.3 Digestion of Metals

Acids are the most common reagents for dissolving metal-containing compounds. Various forms and combinations of these acids can be employed to digest and liberate metals from a variety of compounds. The goal is to disrupt the matrix before the sample is sent to the atomization process applicable in AAS/ICP. Additionally, digestion reduces all the metals to a single oxidation state. The particulate material is lightly adsorbed with extractable metals because digestion is unavoidable for some samples. Strictly controlled conditions should be used to get good findings by maintaining the volume of acid, sample, and contact time constant.

8.3.3.1 Selection of Acid

Most samples are effectively digested by nitric acid (HNO_3). For both electrothermal atomic absorption and flame methods, nitrate is a suitable matrix, and it is the recommended matrix for ICP–MS. For complete digestion, some materials might need the addition of hydrochloric (HCl), perchloric ($HClO_4$), sulfuric (H_2SO_4), or hydrofluoric acid (HF). However, these acids may interfere with the specific metal analysis, and they altogether produce an inferior matrix for both ICP–MS and electrothermal analysis. Nitric acid mixtures, such as HNO_3-H_2SO_4 or HNO_3-HCl, are usually sufficient to clean or oxidize organic materials that are easily oxidizable, while mixtures of HNO_3-$HClO_4$-HF or HNO_3-$HClO_4$ are required for organic materials that are harder to oxidize or contain silicates.

For the samples that are completely digested, acid digestion techniques produce either precise or biased results, depending on how it is carried out. The number of acids used during digestion should be minimized as they will add metals to the number of blanks and samples. This process is not a total digestion method; hence, the microwave digestion procedure can be considered as a substitute.

The microwave method is a closed-vessel procedure that delivers enhanced accuracy in comparison to using a hot plate. The concentration of metal from digestion can be calculated according to Eq. 8.26:

$$\text{Concentration of metal, mg/L} = A*\frac{B}{C} \tag{8.26}$$

where A is the digested solution metal concentration (mg/L), B is the digested solution volume, and C is the sample volume in mL.

8.3.3.2 Nitric Acid Digestion

Though HNO_3 cannot completely degrade organic matrixes on its own, it is still a highly useful component for digestion. Its application can follow two approaches: one for high analyte concentration (>0.1 mg/L) and one meant for trace level concentration (0.1 mg/L). Some of the most typical inorganic dissolutions are:

- 10–15% aqueous dilution – Alkaline earth oxides, lanthanide oxides, actinide oxides, Scandium(III) oxide (Sc_2O_3), and Yttrium(III) oxide (Y_2O_3).
- 1 : 1 HNO_3 + H_2O – Vanadium pentoxide (V_2O_5), copper oxide (CuO), oxides of Mn, Tl, Pb, Bi Zn, Pb, Cd, and Hg.
- Concentrated HNO_3 (69%) – MnO, FeO (hot), CoO, SeO, AsO, AgO, PdO, BiO, and ReO.
- <LI•1 : 3 HNO_3 + HCl – PtO, AuO, Cu alloys, Fe/Ni alloys, Cr/Ni alloys, steel
- 1 : 1 : 1 HNO_3 + HF + H_2O – The metal and oxides of Ti, Zr, Hf, Nb, W, Sn, Al, Si, Ge, Sb, Te, As, Se, Mo, and numerous alloys and oxide mixtures containing one or more of these elements.

The water-soluble alkaline earth metals are the only major group of elements not listed above.

8.3.3.3 Microwave-assisted Digestion

Microwaves are electromagnetic waves with frequencies ranging from 300 MHz (0.3 GHz) to 300 GHz, with a typical frequency of 2450 MHz for both residential and laboratory microwave ovens. Microwaves are very well-suited for sample preparation because of the following factors:

- Fast heating rates – the conversion of electromagnetic energy into heat energy is very rapid and efficient, unlike conventional heating processes. In the microwave, the microwaves pass through the vessel and heat the reaction mixture by directly interacting with molecules.
- Option for instantaneous switching between on or off in terms of direct energy delivery.
- No contact to the heating core required.

The temperature is the most critical factor in acid digestion and acid leaching. High temperatures are beneficial in two ways: they speed up the digestion process and increase digestion quality. Following the Arrhenius equation, as the temperature rises, the rate of reaction increases, leading to the reduced time needed for digestive reactions to take place. A rapid heating impact, immediate on-and-off, and well-designed workflow make microwaves a convenient yet effective tool for assisting in the digestion process [30, 31].

8.3.4 Metals by Atomic Absorption Spectrometry

The element-specific light absorption in AAS determines the concentration of a metal in a solution by aspirating a liquid sample, aerosolizing it, then mixing it with flammable gases. The components of interest are combusted to free atoms that absorb light at a specified wavelength. The elemental composition can be estimated by determining the amount of light absorbed. For combustion, either flame (for larger concentrations) or a graphite furnace (for very low concentrations) is generally employed. When using graphite furnace atomization, the results are obtained at ppm levels and a higher sensitivity of ppb levels. Some of the applications of AAS are:

- Quantitative measurement of metal concentration in solutions.
- Lead in paint analysis.
- Monitoring trace metals in industrial wastewater streams.
- ICP–MS and trace elements in products/raw materials.
- Investigation of additives and purities in steels and other metal alloys.
- Low-level contaminant analysis.

8.3.4.1 Metals by Flame Atomic Absorption Spectrometry

This technique is the most prominent method for detecting metals in samples. The theory behind the approach is that ground-state metals absorb light at a given wavelength. A flame is used to convert metal ions in a solution to their atomic state. When the correct wavelength of light is used, the amount of light absorbed can be measured, thus giving a concentration reading. Flame atomic absorption is effective and accurate as both a quantitative and qualitative technique. The nebulizer, the instrument's most critical component, converts the sample to a mist or aerosol and carries it into the flame. This method is used to determine Ca^{2+}, Mg^{2+}, Na^{+}, and K^{+} at wavelengths 442.7, 285.2, 589.0, and 766.5 nm, respectively [32].

8.3.4.2 Direct Air–Acetylene Flame Method

In this process, the sample is aspirated into an air–acetylene flame and atomized. A beam of light of a specific wavelength is sent through the flame into a monochromator and then finally onto a detector that measures the intensity of light absorbed by the atomized element, providing an indirect measurement of metal concentration. This method is used to determine various heavy metal ions, including Sb, Bi, Cd, Ca, Cs, Cr, Co, Cu, Au, Ir, Fe, Pb, Li, Mg, Mn, Ni, Pd, Pt, K, Rh, Ru, Ag, Na, Sr, Tl, Sn, and Zn.

8.3.4.3 Extraction/Air–Acetylene Flame Method

Low concentrations of Cd, Cr, Co, Cu, Fe, Pb, Mn, Ni, Ag, and Zn can be estimated using the extraction/air–acetylene approach. The technique uses the air–acetylene method but uses chelation by ammonium pyrrolidine dithiocarbamate (APDC) and extraction into methyl isobutyl ketone (MIBK) prior to aspiration.

8.3.4.4 Direct Nitrous Oxide–Acetylene Flame Method

Nitrous oxide is one of the more widely used gases for burning with acetylene, aside from air. Its burning velocity is low, and the flame's temperature is high due to the energy released by the nitrous oxide decomposition. This flame can be safely burned through a slot as long as 10 cm, though most metals prefer a 5 cm slot. Manufacturers of commercial atomic absorption spectrophotometers have adopted this flame as the usual way of atomizing metals that do not atomize successfully in the air–acetylene flame. Heavy metals like Al, Ba, Be, Ca, Mo, Os, Re, Si, Th, Ti, and V can be analyzed by direct aspiration into nitrous oxide– acetylene flare [33].

8.3.5 Cold Vapor Atomic Absorption Spectrometry

Cold vapor atomic absorption spectrometry (CVAAS) is the most often used method for the estimation of mercury. This method is distinguished by its detection boundaries in the single-digit ppt range, a dynamic range of two to three orders of magnitude, and various analytical procedures that enable mercury measurement in any sample matrix. A peristaltic pump is generally employed with CVAAS to inject the sample and stannous chloride ($SnCl_2$) into a gas-liquid separator, which is then bubbled with a stream of pure, dry gas to liberate Hg vapor. The Hg is then transferred via carrier gas into an atomic absorption cell. Hg absorbs 254 nm light in proportion to its concentration in the sample [34].

8.3.6 Electrothermal Atomic Absorption Spectrometry

Most metallic elements may be determined using electrothermal atomic absorption, which eliminates the need for an extraction process and has sensitivity and detection levels 20 to 1000 times lower than traditional flame procedures. This rise in sensitivity is caused by the increasing density of atoms from flame atomic absorption within the furnace. At concentrations lower than 1.0 μg/L, various elements can be analyzed. An added benefit of the electrothermal atomic absorption is that it requires merely small amounts of sample for analysis.

The procedure is essentially similar to direct flame atomization, but instead of a standard burner head, an electrically heated atomizer or a graphite furnace is used. First, a volume of the sampled mixture is discharged into the graphite sampling tube. Final analyzed samples are obtained after a series of heating stages, usually in three steps or more. Initially, a small supply of current is used to heat the tube to dry the sample, followed by an intermediate temperature, which kills the organic materials and volatilizes other matrix components.

Finally, high currents heat the tube to an incandescent state, thereby atomizing the element. Further steps are often added to aid in charring and drying.

Monochromatic radiation from the source is absorbed by the resulting ground-state atomic vapor. The intensity of transmitted radiation is measured by a photoelectric detector. The reciprocal of transmittance is logarithmically correlated to the absorbance across a controlled concentration range [35]. This process is appropriate for determining micro quantities of Sb, Al, Be, Ba, Cd, Co, Mn, Ag, Cr, Fe, Cu, Mo, Pb, Ni, Sn, and Se.

8.3.7 Arsenic and Selenium by Hydride Generation

This section covers two methods: the manual and the continuous-flow methods. The latter of these is especially suggested for measuring selenium concentrations. Hydride production is a method of converting certain elements to volatile metal hydrides by means of a reducing reagent. As a result, atomization efficiency is considerably improved while detection limits are reduced.

Sodium borohydride is used to convert As and Se to hydrides before transporting them into an atomizer. The sodium borohydride reducing agent eliminates hydride dilution by the carrier gas and allows fast, sensitive As and Se readings by rapidly generating the elemental hydrides [36]. For selenium analysis, the hydride generator has the advantage of ease of use, repeatability, lower detection limits, and higher volume throughput.

8.3.8 Inductively Coupled Plasma Optical Emission Spectroscopy

Established in the mid-1960s as a fast, sensitive, and suitable process for determining the metals in wastewater samples, the inductively coupled plasma (ICP) method relies on two components: the ICP torch and the optical spectrometer. The ICP torch is made up of three concentric quartz glass tubes. The output, or the coil of the radiofrequency generator, surrounds part of this quartz torch, and Ar gas is typically used to create the plasma. The ICP can operate in two modes: capacitive (E) mode with low plasma density and inductive (H) mode with high plasma density. The E to H heating mode transition is controlled by external inputs. The torch is operated in H mode [37]. Dissolved metals are analyzed in filtered and acidified samples. Total metal analysis requires proper digestion.

An ICP source is made up of a stream of Ar gas ionized at a radiofrequency of 27.1 MHz. The radiofrequency field is coupled inductively to the ionized gas by a water-cooled coil surrounding a quartz torch that supports plasma through an injector tube situated inside the torch. An aerosol sample is produced in a nebulizer and spray chamber, which is then delivered into the plasma. The aerosol sample is introduced straight into the ICP, exposing the atoms to 6000-8000 °C. The plasma's high temperature stimulates atomic emission, ionizing a large number of atoms to generate an emission spectrum. It is widely used in minerals processing and trace element identification.

8.3.9 Inductively Coupled Plasma–Mass Spectrometry

ICP–MS is used to identify trace elements, metals, and nonmetals in ground, surface, and drinking waters. Although this approach is ideally suitable for freshwater matrixes, it can be used to examine soils, wastewater, sludge, sediments, and biological materials after following the proper dilution, digestion processes, and clean-up to lessen the effects of the matrix. It atomizes the sample, resulting in atomic and tiny polyatomic ions that can then be determined by mass spectrometry alone. ICP–MS is known for its capacity to recognize inorganic and organic materials in liquid samples at extremely low concentrations. It can also distinguish between different isotopes, making it a versatile technique. The ions from the plasma are extracted through a sequence of cones into an MS, commonly a quadrupole, for mass spectrometry coupling. The separation of ions depends on the ratio of mass-to-charge and an ion signal corresponding to the concentration received by a detector.

This analysis uses pneumatic nebulization to aspirate the sample into an Ar-based, high-temperature radiofrequency plasma. The target element dissolves, atomizes, and ionizes as energy is transferred from the plasma to the stream sample. Ions are extracted from the plasma via a differential vacuum interface then separated using a mass spectrometer based on the mass-to-charge ratio. The separated ions are counted by the electron detector, and a data acquisition arrangement processes the essential data.

8.3.10 Anodic Stripping Voltammetry

ASV is currently the most sensitive metal ion detection method. For certain metals, it can be up to 100 times more sensitive than electrothermal AAS. Its detectability can reach the range of nanograms per liter. There is no need for a sample extraction in this method, and it allows for determination of up to six trace elements simultaneously. However, it is limited to forming amalgam metals and requires extensive analytic time compared to spectroscopic approaches. Moreover, the potential interference caused by the high sensitivity may pose serious constraints.

ASV is a two-step electroanalytical procedure. At the negative potential, the metal ions in the sample are reduced and further concentrated onto a Hg electrode during the preconcentration stage. The strength of metal in Hg is 100-1000 times higher than its ion concentration in the solution sample. After the preconcentration stage, a stripping phase measures current by supplying a positive potential. The level of detection relies on the metal ion analyzed, time of deposition, rate of stirring, pH, sample matrix, and type of electrode employed. Cd, Pb, and Zn are efficiently concentrated throughout the pre-electrolysis due to their higher solubility in Hg, and hence they have lower levels of detection ($<1\ \mu g/L$) [38].

8.4 Analysis of Inorganic Anion Constituents

The approaches involved in analyzing anions involve wet chemical procedures and various automatic systems, such as ion chromatography. More recently, there has been increasing demand for estimating the concentration of chlorites, chlorates, and bromites. Though conventional techniques like calorimetric, titrimetric, or electrometric methods are accessible to analyze the anions, ion chromatography (IC) is preferred for fast and subsequent measurements. It discards the usage of dangerous reagents while still being able to differentiate between halides efficiently. These procedures are envisioned for usage in the assessment of water quality, treatment, and measurement operations.

8.4.1 Ion Chromatography with Chemical Suppression of Eluent Conductivity

In IC, the sample is aspirated into the stream and passed through a sequence of ion exchangers. These ions are separated by a lower capacity, strongly basic anion exchanger based on their relative affinity. The separated anions are then passed through a constant suppression process to improve the response of the analyte and reduce the conductivity of the eluent. Though the conductivity of the eluent is significantly reduced, the anions that are separated are transformed into highly conductive forms of acid in the suppressor. Quantification of the ions that are separated in their acid forms is performed by conductivity measurement. They are recognized by comparing their residence time [39]. The obtained peak area or height is used to quantify the measurement.

8.4.2 Single-column Ion Chromatography with Direct Conductivity Detection

In this procedure, the sample is aspirated into an IC, consisting of an injector port, detector, and column. The aqueous sample is mixed with the eluent and driven through the active sites of the column, causing the resulting anions to resolve into separate bands based on their affinity. Direct measurement of conductivity is performed to estimate the concentration of anions without the use of suppression.

8.4.3 Ion Chromatography Determination of Oxyhalides and Bromide

This procedure is used to estimate trace levels of oxyhalides and bromides in the presence of common anions in drinking water, their detection limits, and their respective concentration ranges. This method helps to improve the separation between bromate and chloride ions, allowing lower detection limits of bromate in the presence of high chloride concentrations.

8.4.4 Capillary Ion Electrophoresis with Indirect Ultraviolet Detection

The determination of anions such as fluoride, chloride, bromide, nitrite, nitrate, orthophosphate, and sulfate are critical for assessing water quality. Instrumental measures that can measure many analytes in a single assay, such as IC and capillary ion electrophoresis (CE), save money and time compared to single-analyte wet analysis, e.g. colorimetry and titration. The advantages of CE are the speed at which multi-analyte analytes can be completed and the availability of anion data not offered by isocratic IC.

In CE with ultraviolet (UV) detection, a 25-100 μm internal diameter capillary made of silica is filled with an aqueous electrolyte solution comprising a UV-absorbing anion salt (sodium chromate, Na_2CrO_4) and an electro-osmotic flow modifier. An electric field is created by applying 15 kV where the capillary end of the detector is designated as the anodic side [40]. The sample is aspirated at the cathodic side, and anions are isolated based on their movement in the electric field as they move through the capillary. The cations are undetectable as they move in the reverse path. Organics and water are not drawn to the anode side; instead, they travel after the anions and therefore do not hinder the analysis of anions. The anions are spotted because they shift the UV-absorbing electrolyte chromate anion charge, resulting in a reduction in the UV absorbance in the anion analyte section. Analytes are recognized by their movement, as in chromatography, and quantified by time-corrected peak area comparative to the standards. Once the required analytes are analyzed, a fresh electrolyte solution is sent through the capillary, removing any residue before the performance of the subsequent analysis.

8.5 Analysis of Organic Constituents

Analyses of organic matter in wastewater can be divided into two types: parameterization of organic matter made up of varying constituents and identification of individual organic components. The former can be further classified into four subtypes: (i) substances that require oxygen, (ii) elements that are bound organically, (iii) classes of organic compounds, and (iv) their formation potential (FP). TOC and COD are used to evaluate the overall quantity of existing organic material. BOD measurements signify the presence of biodegradable organics while also denoting the material extracted from a sample with the help of dissolved organic halide (DOX), which can measure organically bound halogens. The FP of trihalomethane is a collective estimate of the concentration of total trihalomethanes resulting from chlorination of the water. The quantitative organics analysis, which can be an indicator for the productivity of treatment processes, is performed to calculate the strength and organic matter composition in wastewater treated effluents and receiving waters.

8.5.1 Biochemical Oxygen Demand

BOD is an estimate of the relative oxygen requirements of polluted water, wastewater, and effluents. Its most common application is in estimating the plants' BOD-removal efficiency. BOD estimates the amount of oxygen utilized throughout a specific period to biochemically break down carbonaceous organic material, oxidize inorganic materials, and estimate the quantity of O_2 utilized to oxidize reduced nitrogen.

The ultimate BOD (UBOD) test determines how much oxygen is needed to completely decompose the ultimate carbonaceous material and oxidize the nitrogen compounds. In water quality modeling investigations, UBOD measurements and adequate kinetic studies are required. The accuracy and precision of BOD calculations are influenced by various aspects like solubility, settleability, floatable solids, organic particulates, and oxidation of reduced Fe and S elements.

8.5.2 Five-Day BOD Test

The BOD test looks for changes in dissolved oxygen (DO) content induced by microorganisms as they break down organic matter in a stoppered container cultured for five days in the dark at 20°C [41]. DO is measured before and after incubation. The difference between the two values is used to calculate BOD as given in Eq. (8.27):

$$BOD_5.\ mg/l = \frac{(D_1 - D_2) - (S)V_s}{P}, \tag{8.27}$$

where D_1 is the diluted sample DO (mg/L), D_2 is the diluted sample DO after 5 d, S is seed oxygen uptake, V_s is the volume of seed in the test container (mL), and P is the fractional volume of the sample utilized, with $1/P$ being the dilution factor.

8.5.3 Ultimate BOD Test

The UBOD test is an extension of the five-day dilution of the BOD examination. The approach entails a single sample dilution sealed in a flask and placed under a lengthy incubation period, which depends on the quality of the final wastewater. DO is estimated using probes at the start and periodically throughout the test. An appropriate statistical technique is used to calculate UBOD from the DO vs. time plot. Supplemental nutrients that quicken the decay process and shorten the duration of the test may be beneficial if just UBOD is sought. The amount of nitrogenous substances that will oxidize throughout the incubation time is determined by the number of relevant oxidizing microorganisms present. These microbes oxidize substantial amounts of N_2 in wastewaters; however, they are not abundantly active in natural surface waters. UBOD can be estimated by Eq. (8.28):

$$BOD_t = UBOD\left(1 - e^{-kt}\right), \tag{8.28}$$

where BOD_t is the measured O_2 uptake at time t, (mg/L) and k is the first-order rate constant.

8.5.4 Chemical Oxygen Demand

COD is described as the quantity of a specific oxidant that reacts with the sample under specified environmental conditions. The amount of oxidant used up during the process is stated in terms of oxygen equivalence. Although both inorganic and organic components of a sample are susceptible to oxidation, their effects on organic components are of larger concern. The open reflux process is appropriate for a wide variety of wastes that require a huge sample. Closed reflux procedures utilize less metallic salt reagents and produce less harmful waste, but then they need homogenized samples containing suspended solids to yield repeatable findings [18].

The open reflux method operates on the fact that a mixture of chromic and sulfuric acids oxidizes most types of organic materials. The sample is refluxed with a strong acid solution containing an excess of potassium dichromate ($K_2Cr_2O_7$). Following digestion, the residual $K_2Cr_2O_7$ is titrated with Mohr salt or ferrous ammonium sulfate to estimate the quantity of $K_2Cr_2O_7$ used up. In addition, the material oxidized is determined in O_2 equivalents. COD is calculated as in Eq. (8.29):

$$\text{COD as } O_2/\text{L} = \frac{(B - A) * M * 8{,}000}{ml\ sample}, \tag{8.29}$$

where B is the ferrous ammonium sulfate used (mL), A is ferrous ammonium sulfate for blank titration (mL), and M is the ferrous ammonium sulfate molarity. 8000 is the milliequivalents of oxygen weight in a 1000 mL sample.

In the closed reflux colorimetric method, the dichromate oxidizes COD content in a sample as it is digested. As a result, chromium (Cr) transitions from the hexavalent (VI) state to trivalent (III). Together, the two Cr types have a distinct color and absorb light in the visible spectrum. At the 400 nm range, the dichromate ion ($Cr_2O_7^{2-}$) absorbs considerable amounts of light, whereas the chromic ion (Cr^{3+}) absorbs much less [18, 42]. While Cr^{3+} absorbs substantial light at a wavelength of 600 nm, the $Cr_2O_7^{2-}$ exhibits zero absorption. The COD for this method is calculated using Eq. (8.30).

$$\text{COD as mg } O_2/\text{L} = \frac{mgO_2 in\ final\ volume * 1000}{ml\ sample} \tag{8.30}$$

8.5.5 Total Organic Carbon

The amount of organic carbon present in wastewater reflects numerous organic mixtures, having many oxidation states. Quite a few of these molecules can additionally be oxidized through chemical or biological means. TOC is unaffected by the oxidation state of the organic matter, and it does not account for the elements that are organically bound, such as nitrogen and hydrogen, or any metal ions that may supply the O_2 demand evaluated by COD and BOD. The quantity of TOC in drinking water ranges from 0.1–25 mg/L, while wastewater values are > 100 mg/L. When performing TOC analysis, the following are estimated:

- TC – Total Carbon
- TIC – Total Inorganic Carbon
- POC – Purgeable Organic Carbon
- NPOC – Non-Purgeable Organic Carbon
- DOC – Dissolved Organic Carbon

- NDOC – Non-Dissolved Organic Carbon
- TOC = TC − TIC
- TOC = POC + NPOC
- TOC = DOC + NDOC

Depending on the strength of TOC and analytical needs (e.g. for speed, sensitivity), a variety of oxidation detection methods are utilized. High-temperature combustion at 1200°C in an O_2 rich atmosphere converts the organic carbon to CO_2. To reduce interferences, the CO_2 produced is routed via scrubber tubes and measured using non-dispersive infrared absorption. This combustion method is primarily employed for higher TOC concentrations.

Procedurally, the test starts with the sample entering the analyzer and passing through a pressure regulator, where the sample is split into two pathways. A portion of the flow is directed into the bypass stream, where resistivity, conductivity, and temperature are estimated. The other portion is directed toward another sensor that measures the conductivity before the oxidation step. As the sample passes through the oxidation chamber, it is exposed to high-intensity UV radiation at 185 nm to oxidize the sample efficiently. After oxidation, the sample is passed through the third sensor, where conductivity and temperature are estimated to evaluate the TOC. The operating flowrate is generally 20 mL/min, resulting in a residence time in the oxidation chamber of less than one minute. As the conductivity measurements are continuous, the response time is directly related to the residence time of the sample in the chamber [43].

8.5.6 Oil and Grease

There is no method for completely measuring the total quantity of oil and grease in a given substance within a solution. A feasible quantitative analysis is done by determining the solubility of some groups of constituents with similar features in the solvent. Knowing the amount of oil and grease aids in the appropriate operation and design of wastewater treatment systems and provides more information to avoid potential treatment issues.

For liquid-liquid samples, the gravimetric partition technique is used. Dissolved or emulsified oil and grease are removed from water using a solvent. When mixed with the samples, organic solvents form an emulsion that is hard to break. This technique offers a way of handling emulsions that are similarly difficult to break [44].

In the partition infrared method, the use of solvent trichlorotrifluoroethane permits the absorbance of C–H bond linkage in the infrared region (IR) to estimate grease and oil. The removal of the evaporation stage allows for the detection of various volatile compounds using infrared technology. Thus, with the exception of gasoline, lighter petroleum distillates can be precisely quantified.

In relatively polar conditions, such as when heavier fractions of petroleum exist, Soxhlet extraction can be implemented. Soluble metallic soaps are hydrolyzed by acidification. A simple filtration process is then used to separate oils, solids, or grease from the liquid samples. Post-extraction, the residue left after evaporation of the solvent is evaluated to assess the content of oil and grease. When the filter is dried, volatilized compounds at or lower than 103°C will have vanished. If the goal is to determine hydrocarbon content instead, then the hydrocarbon solution and fatty materials need to be mixed with Si gel, which removes fatty acids from the solution. The Si gel can adsorb polar materials based on polarity. The constituents that are not removed by the adsorption of Si gel are designated as hydrocarbons.

The solid-phase gravimetric partition method can be used to decrease the volume of solvent and limit matrix problems. In this test, the dissolved or emulsified grease and oil are extracted from water by allowing it to pass through a solid-phase extraction disk where grease and oil are adsorbed by the disk and then eluted with the solvent n-hexane. This technique is not suitable for components that are volatile at temperatures lower than 85°C and heavier fuel oils that contain a certain percentage of insoluble material [18]. This technique might be acceptable as a substitute to the liquid-liquid extraction technique, specifically those that have tough emulsions during the extraction step.

8.5.7 Phenols

Phenols are hydroxy derivatives of benzene that occur in wastewater. Chlorination of such waters produces chlorophenols. Removal of phenols comprises either chlorination, chlorine dioxide or chloramine treatment, or ozonation followed by adsorption using activated carbon. This analytical procedure uses the 4-amino antipyrine colorimetric process that estimates ortho, phenol, and meta substituted phenols under appropriate conditions of pH. There are two types of the 4-amino antipyrine method [18]: direct photometric and chloroform extraction methods. For exceptional sensitivity, the chloroform extraction method can be used in samples comprising lower than 1 mg/L phenol by color concentration in a nonaqueous solution. The aqueous solution color is retained using the direct photometric method.

As it is impossible to predict or estimate the total quantities of several phenolic mixtures in a given sample, colorimetric measures have been chosen as a standard technique. The principle of the photometric methods is that in the presence of potassium ferricyanide ($C_6N_6FeK_3$), steam-distilled phenols react with 4-amino antipyrine to generate a colorful antipyrine dye at pH 7.9 ± 0.1. Their absorbance is estimated at 460 nm after it has been extracted from an aqueous solution with chloroform. This technique covers a wide range of phenol concentrations from 1.0 –250 g/L. When directly extracted without chloroform, the absorbance is estimated at 500 nm in an aqueous solution.

8.5.8 Surfactants

Surfactants enter water and wastewater mainly through the release of wastes from laundry and other home and industrial cleaning processes. A surfactant is a molecule that combines a strongly hydrophobic and hydrophilic group into a single molecule. These molecules tend to group at the boundaries between aqueous media and other phases like air, oily liquids, and particulate material, passing on behavioral properties like suspension of particles, foaming, and emulsification. The hydrophobic end of a surfactant is usually a hydrocarbon of 10 to 20 carbon atoms. There are two kinds of hydrophilic groups. One type ionizes in water, and the other does not. Ionic surfactants are classified based on the charge carrier. Anionic surfactants are negatively charged, whereas a cationic surfactant carries a positive charge [45].

Separation by sublation separates surfactants from a dilute aqueous solution and produces a dehydrated residue that is comparatively devoid of nonsurfactants. A stream of N_2 is bubbled by passing through a column comprising the sample and an ethyl acetate layer. Surfactants are absorbed in the bubbles' gas-water interface and passed into the film of ethyl acetate. The surfactant soluble in ethyl acetate is left behind as the bubbles discharge out. Surfactants are then removed from the solvent, dehydrated, and evaporated, leaving behind a residue.

The methylene blue active substances are commonly used for detecting the presence of anionic surfactants and result in the transfer of the cationic dye methylene blue from an aqueous solution to an immiscible solvent. This happens when the methylene blue anion and cation form an ion pair, where the blue color intensity results as a measure of methylene blue in the organic phase. Similarly, cobalt thiocyanate active substances (CTAS) react with a solution of cobalt thiocyanate to produce a cobalt-comprising product that can be extracted into an organic liquid and quantified. Nonionic surfactants, like other natural and synthetic compounds, have this activity. This approach includes the process of sublation to eliminate interactions of nonsurfactants and the ion exchange process to eliminate surfactants. The separation of CTAS to methylene chloride from aqueous solution by a single extraction process is measured at 620 nm by spectrophotometry.

8.5.9 Tannin and Lignin

Lignin is a key component of vegetation that is frequently discarded as waste during pulp processing. Another plant component, tannin, might enter the water system by the decomposition of vegetable material or through tanning industry wastes. Tannin is also used in the internal treatment of boiler fluids to prevent scale formation by producing sludge. Tannin and lignin comprise aromatic hydroxyl groups that can be analyzed by a colorimetric method, in which the hydroxyl group reacts with the Folin reagent to yield a suitable blue color to estimate the concentrations by up to 9 mg/L. Inorganic and organic compounds respond similarly to this test [46].

8.5.10 Organic and Volatile Acids

As an anaerobic digestion control test, organic acids can be measured by adsorption with the help of a chromatographic column, GC, or distillation process. A chromatographic separation method can be used for organic acids, whereas, for volatile acids, a distillation and a gas chromatographic method may be implemented [47]. Volatile fatty acids are divided into water-soluble fatty acids that can be distilled at atmospheric pressure. These can be eliminated by distillation despite their high boiling points. During the chromatographic separation process, the acidified mixture comprising organic acid is adsorbed onto a column of silicic acid, and the acids are eluted with n-butanol in chloroform ($CHCl_3$). All short-chain organics are eluted by the solvent and are collected as total organic acids. Total organic acids are calculated using Eq. (8.31):

$$\text{Total organic acids} = \frac{(a-b) * N * 60{,}000}{mL\ sample}, \tag{8.31}$$

where a is the sample of sodium hydroxide (NaOH) used (mL), b is the NaOH utilized for blank, and N is the normality of NaOH.

In the distillation method, acids comprising up to six carbon atoms can be recovered. The retrieval of each acid enhances through an increase in the molecular weight. The recovery factor depends on the rate of heating, presence of solids sludge, and volume of final distillate, and it is calculated by Eq. (8.32):

$$\mathrm{f} = \frac{a}{b}, \tag{8.32}$$

where a is the distillate acid concentration (mg/L), and b is the standard solution acid concentration (mg/L).

The GC technique can evaluate the specific concentration of fatty acids such as acetic, butyric, isobutyric, propionic, and isovaleric acids. Heptanoic acid and caproic acids can also be determined, albeit with a reduced amount of precision due to their lower water solubility. All acids must be transformed to a volatile state before introducing into the GC so that the vaporization can take place. The fatty acids need to be separated from solid materials before their introduction into GC to lessen the degradation of the GC column. After separation through centrifugation and filtration process, samples can be safely analyzed with the flame ionization detector.

8.6 Analysis of Radioactive Materials

Radioactivity in wastewater is sourced from both natural and human operations. When drinking water comes in contact with naturally occurring radioactive materials (NORM) bearing rocks, radionuclides can build up to dangerous levels in the water. Radium (Ra), decayed products of thorium (Th), and uranium (U) are the most common radionuclides detected in the water. These radioactive substances are unstable and emit radiation to achieve more stable conditions, termed radioactive decay. Measurements of the type of radiation emitted, specific energy levels of radiation, and rate of decay, can help us identify how much of a radioactive substance is present. It is measured in curies and is used to describe the concentration of radioactivity present in drinking water. U, Th, Ra are NORMs with a long sequence of radioactive daughters that release alpha, beta, and gamma radiations until they form a stable element. These radioactive elements, along with radon (Rn) and thoron (Tn), produce airborne activity and assist in bringing radioactivity through rain and into groundwater.

Gross alpha and beta examinations are used to determine the type and level of radioactivity in wastewater by evaporating the sample and measuring the activity of the residue. As alpha radiation is easily absorbed within a thin layer of solid material, the reliability and sensitivity of the method for alpha determination may be reduced in samples with high TDS content. In the gross beta test, the evaporation method includes the contribution from potassium-40. These two tests serve as an initial screening: if the result is >15 pCi/L for U and >5 pCi/L for Ra, further testing for radioactive materials is required. The coprecipitation test eliminates the contribution due to potassium-40, making its analysis not essential. This process is not related to the assessment of samples comprising fission products, like Cs-137 [48].

The gamma spectroscopic method is applied for measuring the photons released from water containing radionuclides with either germanium (Ge) diodes or thallium-activated sodium iodide [NaI(Tl)] crystals [49]. This technique can be used on samples containing radionuclides that generate gamma rays with energy ranging from 60–2000 keV. Precise quantification with sodium iodide (NaI) is attained for single nuclides or once the gamma emissions are restricted to limited well-separated energies. Cesium (Cs) has been observed as one of the dangerous radioactive nuclides formed during fission. If Cs activity is high, it can be estimated by gamma-counting a liquid sample or the sample that can be evaporated and counted. Radioiodine released during the processing of reactor fuels is a key concern in monitoring radioactivity. The precipitation method is favored since it is simple and involves less time.

In its working mechanism, an iodate carrier is added to an acidified sample, and after being reduced with sodium sulfite (Na_2SO_3) to an iodide (131-I), the 131-I is precipitated with silver nitrate ($AgNO_3$). The precipitate is dissolved and purified with zinc powder and H_2SO_4, and it is reprecipitated as PdI_2 for counting. In the ion exchange technique, iodide is strengthened by absorption onto an anion resin, purified, and counted in a beta-gamma coincidence system, which is sharp and precise. Method-wise, a known quantity of inactive I in the form of KI is added as a carrier, and it is passed through an oxidation–reduction stage utilizing hydroxylamine and sodium bisulfite ($NaHSO_3$) to completely convert iodine to iodide. Initially, the iodine, in iodide form, is concentrated by absorption onto an anion-exchange column followed by a wash with NaCl, and iodine is eluted with sodium hypochlorite (NaOCl). Iodine in the iodate form is reduced to I_2, extracted into carbon tertrachloride (CCl_4), and further reextracted as iodide into water.

8.7 Toxicity Test Systems, Requirements, Evaluation, and Implementation

To have a comprehensive physical and chemical examination of water contents and their potential impacts on aquatic biota, chemical interactions, and effect of composite matrixes, toxicity testing is essential [50]. Toxicity tests help to analyze numerous aspects such as:

- effects on ecological condition for aquatic life.
- properties that dictate eco-friendliness like pH, DO, temperature, turbidity, and salinity.
- toxicity of waste.
- responsiveness of aquatic organisms to the toxicant.

- acceptable effluent discharge rate.
- conformation to water quality and effluent release standards.

Toxicity test results are used in conjunction with other contextual data such as type of water, and site discharge information related to volume, dilution rate, concentration, and time of exposure for regulatory and legal requirements. All parts of the test that affect data quality are covered by the quality assurance practices for toxicity examination. These comprise sample collection and handling, the test organism's source and condition, toxicant reference testing, and assessment strategies. A toxicity test for aquatics is a process wherein the response of aquatic creatures is utilized to notice or assess the outcome of environmental factors associated with it.

8.7.1 Requirements for Toxicity Test

The basic requirements for toxicity valuation are:

- large water reserves
- suitable and powerful water stream consisting of non-polluting materials
- suitable space for well-organized holding, culturing, equipment testing, and services
- appropriate lighting facilities for toxicity tests.

The services, equipment, and water reserves required for successful examination rely on the type of test.

8.7.2 Categories of Toxicity Test: Uses, Pros, and Cons

Toxicity evaluations are classified by:

- Duration (short-term, intermediate-term, or long-term).
- Evaluation solutions process (static, renewal, or flow-through).
- Monitoring objective (quality of effluent, relative toxicity, responsiveness, taste, odor, or growth rate).

Short-period toxicity testing is implemented for continuous tracking and exploratory tests by following the permitted effluent discharge criteria. Acute definitive tests are used to determine the effects due to toxicants. These tests are implemented to show the exact range of concentration for intermediate and long-term tests. Intermediate toxicity examinations are implemented when additional time is required to estimate the consequence of toxicant concentrations in several phases of an organism's life cycle.

Long-period toxicity examinations are normally performed to estimate chronic toxicity. Long-term examination comprises early life cycle, partial life cycle, or full life cycle analysis. Exposures range from a minimum of 7 d to a maximum of 21–28 d, which can be prolonged to several months or even extended to additional partial and full life cycle tests. It is advised to be alert when running static examinations to estimate the solutions comprising higher levels of bacteria, BOD, and COD. These tests can be performed effectively if DO is monitored critically and adequate aeration is included to supply sufficient oxygen. Flow-through examination for toxicity is necessary for samples comprising higher levels of BOD or COD. It is hard to sustain under static exposure situations for organisms with a higher metabolic rate. The flow-through examinations keep the conditions well oxygenized while metabolic wastes are continuously removed. Usage of flow-through toxicity examination is recommended when the solution is likely to degrade quickly.

8.7.3 Short-term Toxicity Tests

8.7.3.1 Range-finding Examination

For an effluent system of unfamiliar toxicity, perform a short-term examination for 24 or 48 hours to estimate a rough range of concentrations. For effluents that are known to have low toxicity, 48- or 96-hour examinations might be essential whereby the test organism is exposed to a wide concentration range, generally in a logarithmic ratio (0.01, 0.1, 1, 10, and 100% of sample). Select a geometrical sequence of concentrations among the highest and lowest concentration ranges that can kill most or all organisms in the test [51].

8.7.3.2 Short-term Definitive Examination

Short-term examinations might be static, renewal, or flow-through. In a static renewal test, the organisms are exposed to a fresh solution of the same concentration of sample every 24 h, either by relocating the test organisms from one chamber to another or by substituting all or a portion of the solution in the test chambers. In the nonrenewal test, the organisms are exposed to the same test solution during the entire test duration.

8.7.3.3 Intermediate Toxicity Examination

There are no strong period differences between intermediate and short or long-term and intermediate examinations. Generally, intermediate examinations might proceed from 11 to 90 days. The length of the organism's life cycle assists in estimating the extent of examination for that species.

8.7.3.4 Long-term Partial or Complete Toxicity Examination

The complete objective of this kind of examination is to evaluate no observed effect concentration (NOECs) or chronic value (ChV) of effluents, toxicants, or wastes [52]. In the life cycle or partial life cycle examination, ensure that the water quality aspects like pH, temperature, salinity, and DO obey the usual periodic sequence.

8.7.3.5 Short-term Examination for Estimating Chronic Toxicity

Examinations are permitted to evaluate the long-term results of a toxicant following seven days of exposure. The endpoints for these examinations are termed chronic estimators, including deadliness, growth, and potential to reproduce. These examinations are intended to assess the effluent toxicity and might not be suitable for other forms of testing.

8.7.4 Toxicity Test Systems

The information gap created by the failure to apply (existing or altered) bioassays, which have been proven to be a potent instrument for understanding the genuine interactions of matrixes discharged into the environment, can be easily closed. Standard toxicological methods like tests on crustaceans, algae, and luminous bacteria have been created to evaluate effluent quality. However, when determining the true impact of wastewater discharge, a variety of biological targets should be considered [53]. In the selection of bioassays, tests are chosen for their repeatability, automation, sensitivity, adequacy, statistical robustness, biological representativeness, and standardization. After selecting endpoints linked to short and long-term effects, criteria of acute and chronic toxicity should be considered. Furthermore, diverse biological complexity (prokaryotes/eukaryotes, unicellular/multicellular, animals/plants) and distinct levels of biological organization (organisms, tissues, cells) should be targeted.

Bioassays can identify both baseline toxicity and specific modes of action, which can be used to link critical events to biological outcomes. Some of the more specific tests in toxicological studies are:

- Baseline toxicity (green alga growth inhibition [54], marine bacteria bioluminescence inhibition [55], plants root growth inhibition).
- Endocrine disruptions (luciferase activity quantification in human breast cancer cell line [56]).
- Genetic toxicity points to reverse mutations in bacteria [18], single-cell gel electrophoresis [18].
- Carcinogenicity (the number of malignant foci [57]).

The detailed methodologies of the above tests are discussed elsewhere [58]. The association between the biological reaction and the degree of dilution (tests based on raw samples) or concentration (tests based on extracts) of the samples is critical information to consider for interpreting results. The computation is relatively easy in the case of dilution of raw wastewater. When a sample pretreatment is required, however, both dilution and enrichment techniques must be considered. The range of sample concentrations investigated in different bioassays was expressed in units of Relative Enrichment Factor (REF) [59], which is the product of the solid-phase extraction process's enrichment factor and the bioassay's actual sample dilution and is given as below in Eqs. (8.21–8.23).

$$REF = EF_{SPE} * DF_{bioassay}$$

$$EF_{SPE} = \frac{V_{water}}{V_{extract}}$$

$$DF_{bioassay} = \frac{V_{extract\ added\ to\ bioassay}}{V_{bioassay}}$$

A REF value of one indicates that the concentration of organic analytes in the bioassay is the same as that of unprocessed water (free of metals, inorganic anions, and a portion of colloidal organic after solid-phase extraction). A value greater than or less than one, on the other hand, denotes a sample concentration or dilution.

8.7.5 Source Evaluation of Toxicity

Once the effluent toxicants have been discovered, a follow-up investigation can be carried out to determine the toxicants' sources. This assessment could include a review of existing pretreatment program data or data gathered and analyzed from additional samples provided by industrial users. In other circumstances, the toxicity identification evaluation (TIE) may not identify the exact compounds producing effluent toxicity. Hence the sources of toxicity must be tracked in the absence of toxicant data. Surfactants and several non-polar chemical molecules are examples of substances that are difficult to identify in the TIE. Surfactants and several non-polar chemical molecules are examples of substances that are difficult to identify in the TIE other than organophosphate insecticides [60, 61].

8.7.6 Toxicity Reduction Evaluation

The goal of toxicity reduction evaluation (TRE) is to identify and implement toxicity control methods and technologies that will allow the effluent toxicity permit limitations to be met. Toxicity control evaluation entails examining prospective control options and deciding on the optimal option(s) for toxicity reduction based on technical and financial factors [60]. At the beginning of the toxicity control evaluation, the criteria for selecting the preferred toxicity control are:

- effluent toxicity limitations
- agreement with other permits
- capital, operating, and maintenance costs
- ease of implementation
- reliability
- environmental effect.

8.7.6.1 Pretreatment Control Evaluation

Public work managers identify pretreatment control options to prevent the pass-through of toxicants, toxicities, and inhibitory material that have been identified as indirect dischargers. The main benefit of pretreatment toxicity control is the reduced volume of waste handled by targeting particular sources. Additionally, expenses are usually borne by industrial users. The toxicants to be controlled are not the same as those already regulated by pretreatment programs like organophosphate insecticides, TDS, biocides, and specialty chemicals used in industry.

8.7.6.2 In-plant Control Evaluation

The objective of the in-plant control evaluation is to select and evaluate feasible treatment options for the reduction of effluent toxicity. Treatability testing can be done to evaluate the operating characteristics of treatment alternatives and toxicity removal efficiencies. These studies should include acute or chronic toxicity tests as well as chemical analysis to evaluate the removal of specific toxicants and toxicities. The results serve as a foundation for finalizing the selection and conceptual design for a viable process modification. To find the best technically viable and cost-effective control alternative, the in-plant control assessment should be undertaken in conjunction with the pretreatment control evaluation wherever possible.

8.7.7 Toxicity Control Implementation

After evaluation and selection of toxicity control measures are complete, the final phases in the TRE are the implementation of the selected pretreatment and in-plant control options, as well as follow-up monitoring to verify permit compliance. The effort required in the implementation step depends on the severity of the effluent toxicity and the difficulty of the approach selected. Based on the results obtained in the TRE process, a toxics control implementation plan (TCIP) should be developed. This plan summarizes the results of the TRE analysis: screening, selection, and justification for the selection of the preferred toxicity control options [62]. In the case of in-plant operations, the TCIP should include a design basis for the selected control options, including capital and operating costs. The TCIP should specify the grounds for selection and technical justification for local limitations and discharge monitoring procedures.

8.7.8 Calculating, Investigating, and Reporting Toxicity Results

This segment lists the statistical approaches for assessing the information based on the acute and chronic toxicity examinations, procedures suited for each specific data, and the benefits and drawbacks of the most commonly used statistical approaches.

Acute toxicity test results are accompanied by simple "yes" or "no" answers to the queries; for example, "did the exposure produce impairment?" or "did the exposure result in demise?" In acute toxicity examinations, continuous measurements are collected in quantifiable data. Data related to the weight, length, or the number of a young generation formed, are normally not utilized as endpoints.

The outcomes of chronic toxicity examinations may require additional numerical approaches, depending on the examination. All such procedures are mathematically intensive, though, in most cases, software programs and easy-to-read materials simplify their use. Regression analysis may be implemented to generate effective concentration and provide confidence intervals for test estimates. Null and alternative hypotheses can be utilized to produce low and no concentration estimates of effect thresholds [63] based on statistical significance. Regression approaches interpolate the points on the concentration modeled response curve to evaluate the effect of dosage, whereas hypotheses outcomes are bound to be one of the dosages tests. In this regard, both examinations profit from partaking in the particular test concentrations to reach as close as possible to the "true" effect. Recently, an alternate hypothesis of bioequivalence has been proposed [64].

Numerous methods are presented for examining quantal toxicity statistics. Lethal concentration 50% (LC_{50}) estimations comprise parametric measures, such as probit [65] and logit [66] investigation, and generalized linear models (GLiM) [67]. The utmost frequently utilized nonparametric measures are the Spearman–Karber technique and the trimmed Spearman–Karber technique [68], whereas numeric interpolation procedures comprise binomial distribution, graphical techniques, and the moving average. Although no individual process is best for all the data sets, binomial distribution and graphical interpolation approaches are preferred for their simplicity. For most data sets, computer programs make it easier to utilize more advanced statistical approaches and models, which provide improved fits.

These approaches have traditionally been employed more effectively for chronic toxicity test findings than acute toxicity examinations. Nevertheless, the emergence of short-term, serious life-stage examination processes (numerous involve fatality as endpoints) has made the distinctions of acute and chronic effects occasionally inaccurate. These numerical approaches do not need to be restricted to one type of examination. The NOEC and low observed effective concentrations can be easily estimated with accessible computerized statistical approaches for hypotheses testing (T-test, Wilcoxon Rank sum, Dunnett's test) [69]. Generally, the practice of Shapiro–Wilk or chi-squared tests is used to find the normality, and Bartlett's test is used for homogeneity of variance where numerical models are appropriate for these tests.

Point estimation techniques for lethal concentration, effective concentration, and inhibition concentration determinations comprise logit, probit, linear interpolation, and GLiM. The benefit of implementing a suitable regression model is that minimal confidence intervals might be estimated for any specified estimation point. This data can be utilized to estimate an appropriate fit from alternate models and provide a statistical equation of statistical confidence in test results.

Outcomes from toxicity tests can be sufficient on their own in use for decision-making if they are appropriately documented. Complete documentation should include:

- Test organism utilized – species, age, food used in culturing, behavior during testing.
- Tested material – its source, storage, characteristics, assortment, process, and period.
- Water dilution – origin, repositing, characteristics, collection method and period, physical and chemical properties of the solution, toxicant concentration.
- Test method – endpoint, deviation from reference, date and period of start and end, the volume of chambers.

While toxicity testing in water quality management programs might seem more problematic to apply and interpret than the commonly employed chemical examinations executed in the lab, the advantages are numerous:

- Bioavailability and the complicated interactions of numerous chemicals are properly conveyed.

- A single unified measurement of an organism's reaction response to chemically complex material is producible.
- They can be a lower cost and simpler implementation compared to a series of chemical parameter quantifications.

References

1 Black, A.P. and Christman, R.F. (1963). Characteristics of colored surface waters. *Journal-American Water Works Association* 55 (6): 753–770.
2 Bennett, L.E. and Drikas, M. (1993). The evaluation of colour in natural waters. *Water Research* 27 (7): 1209–1218.
3 Hongve, D. and Åkesson, G. (1996). Spectrophotometric determination of water colour in Hazen units. *Water Research* 30 (11): 2771–2775.
4 Hardy, A.C. (1936). Handbook of Colorimetry. Cambridge, MA: MIT Press.
5 Shibata, S., Furukawa, M., and Goto, K. (1969). Dual-wavelength spectrophotometry: General method. *Analytica Chimica Acta* 46 (2): 271–279.
6 ASTM E308-95 (1995). *Standard practice for computing the colors of objects by using the CIE system*. West Conshohocken, PA: ASTM International.
7 Allen, W.P., Prescott, W.B., Derby, R.E., et al. (1973). Determination of color of water and wastewater by means of ADMI color values. *Proc. 28th Industrial Waste Conference, Lafayette IN (May 1-3, 1973)*. Lafayette, IN: Purdue University.
8 Hach, C.C., Vanous, R.D., and Heer, J.M. (1985). *Understanding Turbidity Measurements: Technical Information Series-Booklet No. 11*. Loveland, CO: Hach Company.
9 Chambers, E. and Wolf, M.B. (Eds.) (1996). *Sensory Testing Methods*. West Conshohocken, PA: ASTM.
10 Mallevialle, J. and Suffet, I.H. (1987). *Identification and Treatment of Tastes and Odors in Drinking Water*. Denver, CO: American Water Works Association Research Foundation.
11 Bruvold, W.H. and Pangborn, R.M. (1989). A critical review of methods used for the sensory evaluation of water quality. *Critical Reviews in Environmental Science and Technology* 19 (4): 291–308.
12 Faust, S.D. and Aly, O.M. (2018). *Chemistry of Water Treatment*. Boca Raton, FL: CRC Press.

13 ASTM E679-04. (2011). Practice for Determination of odor and taste thresholds by a forced-choice ascending concentration series method of limits. West Conshohocken, PA: ASTM International.

14 Punter, P.H. (1983). Measurement of human olfactory thresholds for several groups of structurally related compounds. *Chemical Senses* 7 (3-4): 215–235.

15 Larson, T.E. and Henley, L.M. (1955). Determination of low alkalinity or acidity in water. *Analytical Chemistry* 27: 851.

16 Thomas, J.F.J. and Lynch, J.J. (1960). Determination of carbonate alkalinity in natural waters. *Journal American Water Works Association* 52: 259.

17 Lowenthal, R.E. and Marais, G.V. (1976). *Carbonate Chemistry of Aquatic Systems*. Ann Arbor, MI: Ann Arbor Science.

18 Baird, R.B., Eaton, A.D., Rice, E.W., and Bridgewater, L. (Eds.). (2017). *Standard Methods for the Examination of Water and Wastewater* (Vol. 23). Washington, DC: American Public Health Association.

19 Balzar, W. (1980). Calcium carbonate saturometry by alkalinity difference. *Oceanolog. Acta* 3: 237.

20 Randtke, S. and Edzwald, J.K. (2011). Precipitation, coprecipitation and precipitative softening. In: *Water Quality & Treatment: A Handbook on Drinking Water* (Ed. J.K. Edzwald). New York: McGraw-Hill.

21 Willard, H.H., Merritt, L.L., Jr., Dean, J.A., and Settle, F.A., Jr. (1988). *Instrumental methods of analysis.*

22 Gray, J.R., Gylsson, G.D., Turcios, L.M., and Schwarz, G.E. (2000). *Comparability of Suspended-Sediment Concentration and Total Suspended Solids Data*. USGS Water-Resources Investigations Report 00-4191. Reston, VA: US Geological Survey.

23 Missouri Department of Natural Resources (2009). Chapter 3: Stream discharge. *In Introductory Level Volunteer Water Quality Monitoring Training Notebook*. Jefferson City, MO: Missouri DNR.

24 Sharp, J.H. (1982). The practical salinity scale 1978: A reply to comments by T.R. Parsons. *Limnology and Oceanography* 27 (2): 387–389.

25 Bradshaw, A. and Schleicher, K. (1980). Electrical conductivity of seawater. *IEEE Journal of Oceanic Engineering* 5 (1): 50–62.

26 Becking, L.B., Kaplan, I.R., and Moore, D. (1960). Limits of the natural environment in terms of pH and oxidation-reduction potentials. *The Journal of Geology* 68 (3): 243–284.

27 Carver, R.E. (1971). *Procedures in Sedimentary Petrology*. New York, NY: Wiley-Interscience.

28 Dick, R.I. and Young, K.W. (1972). Analysis of thickening performance of final settling tanks. In *Proceedings of the 27th Industrial Waste Conference, Lafayette, IN (May 2–4, 1972)*. Lafayette, IN: Purdue University.

29 Mindrup, R. (1978). The analysis of gases and light hydrocarbons by gas chromatography. *Journal of Chromatographic Science* 16 (9): 380–389.

30 Harris, D.C. (2014). Quantitative Chemical Analysis. Springer Publishing House.

31 Gedye, R. (1999). *Microwave-Enhanced Chemistry. Fundamentals, Sample Preparation and Applications* (Eds. H.M. Kingston and S.J. Haswell). Washington, DC: American Chemical Society.

32 Skoog, D.A., Holler, F.J., and Crouch, S.R. (2018). *Principles of Instrumental Analysis* (7th ed.). Boston, MA: Cengage learning.

33 Willis, J. (1965). Nitrous oxide-acetylene flame in atomic absorption spectroscopy. *Nature* 207: 715–716.

34 Hatch, W.R. and Ott, W.L. (1968). Determination of submicrogram quantities of mercury by atomic absorption spectrophotometry. *Analytical Chemistry* 40(14): 2085–2087.

35 Fernandez, F.J. and Manning, D.C. (1971). Atomic absorption analyses of metal pollutants in water using a heated graphite atomizer. *Atomic Absorption Newsletter* 10(3): 65–71.

36 Vijan, P.N. and Leung, D. (1980). Reduction of chemical interference and speciation studies in the hydride generation – Atomic absorption method for selenium. *Analytica Chimica Acta* 120: 141–146.

37 Faires, L.M., Palmer, B.A., Engleman, R., Jr., and Niemczyk, T.M. (1984). Temperature determinations in the inductively coupled plasma using a Fourier transform spectrometer. *Spectrochimica Acta Part B: Atomic Spectroscopy* 39(6): 819–828.

38 Brett, C.M., Brett, A.M.O., and Tugulea, L. (1996). Anodic stripping voltammetry of trace metals by batch injection analysis. *Analytica Chimica Acta* 322(3): 151–157.

39 Bynum, M.A.O., Tyree, S.Y., and Weiser, W.E. (1981). Effects of major ions on the determination of trace ions by ion chromatography. *Analytical Chemistry* 53(12): 1935–1936.

40 Jandik, P. and Bonn, G. (1993). *Capillary Electrophoresis of Small Molecules and Ions*. New York, NY: VCH Verlagsgesellschaft, Weinheim/ VCH Publisher.

41 Lin, S.D. (2014). *Water and Wastewater Calculations Manual*. New York, NY: McGraw-Hill Education.

42 ASTM D1252-06 (2006). Standard test methods for chemical oxygen demand (dichromate oxygen demand) of water.West Conshohocken, PA: ASTM International.

43 Van Hall, C.E., Barth, D., and Stenger, V.A. (1965). Elimination of carbonates from aqueous solutions prior to organic carbon determination. *Analytical Chemistry* 37(6): 769–771.

44 US Environmental Protection Agency. (1995). Report of EPA Efforts to Replace Freon for the Determination of Oil and Grease and Total Petroleum Hydrocarbons: Phase II; EPA-821-R-95-003. Washington, D.C. USEPA.

45 Swisher, R.D. (1987). *Surfactant Biodegradation*, 2nd ed. New York, N.Y: Marcel Dekker.

46 Box, J.D. (1983). Investigation of the Folin-Ciocalteau phenol reagent for the determination of polyphenolic substances in natural waters. *Water Research* 17(5): 511–525.

47 Olmsted, W.H., Whitaker, W.M., and Duden, C.W. (1929). Steam distillation of the lower volatile fatty acids from a saturated salt solution. *Journal of Biological Chemistry* 85(1): 109–114.

48 Eaton, A. D., & Franson, M. A. H. (2005). Standard methods for the examination of water & wastewater. 21st ed. Centennial ed./Washington, DC: American Public Health Association.

49 Hamad, A.M. and Qadr, H.M. (2018). Gamma-rays spectroscopy by using a thallium activated sodium iodide NaI (Ti). *Eurasian Journal of Science and Engineering* 4(1): 99–111.

50 Grothe, D.K., Dicksan, K.L., and Reed-Judkins, D.K. (1996). *Whole Effluent Toxicity Testing: An Evaluation of Methods and Prediction of Receiving Impacts*. Pensacola, FL: SETAC.

51 US Environmental Protection Agency. (2002). Methods for measuring the acute toxicity of effluents and receiving waters to freshwater and marine organisms, 5th ed. EPA-821-R-02-012. Washington DC: EPA Office of Water.

52 ASTM E1022-94R07. (2007). *Standard guide for conducting bioconcentration tests with fishes and saltwater bivalve mollusks*. West Conshohocken, PA: ASTM International.

53 Pedrazzani, R., Bertanza, G., Brnardić, I. et al. (2019). Opinion paper about organic trace pollutants in wastewater: Toxicity assessment in a European perspective. *Science of the Total Environment* 651: 3202–3221.

54 International Organization for Standardization (2012). *Water Quality – Fresh Water Algal Growth Inhibition Test with Unicellular Green Algae*. 8692. Geneva, Switzerland: International Organization for Standardization.

55 International Organization for Standardization (2007). *Water Quality – Determination of the Inhibitory Effect of Water Samples on the Light Emission of Vibrio Fischeri (Luminescent Bacteria Test) – Part 3: Method*

Using Freeze-Dried Bacteria. 11348-3. Geneva, Switzerland: International Organization for Standardization.

56 Bertanza, G., Pedrazzani, R., Dal Grande, M., et al. (2011). Effect of biological and chemical oxidation on the removal of estrogenic compounds (NP and BPA) from wastewater: An integrated assessment procedure. *Water Research* 45 (8): 2473–2484.

57 Urani, C., Stefanini, F.M., Bussinelli, L., Melchioretto, P., and Crosta, G.F. (2009). Image analysis and automatic classification of transformed foci. *Journal of Microscopy* 234 (3): 269–279.

58 Pedrazzani, R., Baroni, P., Feretti, D., et al. (2020). Methodological protocol for assessing the environmental footprint by means of ecotoxicological tools: wastewater treatment plants as an example case. In: *Ecotoxicological QSARs. Methods in Pharmacology and Toxicology*. (ed. K. Roy). New York, N.Y: Humana. doi: 10.1007/978-1-0716-0150-1_14.

59 Escher B.I., Allinson M., Altenburger R., et al. (2014). Benchmarking organic micropollutants in wastewater, recycled water and drinking water with in vitro bioassays. *Environment Science and Technology* 48 (3): 1940–1956. doi: 10.1021/es403899t.

60 Botts, J., Braswell, J.W., Goodfellow, W.L., and Doloff, F.B. (1988). Toxicity Reduction Evaluation at the Patapsco Wastewater Treatment Plant. EPA/600/S2-88/034

61 Burgess, R.M., Charles, J.B., Kuhn, A., Ho, K.T., Patton, L.E., and McGovern, D.G. (1997). Development of a cation-exchange methodology for marine toxicity identification evaluation applications. *Environmental Toxicology Chemistry*. 16: 1203–1211.

62 Fillmore, L.B., Morris, T.L., Champlin, T.L., Welch, M.C., Botts, J.A., Goodfellow, W.L., and Williams, R.T. (1990). *Toxicity Reduction Evaluation at the City of Fayetteville Cross Creek Wastewater Treatment Plant*. Washington DC, USA: US Environmental Protection Agency.

63 Pack, S. (1993). *A Review of Statistical Data Analysis and Experimental Design in OECD Aquatic Toxicology Test Guidelines*. Paris, France: Organization for Economic Cooperation and Development.

64 Erickson, W.P. and McDonald, L.L. (1995). Tests for bioequivalence of control media and test media in studies of toxicity. *Environmental Toxicology and Chemistry: An International Journal* 14 (7): 1247–1256.

65 Finney, D.J. (1971). *Probit Analysis: A Statistical Treatment of the Sigmoid Response Curve* (3rd ed.). London: Cambridge University Press.

66 Berkson, J. (1953). A statistically precise and relatively simple method of estimating the bio-assay with quantal response, based on the logistic function. *Journal of the American Statistical Association* 48 (263): 565–599.

67 Kerr, D.R. and Meador, J.P. (1996). Modeling dose response using generalized linear models. *Environmental Toxicology and Chemistry: An International Journal* 15 (3): 395–401.

68 Hamilton, M.A., Russo, R.C., and Thurston, R.V. (1977). Trimmed Spearman-Karber method for estimating median lethal concentrations in toxicity bioassays. *Environmental Science & Technology* 11 (7): 714–719.

69 Chiang, C.L. (2003). *Statistical Methods of Analysis*. Singapore: World Scientific Publishing.

9

Economic Analysis of Desalination Process

Vinayagam Sivabalan, Jesa Singh, and Bhajan Lal

9.1 Overview

An economic analysis of a desalination process is vital for process selection. In a general desalination process selection, various parameters such as quantity and quality of treated water desired, the quantity and quality of the saline feed water required, operating conditions, raw materials, and the overall energy consumption are considered. The fact that energy consumption alone makes up nearly half the overall cost in water production means that the goal of reducing desalination costs by an order of magnitude likely will not come in the near future. However, any improvements that can bring about lower energy usage in desalination processes are important in ensuring that the goal will eventually become a reality [1].

Although reverse osmosis is the most widely employed desalination technology globally, it is energy-intensive. Hence, there is a need to develop innovative energy-efficient technologies to strengthen the energy–water nexus [2]. The choice of a system is not a matter of competition between the different technologies but is instead based on the most suitable plant design for the given situation. Each case should be studied and analyzed individually to design a plant with the greatest energy efficiency and ensure that the desalinated water is produced at the least cost technically possible, without compromising the quality of the freshwater [3].

The economics of a desalination process is affected by several general factors such as plant capacity, site conditions, energy cost, and plant life [4]. According to Barak [5], the most commonly used criteria in economic analysis of desalination processes are:

Gas Hydrate in Water Treatment: Technological, Economic, and Industrial Aspects,
First Edition. Bhajan Lal and Sirisha Nallakukkala.

- Cost of treated water
- Investment
- Cashflow
- Payback Period
- Rate of Return
- Price of the Treated water
- Value of treated water to the customer.

In addition, several unique techno-economic features also should be considered for desalination processes. These special features are:

- Large quantities of treated water required.
- Very low acceptable price.
- Sensitivity of cost to economic environment (particularly energy prices).
- The national system of long-term production plants.
- Public, political, and media intensive involvement.
- Need for good, effective, and highly reliable transportation.

The value of the desalted water depends on many factors such as water scarcity, specific applications, desalted water quantity, salinity, qualities, and costs of additional or alternative water sources. In many cases, it is difficult to assess the value of the desalted water.

9.2 Cost of Treated Water

The cost of water is an economic parameter that incorporates all project capital and annual operation and maintenance (O&M) expenditures associated with water production. The cost of water is the total expenditure needed to supply the product to the customer per unit product. In practice, two product costs exist. One fits the definition fully and refers to the cost at the location and time of transition from the supplier to the end-user and is termed delivered or gross product cost. The other cost refers to the product leaving the plant/factory and may be termed plant or net product cost. The first includes the expenses that accrue between the plant exit and the point of delivery to the customer, covering the costs of water storage, transportation, possible losses, and distribution. These factors have to be added to the net cost, making the gross cost is more relevant in cost and pricing analyses. However, the net product cost is more often referred to in literature and feasibility studies as the cost of desalted water because the expenses occurring between the plant exit and the point of delivery are highly site-dependent and not process-dependent, so can be analyzed only on a case-by-case basis.

Thus, in pricing the desalted water for the customer, the *gross* product cost is most often the preferred term, whereas for comparing desalination methods or plants, the *net* product cost is more convenient. Typically, this cost parameter is expressed in monetary units per unit volume of desalinated water (US$/m^3). The cost breakdown of treated water is displayed in Figures 9.1 and 9.2.

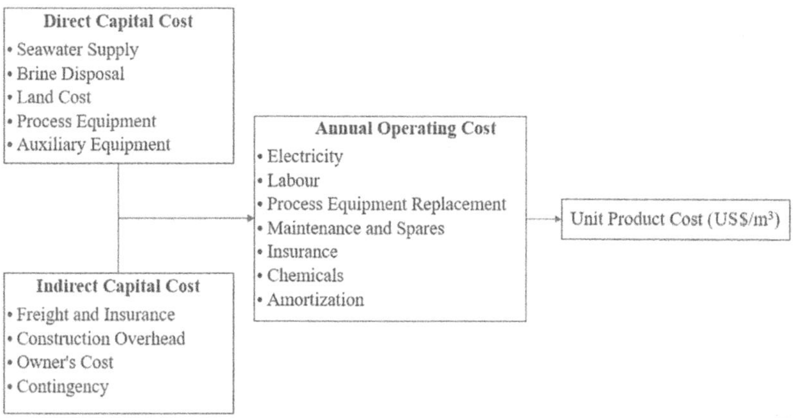

Figure 9.1 Elements of treated water cost. The cost of treated water includes all expenditures associated with project implementation, operation and maintenance, and financing, which can be categorized into fixed and variable costs.

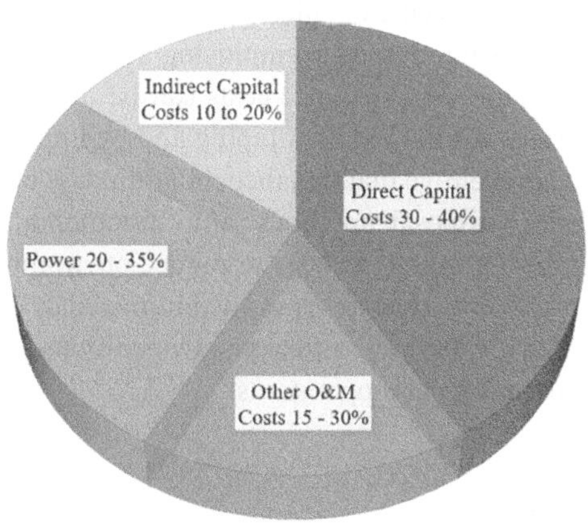

Figure 9.2 Cost breakdown of treated water.

9.2.1 Fixed Cost

The fixed cost component of the total production cost is independent of the actual amount of water produced by the desalination plant. Therefore, this cost must be minimized as much as possible. The major components of fixed cost are the cost recovery factor (CRF), labor costs, maintenance costs, environmental and performance monitoring costs, and indirect O&M costs.

The CRF is a function of the interest rate and time (years) over which the investment is recovered. The capital cost for constructing the plant is usually amortized over the term of repayment (typically a period of 5 to 20 years). The amortized value of the capital cost is calculated by dividing the capital cost by CRF and by the plant's design capacity availability factor (CAF) [6]. CAF is the percentage of the year that the plant is producing freshwater at an average flow rate equal to or higher than the plant's design capacity. If the desalination plant design daily capacity is 10 000 m^3 and plant CAF is at 100%, then the total average annual freshwater production is calculated as 10 000 m^3/d multiplied by 365 days, yielding a total of 3 650 000 m^3/y. The CRF can be calculated as given in Eq. (9.1).

$$\mathrm{CRF} = \frac{(1+i)^n - 1}{i(1+i)^n}, \tag{9.1}$$

where n is the period of repayment of capital expenditures, and i is the interest rate of the amortized investment (%). The CRF cost can be reduced by increasing the CAF, which is achieved by optimizing the desalination process.

Labor costs can be reduced by building the desalination plant with high levels of automation in mind and by employing a limited number of highly qualified operations staff. Maintenance costs can be minimized by selecting high-quality materials, equipment, and piping and by implementing a proactive and systematic preventive maintenance program. Use of environmentally safe, low-cost brine disposal methods and automation of most plant performance monitoring functions help to reduce environmental and monitoring costs. Indirect O&M costs are typically minimized by using highly qualified operations staff or subcontracting plant operations to a private party specialized in desalination plant operations.

9.2.2 Variable Cost

The varying component of the cost of treated water consists of power consumption, chemical usage, replacement of process equipment parts, and waste stream disposal. Power expenditure is the largest variable cost component and usually accounts for 25–35% of the total cost of water. The power consumption varies according to the specifications within the desalination process itself and can be reduced through process optimization and by selecting the appropriate process to fit the supply and demand of the desalinated water.

In terms of chemical usage, materials used vary largely based on what is required within a process. For example, reverse osmosis (RO) requires chemicals for anti-scaling and cleaning the membrane, while gas hydrate-based desalination requires chemicals to promote hydrate formation. In some circumstances, the treatment of raw seawater to a less harsh quality requires a lower amount of pretreatment chemicals and less frequent membrane cleaning, which, in turn, yields lower plant chemical costs. The difference between the chemical pretreatment and membrane cleaning costs for good and below-average quality source water could be significant – often two to four times lower.

Replacement of process equipment parts, often considered a fixed maintenance cost (i.e. preventive maintenance costs), may be accounted for as a variable cost. For instance, if the source water quality is poor, the membranes in a membrane-based desalination process (RO, electrodialysis, and nanofiltration) will require frequent cleaning and replacement due to fouling. Among thermal-based desalination technologies, multistage flash (MSF) and multi-effect distillation (MED) use direct heat exchange from steam, while vapor compression (VC) uses heat from compression of vapor as the energy source for evaporation [7]. For these thermal processes, corrosion and degradation of heat transfer fluids are common and necessitate frequent replacement. In gas hydrate-based desalination using porous media, the porous media may need to be replaced to avoid salt occlusion. In general, most processes will require replacing process equipment parts to maintain high efficiency.

Waste stream disposal costs are usually relatively small. However, in some cases, the operation of the concentrate disposal facilities could constitute a significant portion of the plant water production costs. Malfunction of these facilities could significantly reduce the plant capacity factor (e.g. can increase downtime). Therefore, simple and environmentally safe methods of concentrate disposal such as co-discharge with power plant cooling water, sanitary sewer discharge, or direct open ocean discharge, when viable, are recommended over deep-well injection discharge, evaporation pond disposal, or zero liquid discharge [6].

9.3 Factors Affecting the Product Cost

As discussed by Dessouky and Ettouney [8], there are many contributors to the cost of treated water, several of which are listed in Table 9.1.

Besides these, there are two other paradigms to the cost factors: those costs within the owner's control and those outside of the owner's control. Voutchkov has elaborated thoroughly on the cost factors in each category, which are briefly laid out in Table 9.2 [6].

One way to reduce the cost of desalinated water is to improve upon the desalination technology, particularly with enhancements that would increase the performance ratio (the ratio of freshwater produced to the amount of energy consumed) [10]. Desalination costs have decreased over the years both due to technical improvements and through leveraging the increase in fossil fuel prices. For conventional systems, the cost per cubic meter of seawater ranges from €0.40 to more than €3.00, while the cost is almost half that for brackish water desalination. When renewable energy sources are used, the cost is much higher, and in some cases can reach up to €15 per cubic meter due to the expensive supply systems required. However, this cost is counterbalanced by the environmental benefits. Under special conditions, hybrid systems can offer increased and more stable production of freshwater [9]. The use of renewable energy as a power supply is further discussed in Chapter 10.

Table 9.1 Factors affecting the product cost.

Factors	Description
Salinity and quality of feed water	Lower feed salinity allows for higher conversion rates. Thus, the plant can operate at lower specific power consumption and chemical dosage [9].
Plant capacity	Larger plant capacity reduces the capital cost for unit product. Although, the increase in the plant capacity implies higher capital.
Site conditions	Installation of new units as an addition to existing sites, would eliminate costs associated with facilities for feed water intake, brine disposal, and feed water pretreatment.
Qualified manpower	Availability of qualified operators, engineers, and management would result in higher plant availability, production capacity, and lower downtime caused by trips of devices.
Energy cost	Availability of inexpensive sources for low-cost electric power and heating steam have a strong impact on the unit product cost.
Plantlife and amortization	An increase in plant life reduces the capital product cost.

Table 9.2 Cost factors within and beyond owner's control.

Factors		Descriptions
Within owner's control	Project size	The cost of water production by seawater desalination can be reduced significantly by building fewer large-scale desalination plants rather than a large number of small facilities. This is because the cost of water production reduces with larger plant capacity
	Capacity availability factor (CAF)	Discussed in Section 4.2.1
	Source water quality	Pretreatment cost for source water with high total dissolved solids (TDS), temperature, turbidity, silt density index (SDI), organic content, nutrients, algae, bacteria, boron, silica, barium, calcium, and magnesium can be reduced by selecting efficient processes
	Target product water quality	Incremental expenditures are needed to achieve more stringent product water quality goals, which usually require lesser TDS count and lower amount of chloride, boron, and bromide
	Concentrate disposal method	Depending on the site-specific conditions of a given project, concentrate disposal expenditures may have a measurable contribution to the total plant construction, O&M, and the overall water costs.
	Power supply and unit power cost	The selection of suitable desalination processes according to the Feed and Output conditions can help in reducing the overall energy usage. Power can be obtained via steam turbine generation or solar power, when and wherever possible, to reduce energy costs.
Beyond owner's control	Environmental regulations	These cost factors are very site- and project-specific and are usually outside of the control of the desalination plant owner but can have a very significant impact on the overall water production costs.
	Climate conditions	
	Seasonal water demand and power tariff variations	
	Local labor and material costs	
	General economic environment	

9.4 Case Studies

While cost remains the primary factor in selecting a particular desalination technique for water treatment, desalination costs have decreased markedly over the past few decades. For example, typical desalination costs in the 1940s and 1950s ranged from 15.00–US$20.00 per thousand gallons. By the early 1960s, desalination costs had dropped to about 5.50–US$9.00 per thousand gallons. Recent cost analyzes indicate that desalination now has comparable costs of approximately 0.21–US$5.00 per thousand gallons under near-optimum operating conditions. If the desalination equipment is not operated efficiently, the latter cost can increase to as much as US$10.00 per thousand gallons. Some marginal reductions in distillation costs may be realized from improvements in plant designs, fabrication techniques, heat exchange materials, scale control techniques, and plant automation.

Detailed evaluations are required to determine the cost-effectiveness of a certain process for a specific location. A particular desalination process should neither be selected nor rejected based on the costs provided in this report alone. Also, when comparing the costs of desalination plants with the costs of conventional treatment plants, it is important to recognize that conventional desalination plants typically have an operating lifetime of 15 to 20 years. Conventional water treatment plants often last 30 to 40 years, and sometimes longer [11].

This section presents the calculations for well-documented major desalination processes [8]. Similar calculations can be done for gas hydrate desalination, but no such case study is presented here due to the lack of established field data and design studies. Not enough experimental data are available on the hydrate desalination process [12]. Based on the results of the latest studies on gas hydrates-based desalination, common assumptions used in the calculations are:

- Plantlife (n) = 30 y
- Electric cost (c) = US$0.05/m^3
- Average latent heat of heating steam (X) = 2200 kJ/kg
- Steam heating cost (s) = US$1.47/GJ
- Performance ratio (PR) of MSF and multi-effect evaporation (MEE) = 8 kg/kg
- Specific cost of operating labor = US$0.10/m^3
- Interest rate (i) = 5%
- Plant availability (f) = 0.9

9.4.1 Multistage Flash Distillation

The MSF cost data includes:

- Direct capital cost (DC) = US$64 x 10^6
- Plant capacity (m) = 32 731 m^3d^{-1}
- Specific consumption of electric power (w) = 5 kWh/m^3
- Specific chemicals cost (k) = US$0.025/$m^3$

The calculations proceed as follows.

1. **Calculate the amortization factor.**

$$a = \frac{i(1+i)^n}{(1+i)^n - 1} = \frac{0.05(1+0.05)^{30}}{(1+0.05)^{30} - 1} = 0.065051/\text{year}$$

2. **Calculate the annual fixed charges.**
   A_1 = (a) (DC) = (0.065051) (64 x 10^6)
   = US$4.163264 x 10^6/yr
3. **Calculate the annual heating steam cost.**
   A_2 = (s) (λ) (f) (m) (365)/ ((1000)(PR))
   = (1.466) (2200) (0.9) (32 731) (365)/(1000)(8))
   = US$4.334723 x 10^6/y
4. **Calculate the annual electric power cost.**
   A_3 = (c) (w) (f) (m) (365)
   = (0.05) (5) (0.9) (32 731) (365)
   = US$2.688033 x 10^6/y
5. **Calculate the annual chemicals cost.**
   A_4 = (k)(f)(m)(365)
   = (0.025) (0.9) (32 731) (365)
   = US$2.68803 x 10^5/y
6. **Calculate the annual labor cost.**
   A_5 = (l)(f)(m) (365)
   = (0.1) (0.9) (32 731) (365)
   = US$1.075213 x 10^6/y
7. **Calculate total annual cost.**
   $A_t = A_1 + A_2 + A_3 + A_4 + A_5$
   = 4.334723 x 10^6 + 4.334723 x 10^6 + 2.688033 x 10^6 + 2.68803 x 10^5 + 1.075213 x 10^6
   = US$12.530065 x 10^6/y
8. **Calculate unit product cost.**
   $A_s = A_t$/((f)(m)(365))
   = (12.530065 x 10^6)/((0.9)(365)(32 731))
   = US$1.165/$m^3$

$A_s = A_t/((f)(m)(365))$
$= (12.530065 \times 10^6)/(32\ 731)$
$= US\$383/m^3d^{-1}$

9.4.2 Multi-Effect Evaporation

MEE cost data includes:

- Direct capital cost (DC) = US$20 x 10^6
- Plant capacity (m) = 12 000 m^3d^{-1}
- Specific consumption of electric power (w) = 3 kWh/m^3
- Specific chemicals cost (k) = US$0.025/$m^3$.

The calculations proceed as follows.

1. **Calculate the amortization factor.**

$$a = \frac{i(1+i)^n}{(1+i)^n - 1} = \frac{0.05(1+0.05)^{30}}{(1+0.05)^{30} - 1} = 0.065051\,/\,\text{year}$$

2. **Calculate the annual fixed charges.**
   $A_1 = (a)\,(DC) = (0.065051)(22 \times 10^6)$
   $= US\$1\ 301\ 029/y$
3. **Calculate the annual heating steam cost.**
   $A_2 = (s)\,(\lambda)\,(f)\,(m)\,(365)/((1000)(PR))$
   $= (1.466)\,(2200)\,(0.9)\,(12\ 000)\,(365)/((1000)(8))$
   $= US\$1\ 589\ 217.30/y$
4. **Calculate the annual electric power cost.**
   $A_3 = (c)\,(w)\,(f)\,(m)\,(365)$
   $= (0.05)\,(3)\,(0.9)\,(12\ 000)\,(365)$
   $= US\$591\ 300/y$
5. **Calculate the annual chemicals cost.**
   $A_4 = (k)(f)(m)(365)$
   $= (0.025)\,(0.9)\,(12\ 000)\,(365)$
   $= US\$98\ 550/y$
6. **Calculate the annual labor cost.**
   $A_5 = (l)\,(f)\,(m)\,(365)$
   $= (0.1)\,(0.9)\,(12\ 000)\,(365)$
   $= US\$394\ 200/y$
7. **Calculate total annual cost.**
   $A_t = A_1 + A_2 + A_3 + A_4 + A_5$
   $= 1\ 301\ 029 + 1\ 589\ 217.3 + 985\ 500 + 98\ 550 + 394\ 200$
   $= US\$3\ 974\ 296/y$

8. **Calculate unit product cost.**

$A_S = A_t/((f)(m)(365))$
$= (3\ 974\ 296)/((0.9)(365)(12\ 000))$
$= US\$1.008/m^3$

$A_S = A_t/((f)(m)(365))$
$= (3\ 974\ 296)/(12\ 000)$
$= US\$331.2/m^3d^{-1}$

9.4.3 Mechanical Vapor Compression (MVC)

The MVC cost data includes:

- Direct capital cost (DC) = US$400 000
- Plant capacity (m) = 3000 m^3d^{-1}
- Specific consumption of electric power (w) = 7 kWh/m^3
- Specific chemicals cost (k) = US$0.025/$m^3$.

The calculations proceed as follows.

1. **Calculate the amortization factor.**

$$a = \frac{i(1+i)^n}{(1+i)^n - 1} = \frac{0.05(1+0.05)^{30}}{(1+0.05)^{30} - 1} = 0.065051\,/\,\text{year}$$

2. **Calculate the annual fixed charges.**

$A_1 = (a)\,(DC) = (0.065051)(400\ 000)$
$= US\$26\ 020/y$

3. **Calculate the annual electric power cost.**

$A_3 = (c)(w)(f)(m)(365)$
$= (0.05)(7)(0.9)(3000)(365)$
$= US\$344\ 925/y$

4. **Calculate the annual chemicals cost.**

$A4 = (k)(f)(m)(365) = (0.025)(0.9)(3000)(365) = US\$24\ 638/y$

5. **Calculate the annual labor cost.**

$A_5 = (l)(f)(m)\,(365) = (0.1)(0.9)(3000)(365) = US\$98\ 550/y$

6. **Calculate total annual cost.**

$A_t = A_1 + A_3 + A_4 + A_5$
$= 26\ 020 + 344\ 925 + 24\ 638 + 98\ 550$
$= US\$494\ 133/y$

7. **Calculate unit product cost.**

$A_S = A_t/((f)(m)(365))$
$= (494\ 133)/((0.9)(365)(3000))$
$= US\$0.501\ /m^3$

$A_s = A_t/((f)(m)(365))$
$= (494\ 133)/(3000)$
$= US\$165/m^3d^{-1}$.

9.4.4 Reverse Osmosis

The RO cost data includes:

- Direct capital cost (DC) = US\$98 x 10^6
- Membrane purchase cost (@60% DC) = US\$58.8 x 10^6
- Membrane annual replacement cost (@10% of membrane purchase cost) = US\$5.88 x 10^5
- Plant capacity (m) = 94 625 m^3/d
- Electric cost (c) = US\$0.04/m^3
- Specific consumption of electric power (w) = 5 kWh/m^3
- Specific cost of operating labor (t) = US\$0.05/m^3
- Specific chemicals cost (k) = US\$0.033/m^3.

The calculations proceed as follows.

1. **Calculate the amortization factor.**

$$a = \frac{i(1+i)^n}{(1+i)^n - 1} = \frac{0.05(1+0.05)^{30}}{(1+0.05)^{30} - 1} = 0.065051 / \text{year}$$

2. **Calculate the annual fixed charges.**
   $A_1 = (a)(DC)$
   $= (0.065051)(98 \times 10^6)$
   $= US\$6\ 375\ 041/y$
3. **Calculate the annual electric power cost.**
   $A_2 = (c)(w)(f)(m)(365)$
   $= (0.04)(5)(0.9)(94\ 625)(365)$
   $= US\$6\ 216\ 863/y$
4. **Calculate the annual chemicals cost.**
   $A_3 = (k)(f)(m)(365)$
   $= (0.033)(0.9)(94\ 625)(365)$
   $= US\$1\ 039\ 129/y$
5. **Calculate the annual membrane replacement cost.**
   $A_4 = US\$5\ 880\ 000/y$
6. **Calculate the annual labor cost.**
   $A_5 = (l)(f)(m)(365)$
   $= (0.05)(0.9)(94\ 625)(365)$
   $= US\$1\ 554\ 216/y$

7. Calculate total annual cost.

$A_t = A_1 + A_2 + A_3 + A_4 + A_5$

$= 6\,375\,041 + 6\,216\,863 + 1\,039\,129 + 5\,880\,000 + 1\,554\,216$

$=$ US\$21 065 248/y

8. Calculate unit product cost.

$A_s = A_t/((f)(m)(365))$

$= (21\,065\,248)/((0.9)(365)(94\,625))$

$=$ US\$0.678/m^3

$A_s = A_t/((f)(m)(365))$

$= (21\,065\,248)/(94\,625)$

$=$ US\$223.10/m^3d^{-1}

9.4.5 Gas Hydrate-Based Desalination

Though improvements in hydrate-based desalination reactor designs for better commercial performance are constant, the actual adoption of the technology in commercialized applications has so far been unsuccessful. The economic viability of the process can be analyzed for many avenues for savings, as the performance variables that affect the cost of operation, such as the source brine temperature, suitable pressure-temperature conditions, percent salt mobility, and water recovery, can certainly be explored for optimization [13].

In recent years, gas hydrate formation technology has been given serious consideration as a method for brine water desalination [14]. From discussions in other chapters, it can be concluded that the gas hydrate-based desalination method has been measured as an attractive alternative to RO or other conventional distillation methods. However, the hydrate-based desalination process, which displays low-pressure requirements and high rejection of salt, has not been evaluated on a continuous basis and needs to undergo further optimization for separation of hydrates from brine and elimination of salt molecules from hydrate cages [15].

The economic study is an essential step for checking the feasibility of a process. The operation cost of seawater desalination by hydrate technology depends on many factors such as seawater temperature, salt content and mobility, and yield [16]. The removal efficiency of desalination plants also depends on the ionic radius and ionic charge of the mineral cations present. As larger cation sizes are involved, the efficiency for salt removal using hydrates is improved, though cations with low ionic charge can also risk efficiency reductions [17].

This section presents the latest techno-economic analyzes for gas hydrate-based desalination. Capital cost is the cost required to make a project commercially operable. For this project, the equipment cost focuses on determining the capital cost. The main factors contributing to the production costs are the operating costs required to run pumps for the cooler and reactor, operate the cooler, obtain carbon dioxide (CO_2) gas to serve as a hydrate former, and power consumption. Total capital investment is made up of total fixed investment and working investment.

Table 9.3 presents the assumptions used in the gas hydrate-based desalination calculations.

Determine the Reynold's number (Re).

$$Velocity = \frac{Volumetric\ flowrate}{Area} = \frac{0.00183}{0.00785} = 0.233\frac{m^3}{hr}$$

$$Reynolds\ Number(Re) = \frac{D\nu\rho}{\mu} = \frac{0.1 * 0.233 * 1025}{8.9 * 10^{-4}} = 26\ 489$$

The Reynolds Number value obtained is very high, which means that the inertial forces of the fluid dominate the viscous forces. The obtained value is indicative of turbulent flow.

$$Relative\ roughness = \frac{Absolute\ roughness}{Diameter} = \frac{0.046}{100} = 0.00046$$

For this purpose, a commercial steel pipe with an absolute roughness (ε) value of 0.046 mm is selected.

Table 9.3 Properties of inlet and outlet streams at pump and piping diameter specifications used in gas hydrate desalination calculations.

Specification	Inlet	Outlet
Pressure, kPa	101.325	4000
Temperature, °C	24	24
Density, kg/m^3	1025	
Volumetric flow rate, m^3/h	6.588	
Mass flow rate, kg/h	6752.7	
Viscosity, Pa.s or Ns/m^2	8.9×10^{-4}	
Piping diameter		
Length	100 m	
Diameter	0.1 m	
Cross-sectional area	$0.00785\ m^2$	

Using a Moody Diagram [18] with a *Re* value of 26 489 and a relative roughness of 4.06 *10^{-4}, the friction factor (*f*) value of 0.026 is obtained.

Calculate the pressure drop in the pipe.

$$\Delta P_f = 8f\left(\frac{L}{D}\right)\left(\frac{1}{2}\right)\rho V^2$$

$$= 8(0.026)\left(\frac{100}{0.1}\right)\left(\frac{1}{2}\right)(1025)(0.23^2) = 5639.14 \text{ Pa.}$$

Head loss and head rise (h^a) are then calculated as:

$$\text{Head loss} = \frac{\Delta P}{\left(\frac{1}{2}\right)\rho V^2} * \frac{V^2}{2g} = \frac{5639.14}{\left(\frac{1}{2}\right)(1025)(0.00183)^2} * \frac{(0.00183)^2}{2*9.81} = 0.56\text{m},$$

and

$$\text{Actual headrise } h_a = \frac{\Delta P}{g\rho} = \frac{(400000 - 101325)}{1025*9.81} = 387.73\text{m},$$

where g reflects gravitational acceleration in m/s.

The total energy required, assuming $\Delta z = 390$ m, then is expressed as:

$$W = \left(\frac{\Delta P}{\rho}\right) - \left(\frac{\Delta P_f}{\rho}\right) - (g\Delta Z)$$

$$= \left(\frac{3898675}{1025}\right) - \left(\frac{5639.14}{1025}\right) - (9.81*390) = -27.82\frac{J}{kg}.$$

The negative value for W indicates that a pump is required.

The total head required by a pump (hr) is calculated as:

$$h_r = \left(\frac{\Delta P_f}{\rho}\right) + \left(\frac{\Delta P}{\rho}\right) + (g\Delta Z) = \left(\frac{5639.14}{1025}\right) + \left(\frac{3898675}{1025}\right) + (9.81*390) = 3749.03\text{m}$$

$$\text{Power} = \frac{W}{\textit{Efficiency}} = \frac{27.82*6752.7*\frac{1}{3600}*\frac{1}{1000}}{0.85} = 0.061 \text{ kW}$$

$$W_{shaft} = \frac{\textit{Power}}{\textit{Efficiency}} = \frac{0.061}{0.85} = 0.072 \text{ kW}$$

Net Positive Suction Head

$$(\text{NPSH}) = \frac{P}{\rho} + \text{H} - \frac{P_f}{\rho} - \frac{P_v}{\rho}$$

$$= \frac{4000*10^3}{1025} + (387.73) - \frac{5639.14}{1025} - \frac{101325}{1025}$$

= 4185.81 Pa = 4.186 kPa

$$\text{Brake horsepower}(BHP) = \frac{\textit{volumetric flowrate} * \textit{operating pressure}}{1714 * n}$$

$$= \frac{0.00183 * 580.151}{1714 * 0.85} = 0.000729 \text{ bhp}$$

Equipment cost for the reactor pump is calculated as:

$$C_E = C_B\left(\frac{Q}{Q_B}\right)M,$$

where C_E is the equipment cost with capacity Q, C_B is the known base cost for equipment with capacity Q, and M = constant depending on equipment type. Therefore:

$$C_E = \left(1.97 * 10^3\right)\left(\frac{0.072}{1}\right)^{0.35} = \$784.39.$$

According to Jenkins et al. [19], the Chemical Engineering Plant Cost Index (CEPCI) for the year 2020 experienced a 3.5% decline compared to the same time in 2019. Hence, the CEPCI for 2020 will be 597.53, with the cost index of the base year 2000 being 435.8.

$$C_E = \text{US\$784.39} * \left(\frac{597.53}{435.8}\right) = \text{US\$1075.49}$$

By implementing the material correction cost ($f_m = 1.0$), design pressure correction factor ($f_p = 1.38$), and design temperature correction factor ($f_t = 1.0$), the final costing value for this pump can be obtained.

$C_F = C_E \, f_m \, f_t \, f_p$ = (US$1075.49)(1.0)(1.0)(1.38) = US$1 484.18 = RM6 45 2.94

The properties of the inlet and outlet streams of the cooler pump are listed in Table 9.4.

The calculations proceed in a manner similar to that of the reactor pump.

$$Velocity = \frac{0.144}{0.071} = 2.03\frac{m}{s}$$

$$\text{Re} = \frac{1000 * 2.03 * 0.3}{1.68 * 10^{-3}} = 362500$$

2. Calculate the relative roughness of the pipe.

$$Relative\,roughness = \frac{Absolute\ roughness}{diameter} = \frac{0.046}{300} = 1.53 * 10^{-4}$$

Table 9.4 Properties of inlet and outlet streams at pump and piping diameter specifications for the cooler pump

Specification	Inlet	Outlet
Pressure, kPa	101.325	4000
Temperature, °C	24	24
Density, kg/m^3	1000	
Volumetric flow rate, m^3/h	518.4	
Mass flow rate, kg/h	518 400	
Viscosity, Pa.s or Ns/m^2	$1.68x10^{-3}$	
Piping parameter		
Length	100 m	
Diameter	0.3 m	
Cross-sectional area	$0.071\ m^2$	

Using a Moody Diagram [18] with the values of Reynolds Number of 36 2500 and the relative roughness of $1.53x10^{-4}$, the friction factor value of 0.0175 is obtained.

Pressure drop in the pipe,

$$\Delta P_f = 8f\left(\frac{L}{D}\right)\left(\frac{1}{2}\right)\rho V^2 = 8(0.0175)(\frac{100}{0.3})(\frac{1}{2})(1000)(2.03^2)$$

$$=96\ 154.33\ \text{Pa}$$

Determine the loss coefficient, K_L, to calculate head losses.

$$\text{Head loss} = \frac{\Delta P}{\left(\frac{1}{2}\right)\rho V^2} * \frac{V^2}{2g} = \frac{96154.33}{\left(\frac{1}{2}\right)(1000)(0.144)^2} * \frac{(0.144)^2}{2*9.81} = 9.8\ \text{m}$$

The total head required assuming $\Delta z = 10$ m.

The total energy required to run the cooler pump then is calculated as:

$$\text{W} = (\frac{\Delta P}{\rho}) - (\frac{\Delta P_f}{\rho}) - (g\Delta Z),$$

$$= -\left(\frac{96154.33}{1000}\right) - (9.81*10) = -194.25\frac{J}{kg}.$$

Again, the negative value for W indicates the use of a pump is required.

The total head required (h_r) by the cooler pump is calculated as:

$$h_r = \left(\frac{\Delta P_f}{\rho}\right) + (\frac{\Delta P}{\rho}) + (g\Delta Z) = \left(\frac{96154.33}{1000}\right) + (9.81*10) = 194.25\ \text{m}.$$

$$\text{Power} = \frac{W}{Efficiency} = \frac{194.25 * 518400 * \frac{1}{3600} * \frac{1}{1000}}{0.9} = 31.08\ \text{kW}.$$

$$\text{Shaft work } W_{shaft} = \frac{Power}{Efficiency} = \frac{31.08}{0.9} = 34.53\ \text{kW}.$$

$$Brake\,horsepower\,(BHP) = \frac{Volumetirc\ flowrate * Operating\ pressure}{1714 * \eta}.$$

$$= \frac{0.144 * 14.7}{1714 * 0.9} = 0.00137\ \text{bhp}$$

Based on the centrifugal pump selection guide [20], the rotational speed of the pump can be determined using the flow rate and total head of the pump. Based on those values, the revolution of the pump can be said to be at 3500 rpm at a single stage.

$$\text{Specific rotational speed, } NS = \frac{NQ^{0.5}}{(gH)^{0.75}} = \frac{(3500)(0.144)^{0.5}}{(9.81 * 194.25)^{0.75}}$$

$$= 4.60\ \text{rps} = 276\ \text{rpm}.$$

The equipment cost for the cooler pump is calculated as:

$$C_E = C_B\left(\frac{Q}{Q_B}\right)M,$$

where C_E represents the equipment cost with capacity Q, C_E is the known base cost, and, for equipment with capacity Q, M represents the constant depending on equipment type. Therefore:

$$C_E = (1.97 * 10^3)(\frac{34.53}{1})^{0.35} = \$6805.11.$$

According to Jenkins et al. [19], the CEPCI for 2020 experienced a 3.5% decline compared to the same time in 2019. Hence, the CEPCI for 2020 will be 597.53, with the cost index of the base year 2000 being 435.8.

$$C_E = \$6805.11 * (\frac{597.53}{435.8}) = \$9330.56$$

By implementing the material correction cost ($f_m = 1.0$), design pressure correction factor ($f_p = 1.38$), and design temperature correction factor ($f_t = 1.0$), the final costing value for this pump can be obtained.

$C_F = C_E f_m f_t f_p = (\text{US\$}9330.56)(1.0)(1.0)(1.0) = \text{US\$}9330.56 = \text{RM } 40\,567.65$

Coolers are a type of heat exchanger used to reduce temperature. The fluid used to cool the other stream is in the tubes within the cooler while the fluid to be cooled fills the shell. In this context, the cooler is used to reduce the water temperature from 24°C to 2°C, achieving the prescribed optimal low-temperature condition. The log mean temperature difference is calculated as:

$$\Delta T_{lm} = \frac{(T_1 - t_2) - (T_2 - t_1)}{ln\frac{T_1 - t_2}{T_2 - t_1}} = \frac{(24-5) - (2-(-10))}{ln\frac{24-5}{2-(-10)}} = 15.23^{o}C,$$

where ΔT_{lm} = log mean temperature difference, T_1 = hot fluid temperature inlet (24°C), T_2 = hot fluid temperature outlet (2°C), t_1 = cold fluid temperature inlet (−10°C), *and* t_2 = cold fluid temperature outlet (5°C).

The type of heat exchanger used is the one shell and two tube type. Effectiveness (*S*) is the ratio of the attained heat transfer rate to the maximum feasible heat transfer rate. The ratio of the thermal capacities of the two streams is designated *R*. The *R* and *S* values have to be obtained from the T_1, T_2, t_1, and t_2 temperatures to calculate the correction factor, F_t, which will then be used to calculate the true temperature difference, T_m.

$$R = \frac{T_1 - T_2}{t_2 - t_1} = \frac{24-2}{5-(-10)} = 1.466667$$

$$S = \frac{t_2 - t_1}{T_1 - t_1} = \frac{5-(-10)}{24-(-10)} = 0.441176$$

Correction Factor, F_t is calculated as:

$$F_t = \frac{\sqrt{R^2+1}\ \ln[\frac{(1-S)}{(1-RS)}]}{(R-1)\ln\left[\frac{2-S(R+1-\sqrt{R^2+1}}{2-S\left(R+1+\sqrt{R^2+1}\right)}\right]}, \text{ or}$$

$$= \frac{1.775137\ \ln[\frac{(0.558824)}{(0.352941)}]}{(0.466667)\ln\left[\frac{2-0.441176(2.466667-1.775137}{2-0.441176(2.466667+1.775137)}\right]} = 0.68.$$

The purpose of the temperature correction factor being applied to the log mean temperature difference to calculate the true temperature difference is to allow for the departure from true counter-current flow [21]. True Temperature difference, T_m, is calculated as:

$$T_m = F_t \Delta T_{lm} = (0.68)(15.23288) = 10.32635^{o}C,$$

and

$$\text{Heat transfer area} \quad A = \frac{Q}{UT_m} = \frac{980000}{740 * 10.33} = 128.2\text{m}^2,$$

where Q = heat transferred per unit time, W (980 000 W), U = overall heat transfer coefficient, W/m^{2}°C (740 W/m^{2}°C), A = heat transfer area, m^2, and T_m = true temperature difference, in°C (10.33°C).

Values for the area of heat transfer can be calculated or obtained by manual calculations, except for the overall heat transfer coefficient (U). The overall heat transfer coefficient obtained is in the range of 690 –830 W/m^{2}°C for coolers. For this purpose, an overall heat transfer coefficient value of 740 W/m^{2}°C was selected [21]. Therefore:

$$\textit{Cost Estimation}, C_E = C_B \left(\frac{Q}{Q_B} \right) M,$$

where C_E = equipment cost with capacity Q, C_E = known base cost for equipment with capacity Q, *and* M = constant depending on equipment type.

$$C_E = (3.28 * 10^4)(\frac{128.2}{80})^{0.68} = \$45200.22$$

According to Jenkins et al. [19], the corresponding CEPCI for the year 2000 (base year) for this calculation is 435.8.

$$C_E = \$45200 * (\frac{597.53}{435.8}) = \$61974.51$$

By implementing the material correction cost ($f_m = 2.9$), design pressure correction factor ($f_p = 1.38$), and design temperature correction factor ($f_t = 1.0$), the final costing value for this pump can be obtained.

The capital cost required for this pump C_E is calculated as:

$C_F = C_E \, f_m \, f_t \, f_p$

= (US\$69 635.21)(2.9)(1.38)(1.0)

= US\$248 022 = RM1 078 356.5.

The cost of the reactor is determined as:

$$C_E = C_B \left(\frac{Q}{Q_B} \right) M, \text{ or}$$

$$C_E = (1.15 * 10^4)(\frac{3}{1})^{0.45} = \$18{,}853.93.$$

Adjusting for the change in the CEPCI then gives:

$$C_E = \$18\,853.95 \text{ x } (\frac{597.53}{435.8}) = \text{US}\$25\,850.85.$$

By implementing the material correction cost ($f_m = 1$), design pressure correction factor ($f_p = 1.38$), and design temperature correction factor ($f_t = 1.0$), the final costing value for this pump can be obtained. The capital cost incurred by the pump is given by C_F.

$C_F = C_E f_m f_t f_p$ = (US$25 850.85)(3.2)(1)(1.38) = US$114 157.37 = RM496 3 36.40

Total Equipment cost is the sum of the reactor pump, cooler pump, cooler, reactor, and 13 butterfly valves, or:

= 1 484.18 + 9 330.56 + 248 022 + 95 118.04 + 3 547.64

= US$357 502.42.

Fixed capital investment (FCI) is made up of direct costs and indirect costs. FCI incurred by a plant can be calculated using the factors to estimate project fixed capital cost [21]. Table 9.5 provides the factors for project fixed capital cost, which encompasses direct costs. The total purchase cost of equipment is multiplied by the direct cost factorial, listed in Table 9.5, to obtain the total physical plant cost.

Once the total physical plant cost is obtained, the FCI can be calculated by multiplying the total physical plant cost by the indirect cost factorial, as shown in Table 9.6.

Table 9.5 Calculation for total physical plant cost.

Factors	Process Type: Fluids
f_1 Equipment erection	0.4
f_2 Piping	0.7
f_3 Instrumentation	0.2
f_4 Electrical	0.10
f_5 Buildings, process	0.15
f_6 Utilities	0.50
f_7 Storages	0.15
f_8 Site development	0.05
f_9 Ancillary buildings	0.15
Total	3.4
Total purchase cost of equipment (PCE)	US$357 502.42
Total physical plant cost (PPC)	US$357 502.42 (PCE) x 3.4 (Direct Cost Factorial) = US$1 215 508.23

Table 9.6 Calculation for fixed capital investment.

Factors	Process Type: Fluids
f_{10} Equipment Erection	0.30
f_{11} Piping	0.05
f_{12} Instrumentation	0.10
Indirect cost factorial	1.45
Fixed capital investment (FCI)	US$1 215 508.23 (PPC) x 1.45 = US$1 762 486.93

It is now known that total capital investment (TCI) is made up of FCI and working investment. Since FCI has already been determined, working investment can be calculated. Working investment is defined as the cost required for the day-to-day operation of a business. It can vary from as low as 5% of fixed capital investment for a simple, single product process with little or no storage of finished product to 30% of FCI for processes with a diverse range of products [21]. For the purpose of this project, 10% of FCI is taken as the working capital. Table 9.7 presents the TCI calculation.

The purpose of the cooler pump is to pump water into the chamber that houses the reactor in order to create an environment at a temperature of approximately 2°C, which will cater to the formation of gas hydrates in the presence of high pressures and CO_2 as hydrate former.

The operating cost for the cooler pump and reactor pump is calculated based on the information in Table 9.8 and expressed as:

$$\text{Power, P} = \frac{Q\rho gH}{\epsilon} = \frac{0.000144 * 1000 * 9.8 * 0.998017}{0.9} = 1.56\,\text{watt}$$

Converting this value to kWh yields:

$$\text{Power, kWh} = \frac{(1.56\text{watt})(4\text{hr})}{1000} = 0.0062\ \text{kWh}.$$

Table 9.7 Calculation for total capital investment.

Capital Cost	Value
Fixed capital investment	US$1 762 486.93
Working capital	10% of FCI = US$176 248.69
Total capital investment	US$1 938 735.62

Table 9.8 Information pertinent to the requirement of pump for cooler.

Property	Value
Flow Rate (m^3/s)	0.000144
Fluid Density (kg/m^3)	1000
Mass Flow Rate (kg/s)	0.143797
Volumetric Flow Rate (m^3/s)	0.000144
Differential Pressure Head (bar)	0.998017
Pump Efficiency	0.9
Operation Hours	4
MYR to US$ conversion	0.23

The pump will require 0.0062 kW for every hour it is operated. To translate this value to incurred cost requires the industrial tariff value, which for Malaysia is RM0.38/kWh, or roughly US$0.09/kWh. Given this information, the cost for operating the pump is:

$$Cost = (0.0062\ \text{Wh})\left(\frac{0.087}{\text{kWh}}\right) = \text{US\$}0.00054.$$

The cost value obtained is the cost required for one run. The pilot reactor has a capacity of 2000 mL and a yield factor of 0.5, or 1000 mL desalinated water per run. Converting to m^3 of desalinated water gives a value of US$0.54/$m^3$ of desalinated water.

The power of the reactor pump is calculated using the factors in Table 9.9 and is expressed as:

$$\text{Power, P} = \frac{Q * \rho * g * H}{\epsilon} = \frac{0.0000018333 * 1000 * 9.81 * 0.499009}{0.85} = 0.010549 \text{ watts}$$

Converting this value to kWh yields:

$$\text{Power, kWh} = \frac{0.010549 * 1}{1000} = 0.0062 \text{ kWh}$$

The pump will require 1.05 x 10^{-5} kW for every hour it is operated. This value is translated to incurred cost in the same manner as for the cooler pump (tariff rate of US$0.087). Thus, the cost for operating the pump is:

$$\text{Cost} = (1.05 * 10^{-5})\left(\frac{US\$0.087}{\text{kWh}}\right) = \text{US\$}0.0000009$$

As with the cooler pump calculation, this result reflects costs per 1000 milliliter. Converting to m^3 of desalinated water gives a value of US$0.0009/$m^3$ of desalinated water.

Table 9.9 Information pertinent to the requirement of pump for reactor.

Property	Value
Flow Rate (m^3/s)	0.0000018333
Fluid Density (kg/m^3)	1000
Mass Flow Rate (kg/s)	0.001833
Volumetric Flow Rate (m^3/s)	0.0000018333
Differential Pressure Head (bar)	0.499009
Pump Efficiency	0.85
Operation Hours	4
MYR to US$ conversion	0.23

The cooler reduces water temperature from approximately 24°C to 2°C, thus allowing hydrate formation. The cooler is of the immersed coil type with refrigerant −141b flowing in the coils and a pool of water immersing said coils. The energy required to cool the water pool is calculated as:

$$Q = mC_p\Delta T = (168\ \text{kg})\left(4200\frac{J}{\text{KgK}}\right)(20\text{K}) = 14112000\ \text{J}$$

To calculate the cost required to run the cooler, energy consumption needs to be calculated on a kWh basis. The cooler runs for four hours. Therefore:

$$Q = \frac{14112000J}{14400s} = 900\frac{J}{s} = 980\ \text{W} = 9.8\text{kW}.$$

$$\text{In kWh, } Q = \frac{(980\,\text{W})(4\text{h})}{1000} = 3.92\ \text{kWh}.$$

This indicates that for every hour the cooler is operational, it requires 3.92 kW for that corresponding hour. The cost can be calculated from this value, using the industrial tariff value, which in Malaysia is RM0.38 (US$0.087).

$$\text{Cost} = (3.92\ \text{kWh})\frac{\text{RM0.38}}{\text{kWh}}) = \ \text{RM1.49} = \$0.34$$

The cost value obtained is the cost required for one run, which produces 1000 mL of desalinated water. The pilot reactor has a capacity of 2000 mL and with a yield of 0.5 being utilized, 1000 mL desalinated water is produced during each run. However, it is assumed that the energetic demand of the cooler will not be constant because, while the initial cooling to bring the warmer

(24°C) water to 2°C is energy-intensive, maintaining the water at 2°C requires less energy.

In this case, CO_2 gas is used as the hydrate former with water molecules. While using the pilot reactor for gas hydrate-based desalination, a 50 L CO_2 cylinder was used. The number of moles of CO_2 present in the cylinder is determined as:

$$PV = ZnRT,$$

where $P = 7\,000\,000$ Pa, $V = 0.05$ m^3, $R = 8.314$ J/mol.K, $T = 297.15$ K, $Z = 0.18$.

Plugging these values into the equation yields:

$$n = \frac{PV}{ZRT} = \frac{(7000000)(0.05)}{(0.18)(8.314)(297.15)} = 787.0635 \text{ mol.}$$

The cost of 787 moles of CO_2 is then calculated as:

$$\text{Cost of one mol of } CO_2 = \frac{RM\ 694}{787.0635 \text{ mol}} = \frac{RM\ 0.88}{\text{mol}} = \frac{US\$\,0.20}{\text{mol}}.$$

The amount of CO_2 required for one run can be calculated using the same equation used to determine the number of moles in the cylinder of CO_2.

$$n = \frac{PV}{ZRT}$$

Where:

$P = 2\,000\,000$ Pa and $4\,000\,000$ Pa

$V = 0.0005$ m^3

$R = 8.314$ J/mol.K

$T = 275.15$ K.

Based on the information in Table 9.10, the number of moles required for one run of gas hydrate-based desalination is 0.43714 mol for a run at 20 bar and 0.87428 mol for a run at 40 bar. However, a safety factor for CO_2 has to be considered. For this instance, it is assumed that 115% mole of CO_2 is required to ensure that there is a sufficient amount of CO_2 for the run inclusive of losses that may or may not happen. In the event where this is a possibility, the number of moles of CO_2 required for a run inclusive of the safety factor will be:

With the above information, the cost required for the utilization of CO_2 can be calculated. The cost per mol of CO_2 has already been calculated above for US\$0.20/mol. The cost of CO_2 required at 20 bar and a run at 40 bar is determined as:

$$\text{Cost at 20 bar}: \frac{US\$0.20}{mol} * 0.50271 \text{ mol} = \text{USD}0.10.$$

Table 9.10 Number of moles required for exactly one run at 20 and 40 bar.

20 bar	40 bar
$n = \frac{(2000000)(0.0005)}{(1)(8.314)(275.15)}$	$n = \frac{(4000000)(0.0005)}{(1)(8.314)(275.15)}$
$= 0.43714$ mol	$= 0.87428$ mol

$$\text{Cost at 40 bar}: \frac{US\$0.20}{\text{mol}} * 1.005422\text{ mol} = \text{ US\$0.20.}$$

CO_2 is not consumed at any point throughout the process and can therefore be reused. Hence, the cost of CO_2 is capped at US$0.10 at 20 bar and US$0.20 for the process operating at 40 bar.

In this instance, the total product cost for gas hydrate-based desalination includes operating costs of the pump and cooler and the cost for CO_2 used as the hydrate former. The incurred costs for the operation at 20 bar and 40 bar are given in Tables 9.12 and 9.13, respectively.

The product costs obtained here are comparable to those obtained by [22], which reports cost/m^3 of US$4.23 and US$2.76 with yields capped at 0.2 and 0.3, respectively.

Because hydrate-based desalination requires refrigeration to maintain low temperatures, using liquefied natural gas (LNG) as a coolant or source of cold energy could be a viable option. A design of using LNG cold energy was recently proposed [23]. The hydrate formation reactor is cooled with LNG to provide the low-temperature environment required for hydrate formation and remove the exothermic heat emitted during hydrate dissociation. When LNG cold energy is used to replace the external refrigeration cycle, the specific energy consumption of hydrate desalination utilizing propane as the guest gas was lowered from 65.15 kWh/m^3 to 0.84 kWh/m^3. Chong et al. [24] further assessed the economic viability of the ColdEn facility in Singapore. The ColdEn facility simulated 260 m^3/h water generation at a regasification rate of 200 t/h. Integrating cold energy into the economic analysis resulted in a considerable

Table 9.11 Number of moles required for exactly one run at 20 and 40 bar.

20 bar	40 bar
$n = 0.43714 \text{ x } 1.15 = 0.502711$ mol	$n = 0.87428 \text{ x } 1.15 = 1.005422$ mol

Table 9.12 Product cost for process at 20 bar.

Aspect	**Cost/m^3**
Reactor Pump	US$0.00092
Cooler Pump	US$0.54
Cooler	US$0.34
Carbon Dioxide	US$0.10
	US$0.98

Table 9.13 Product cost for process at 40 bar.

Aspect	**Cost/m^3**
Reactor Pump	US$0.00092
Cooler Pump	US$0.54
Cooler	US$0.34
Carbon Dioxide	US$0.20
Total Cost	**US$1.08**

reduction in the levelized cost of water: US\$9.31/m^3 reduced to US\$1.11/m^3. The specific energy consumption utilizing cyclopentane as guest molecules integrated with LNG was 0.35 kWh/m^3 [25]. These results show that combining LNG cold energy with hydrate-based desalination is a promising method in terms of both economic feasibility and long-term energy use. Moreover, they reveal that when regasification capacity is higher, as in the case of LNG cold energy use, hydrate-based desalination is more cost-effective than other desalination technologies.

Hydrate-based desalination has not been used in industrial or pilot-scale applications thus far, partly due to the relatively large energy consumption of the process. The major contributor to the total energy requirement is the compressors, as part of the external refrigeration cycle. Therefore, to minimize the total energy and specific energy consumption, it is imperative to find a suitable hydrate former that can enable the process to operate at a higher temperature [26]. The other main issue is the inverse variation between water recovery and salt rejection. Guangzhou Institute of Energy Conversion sought to overcome this issue by developing a pilot-scale hydrate-based seawater desalination setup that produced hydrate pellets and then dissociated the hydrates into pure water. However, the salt removal efficiency was low, and issues with hydrate flowability remained unresolved. Though the implementation of hydrate-based methods is still confined to lab-scale applications [27, 28], developments continue to show rapid and optimistic progress [13]. As discussed in Chapters 3–5 and Chapter 7, the gas hydrate-based desalination process can be made more efficient with the proper equipment design and selection of hydrate promoters.

An extensive economic analysis can only be properly done once a suitable set of parameters for an optimized gas hydrate-based desalination has been established. Moving forward, more research into the kinetics of hydrate formation, including the search for new promoter species capable of improving both the kinetics and water recovery rates, is required for comprehensive process design.

References

1 Kucera, J., (Ed.). *Desalination Water from Water*. Scrivener Publishing.
2 Babu, P., Nambiar, A., He, T., Karimi, I.A., et al. (2018). A review of clathrate hydrate based desalination to strengthen energy–water nexus. *ACS Sustainable Chemistry & Engineering*. 6 (7): 8093–8107.
3 Mesa, A.A., Gómez, C.M., and Azpitarte, R.U. (1997). Energy saving and desalination of water. *Desalination* 108 (1–3): 43–50.
4 Micale, G., Rizzuti, L., and Cipollina, A. (Eds.) (2009). *Seawater Desalination. Green Energy and Technology*. Berlin, Heidelberg: Springer.
5 El-Sayed, M.Y., Lattemann, S., Rodriguez, S.G.S., Kennedy, D.M., Schippers, C.J., Amy, L.G. et al. (2013). *Advances in Water Desalination*, 1e (Eds. N. Lior, M. Balaban, A.M. Darwish, O. Miyatake, S. Wang, and M. Wilf), John Wiley & Sons, Inc.
6 Voutchkov, N. (2019). *Desalination Project Cost Estimating and Management*. Taylor & Francis.
7 Wetterau, G., Moch, I., Frenkel, V., Huehmer, R., Hunt, H., Kiefer, K. et al. (2011). *Desalination of Seawater*. AWWA MANUAL. American Water Works Association, 120.
8 El-Dessouky, H.T. and Ettouney, H.M. (2002). *Fundamentals of Salt Water Desalination*. Elsevier Science.
9 Karagiannis, I.C. and Soldatos, P.G. (Mar 2008). Water desalination cost literature: Review and assessment. *Desalination* 223 (1–3): 448–456.
10 Winter, T., Pannell, D.J., and McCann, L.M.J. (2002). The economics of desalination and its potential application to Australia. In: 46th Conference of Australian Agricultural and Resource Economics Society (AARES). 1–13.
11 Assessment USCO of T. (1988). Using desalination technologies for water treatment: background paper. Congress of the United States, Office of Technology Assessment.

12 Khan, M.S., Lal, B., Sabil, K.M., and Ahmed, I. (2019). Desalination of seawater through gas hydrate process: An overview. *Journal of Advanced Research in Fluid Mechanics and Thermal Sciences* 55 (1): 65–73.

13 Nallakukkala, S. and Lal, B. (Apr 2021). Seawater and produced water treatment via gas hydrate: Review. *Journal of Environmental Chemical Engineering* 9 (2): 105053.

14 Fakharian, H., Ganji, H., Naderifar, A., Mofrad, H.R., and Kakavand, M. (2019). Effect of gas type and salinity on performance of produced water desalination using gas hydrates. *Journal of Water Reuse and Desalination* 9 (4): 396–404.

15 Subramani, A. and Jacangelo, J.G. (May 2015). Emerging desalination technologies for water treatment: A critical review. *Water Research* 75: 164–187.

16 Javanmardi, J. and Moshfeghian, M. (May 2003). Energy consumption and economic evaluation of water desalination by hydrate phenomenon. *Applied Thermal Engineering* 23 (7): 845–857.

17 Sangwai, J.S., Patel, R.S., Mekala, P., Mech, D., and Busch, M. (2013). Desalination of seawater using gas hydrate technology – Current status and future direction. In: XVIII Conference on Hydraulics, Water Resources, Coastal and Environmental Engineering.434–440.

18 Brown, G.O. (2017). The history of the darcy-weisbach equation for pipe flow resistance. *ResearchGate*, 34–42.

19 Jenkins, S. (2019). CEPCI UPDATES: NOVEMBER (PRELIM.) AND OCTOBER (FINAL); Chemical Engineering Essentials for the CPI Professional. Available online: https://www.chemengonline.com/2019-cepci-updates-november-prelim-and-october-final (accessed on 27 January 2020).

20 Kumareswaran, S. (2014). *Design of a Shell and Tube Heat Exchanger* (Doctoral dissertation, University of Moratuwa).

21 Sinnott, R.K., Coulson, J.M., and Richardson, J.F. (2005). *Chemical Engineering Design*. Oxford: Elsevier Butterworth-Heinemann. (Vol. 6, No. 4).

22 Javanmardi, J. and Moshfeghian, M. (2003). Energy consumption and economic evaluation of water desalination by hydrate phenomenon. *Applied Thermal Engineering* 23 (7): 845–857.

23 He, T., Nair, S.K., Babu, P., Linga, P., and Karimi, I.A. (2018). A novel conceptual design of hydrate based desalination (HyDesal) process by utilizing LNG cold energy. *Applied Energy* 222: 13–24.

24 Chong, Z.R., He, T., Babu, P., Zheng, J.N., and Linga, P. (2019). Economic evaluation of energy efficient hydrate based desalination utilizing cold energy from liquefied natural gas (LNG). *Desalination* 463: 69–80.

25 He, T., Chong, Z.R., Babu, P., and Linga, P. (2019). Techno-economic evaluation of cyclopentane hydrate-based desalination with LNG cold energy utilization. *Energy Technology*. doi:10.1002/ente,201900212.

26 Babu, P., Bollineni, C., and Daraboina, N. (Feb 2021). Energy analysis of methane-hydrate-based produced water desalination. *Energy & Fuels* 35 (3): 2514–2519.

27 Xu, C., Li, X., Yan, K., Ruan, X., Chen, Z., and Xia, Z. (Sep 2019). Research progress in hydrate-based technologies and processes in China: A review. *Chinese Journal of Chemical Engineering* 27 (9): 1998–2013.

28 Lv, Q., Li, X., and Li, G. (Feb 2019). Seawater desalination by hydrate formation and pellet production process. *Energy Procedia* 158: 5144–5148.

10

Renewable Energy in Desalination and Hybrid Technologies

Jai Krishna Sahith and Bhajan Lal

10.1 Introduction

The alternative way to address the rising need of larger populated areas is to build additional options for freshwater alongside other conventional water reserves such as dams, rivers, and shallow wells. These new sources include seawater, deep wells, and aquifers for groundwater [1]. The primary downside of water from these sources is the high percentage of salts in minerals which has a detrimental effect on the consistency of water. Widely known conventional desalination processes have been enhanced over the past decades and continue to improve as knowledge of the field expands. Current advancements primarily focus on lowering freshwater processing costs, enabling clean water extraction from low-quality sources.

Desalination is a method of removing or separating dissolved salts and minerals. The first applications of desalination go back to the sixteenth and seventeenth centuries when solar evaporation was carried out to supply sailors with clean seawater. Modern desalination technology was first seen on the island of Curaçao and the Arabian Peninsula in the early twentieth century. Research and production of desalination plants intensified midcentury, and the manufacturing and use of this technology have steadily developed over the past 30 years [2]. Today, the method of desalination, in certain aspects, has important and undeniable effects on human development, which may be attributable to:

- Increased demand for drinking water with rapid population growth and poor access to drinking water in arid regions.
- Rising water use per capita in manufacturing areas and urban neighborhoods.

Gas Hydrate in Water Treatment: Technological, Economic, and Industrial Aspects,
First Edition. Bhajan Lal and Sirisha Nallakukkala.

The use of desalination equipment has increased dramatically, particularly in the Persian Gulf, Algeria, Australia, and Spain. The use of seawater is only feasible for marine areas and offshore areas, while underground saltwater aquifers must be used for areas far from the sea [3, 4].

According to figures from the International Desalination Association (IDA), the standard operational capacity for a desalination plant built in 1980 was five million cubic meters per day. By 2013, nearly 150 countries worldwide used this process to access freshwater, and standard capacity had grown to 80 million cubic meters per day. The capital cost of desalination machinery installed in 2014 was valued at US$12 billion, which rose to US$21 billion by 2019, carrying a global desalination capacity surpassing 23 million cubic meters a day [4, 5].

Desalination methods have given a path to obtaining clean water for many arid areas of the world, such as the Middle East, the Mediterranean, and the Caribbean. The Middle East has the largest built capacity of desalination plants, accounting for 65% of the overall desalination capacity, owing to the scarcity of freshwater supplies and widespread access to fossil fuels. The highest desalination potential is seen in Saudi Arabia, followed by the United Arab Emirates, the USA, Spain, and China. India has had the highest growth since 2002 in this sector [3, 6]. A diagram of desalination capacity (cubic meter per day) by region and source (percent) is shown in Figure 10.1. The Middle East has the largest share of seawater use. North America leads in brackish water desalination capacity.

Freshwater demand has already soared above sustainable levels. With usage continuing to rise, the current aim in desalination technology is to contend with existing and new water supply shortages. In 2015, roughly 5.5% of the world's population used desalinated water, and an estimated 14% will need to use desalinated water by 2025. More and more sources have been considered for obtaining freshwater, such as spent water from agricultural uses, which contains high amounts of dissolved salts and minerals that can be reduced or removed by methods of desalination [7].

10.2 Renewable Energy

Humanity has come to experience many challenges in recent years. These issues are primarily connected to the growing demand for energy and water. Unfortunately, energy and water strategies in most countries are focused on the widespread usage of fossil fuels and accessible water supplies, which is disrupting resource security and will contribute to future demand for access to

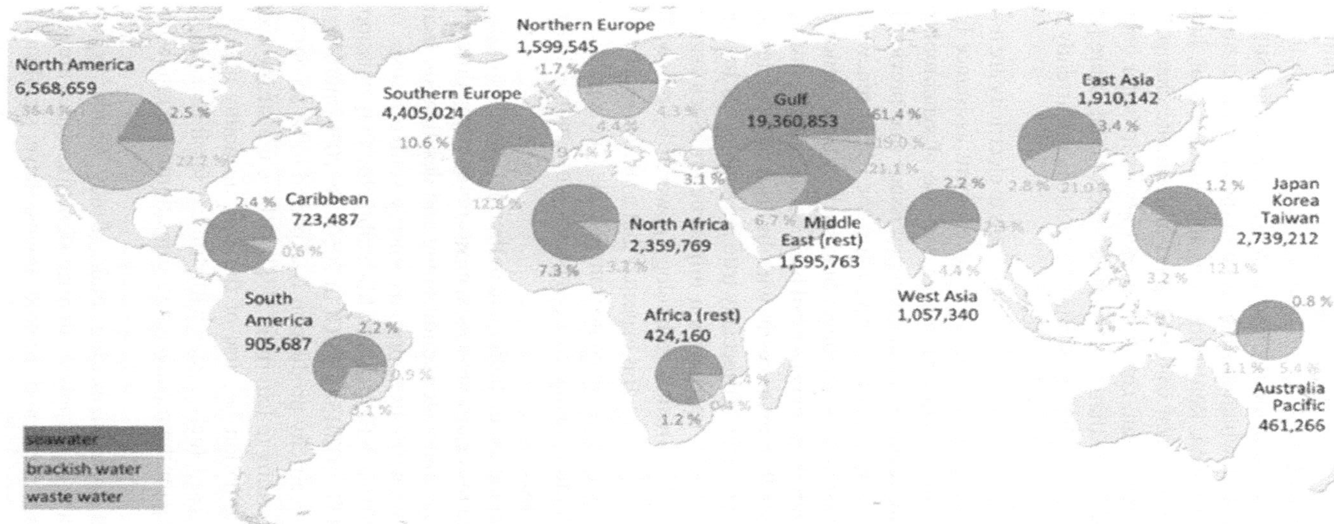

Figure 10.1 International desalination capacity [m^3/d] by water source [%] of which the Middle East has the largest share [8].

new sources [9]. Most countries within the Organisation for Economic Co-operation and Development (OECD) have recognized that new facilities are needed for the sustainable use of energy resources. However, water supply infrastructure remains outdated, which in some areas of the globe has contributed to extreme water restrictions. Koutroulis et al. and Manios et al. have investigated the connection between water supplies and climate change in South Africa and the Mediterranean, concluding that access to water could become more of a significant issue than access to fossil fuels in the future [10, 11]. Therefore, energy and water shortages must be tackled concurrently by establishing solutions that overcome both factors.

While advanced technologies for converting fossil fuels to electricity are now cost-effective, the use of renewable sources with government assistance has seen sustainable and economic results. Moreover, the use of synchronistic multi-source systems has overcome the intermittency issues that previously challenged a transition away from fossil energy. Thus, the use of renewable resources as primary energy sources is no longer held back by dependence on fossil fuels for backup generation. So long as fossil fuels are used for freshwater processing, these processes will remain costly with indirect and deleterious impacts on the air and climate. This is especially true for regions low in fossil fuel supplies or lacking electricity transmission infrastructure. Such regions would benefit greatly from desalination technology powered by renewable energy sources and water storage sites. It is expected that a large proportion of these sustainable energy resources would come from a single type source, as it did with fossil fuels, though multi-source energy systems prove more efficient and reliable in some locations.

Renewable energy technologies, including solar thermal, wind, and geothermal energy, can be used in various processes, most commonly in different thermal schemes, such as air and water conditioning and desalination facilities. Advancing renewable energy as a primary energy source focuses on expanding it from 14% in 2015 to 63% in 2050 while stabilizing CO_2 emissions from the energy sector (33 Gt in 2015 to 35 Gt in 2050). By 2050, more than 70% of overall renewable electricity production is expected to be accounted for by wind and solar. According to Latorre et al. [12], there is a strong need for green energies in desalination because of the cost instability and the decreasing abundance of fossil fuels. Areas with limited fossil fuels and a shortage of water supplies gain the most value in transitioning to renewable resources.

Population growth and industrialization make Saudi Arabia heavily dependent on seawater as a water supply. Water transfers to crowded areas in central Saudi Arabia, and rural and remote towns, have significantly increased, generating new opportunities for renewable energy technologies in desalination [13, 14]. Like several other countries, Saudi Arabia has areas with high solar radiation and has considered the use of solar energy technologies. Additionally, the use of wind energy has increased, especially in coastal areas [15, 16]. However, the use of renewable energy for desalination is not very widespread in Saudi Arabia's desalination industry. Only 0.02% of the total desalination capacity is powered by renewable energy resources. Nevertheless, renewable energy technologies can be ideal for seawater desalination due to the wide variety of clean energy options available and decreasing cost.

The logistics for creating, sustaining, and managing renewable energy systems can be more convenient than conventional energy sources, making them suitable for remote areas. Moreover, most desalination units are situated in coastal areas where renewable energy is plentiful. Dependence on fossil fuels requires transport of fuel materials and transmission of generated power. The lack of established electricity distribution and transmission networks in these regions, therefore, makes fossil fuel power very expensive. Conversely, power from renewable energy sources may connect to a power distribution grid or be generated and used onsite (off-grid), thus eliminating the need to link to a power grid. This latter approach may require site distribution systems (microgrid) and energy storage to satisfy the needs of the complex throughout the year, as some renewable sources, such as solar, wind, or wave, fluctuate over time. It is possible to view microgrid generation units as a small-scale power grid that incorporates the sectors of manufacturing, transmission, and storage and can fulfill local needs [17, 18]. The main advantage to generating green energy on-site is primarily in the power-saving capabilities due to the efficiency brought about by on-site energy generation.

In several countries around the world, including China, Mexico, Kenya, and Bangladesh, standalone units have been built to provide electricity to remote areas. Such energy systems also contain photovoltaic storage cells. While most of these units are installed in non-industrial rural areas, a range of interconnected micronetworks is now available to meet the needs of residential and commercial complexes. In Japan, for instance, the microgrid is capable of providing energy by photovoltaic cells, fuel cells, and storage batteries to different sectors. The use of energy supplies as a prime alternative is a challenge for the sustainable growth of freshwater supply. For starters, it would make sense to use hybrid wind–solar systems for areas with high radiation because of the difficulties and costs of delivering power to remote areas. In urban and rural areas, the use of standalone hybrid systems can be used for several purposes, such as lighting, water transport, and water treatment. In order to include desalination electricity, solar and wind energy must reach a position of

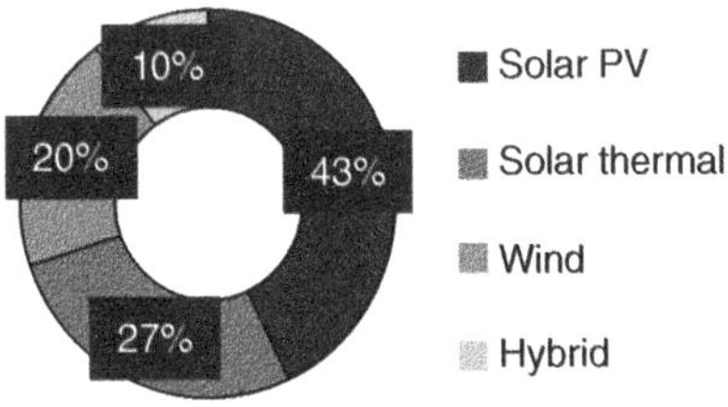

Figure 10.2 The contribution of renewable energy resources in desalination.

having the largest share of all renewable energy sources. Hydropower and ocean resources can also be used to produce electricity for smaller desalination plants [19]. In the case of solar and wind energy, batteries are generally needed to overcome the most notable drawbacks or the resources: unpredictability and intermittency. However, the prevalence of batteries with short storage life, high maintenance costs, and environmentally threatening materials has become a growing issue. Experiments in combining electrolysis with fuel cells have shown initial success, showing decent efficiency and environmental friendliness. Solar, wind, and geothermal energies are renewable energies that have potential roles to play in enabling a smoother transition for renewables in the desalination processes. The size and capability of the processes, however, need to be fine-tuned to allow for better efficiency in utilizing these alternative sources [20]. The contribution of renewable energy supplies to desalination processes [21] for freshwater production is shown in Figure 10.2.

10.3 Renewable Energy Usage in Desalination

10.3.1 Solar Energy

The utilization of solar energy for desalination has the largest share of all forms of green energy at the moment. The human energy needs of today can be fulfilled by harvesting just 1% of the surface sunlight from semi-arid lands. In the Middle East and North Africa (MENA) region, for instance, each square meter of land receives between 5–7 kWh of solar energy per day. In this area of the world, many areas produce 1.7–2.2 MWh/m^2 annually [22, 23]. More of these areas have seen increased use of solar energy for desalination methods. In rural and remote areas, using this approach may have a significant positive effect on people's health, education, and economic growth. However, there are numerous complications in the performance of the complex, variable amount of energy obtained from the sun and a lack of resources, all of which hinder the utilization of solar energy in large-scale desalination capacities [24]. The MENA region is largely situated in semi-arid areas, which are naturally prone to water supply issues. In these areas, solutions to the water shortages are implemented by either importing potable water from other regions or by generating airborne water, both of which are very expensive. That being the case, implementing the most cost-effective desalination approach is dependent on the selection of available techniques, economic parameters, and water conditions, i.e. total dissolved solids (TDS), humidity, temperature, etc. Current research is focused on increasing the performance of solar energy conversion systems, desalination technology, and the optimum combination of the two to further lower the economic barrier to medium and small-sized desalination units. Two types of solar energy systems commonly exist;

- Photovoltaic (PV) systems receive and convert the energy of the sun into electricity. It has been used in applications for reverse osmosis (RO) and electrode ionization.
- Solar thermal systems transform the collected solar energy into heat energy that can be used to provide energy directly or indirectly. It is possible to use the thermal form in solar distillers, solar collectors, and solar ponds.

10.3.2 Wind Energy

The kinetic energy captured from wind flowing through aerodynamic windmill blades to power turbines is one of the cleanest sources of energy due to its relative simplicity. The mechanical energy acquired is directly converted into electrical energy by a generator. Based on the orientation of their axis, wind turbines are commonly classified into vertical and horizontal axis classes. For the production of electricity, horizontal axis turbines are used [25]. With individual wind energy converters, small desalination units can be integrated and have a high capacity for turning salt and brackish water into freshwater. In this situation, the provision of wind powered freshwater to the local community surrounding the area is technically viable. In addition, an on-grid link can be used to run wind turbines [26].

A good example of an RO process using wind turbines is in the Kwinana desalination unit in South Perth, Western Australia, where approximately 140 ML/d of freshwater is provided by the site. An 80 MW wind power plant supplies the electricity needed for the unit [27]. The integrated wind powered desalination device by ENERCON was designed primarily with the locale in mind for utilizing the plentiful source of wind. Their machine runs at variable power, coupled to an RO desalination unit to fit a range of 12.5–100% output.

The variable power design was made to address the changing environmental patterns and wind movement [28]. Similar desalination units have been deployed around the world. As with solar energy, one of the more pervasive challenges is variations in power output over time [29].

The unstable conditions of this supply can be remedied by integrating the system with other clean energy options and by utilizing energy storage reservoirs, such as batteries. The quality and cost of water output were explored by Park et al. through several small wind turbine models linked to Ghana's RO system [30]. The findings showed that the most powerful and the least expensive unit is a 1 kW wind turbine. The effect on the pricing of each freshwater unit on different variables, such as environmental conditions, site layout, turbine strength, saltwater concentration, operating conditions, output capability, RO, and wind turbine prices, was further analyzed by García-Rodríguez et al. [31].

Peñate et al. developed a method for RO's use of wind energy [32]. Their configuration is capable of various modes of operation, with a nominal output capacity of 1000 m^3/d relative to the constant production mode. The results revealed that the degassing units would eventually respond to the variable energy provided and optimize the amount of freshwater. The findings demonstrated that the freshwater capacity is better suited to be delivered by smaller desalination systems. Of course, depending on the variables surrounding the desalination system, this may not always be the optimal case. By increasing the size of the storage tank, it is possible to increase the share of wind power in the scheme (with the same installed capacity of the wind turbine) and to increase the desalination capacity by 25%, from its original projection. Its use of appropriate wind turbines has also managed to deliver an optimized increment of 56% more freshwater, minimizing wind energy usage and total expected costs by around 7% [33].

Gökçek et al. evaluated the technological and economical use of small-scale wind energy to move a RO unit [34]. Based on their analysis, the cost per cubic meter of produced water for a standalone installation ranged from US$2962.00–6457.00 depending on turbine size (6 kW–30 kW). With a 30kW turbine, CO_2 emission reductions amounted to 80×10^3kg/y.

Solar energy is available during the day, but wind speeds are higher during the night. A hybrid solar–wind power system is thus proposed for both optimized and sustained energy usage. Ismail et al. modeled the efficiency of desalination approaches for multi-effect distallation (MED) and MVC and powered by renewable energy for a desalination capacity of 100 m^3/h. The findings revealed that the location, solar radiation strength, wind speed, atmospheric temperature, and water salinity of the desalination device are significant factors in the efficiency of the system [35].

The techno-economic feasibility of using green energy systems for industrial desalination in the Netherlands has been explored by Soshinskaya et al. Modeling results indicate that there is a great opportunity in the area for generating energy from renewable sources, enough to satisfy 70–96% of electricity needs of the potable water production unit using photovoltaic (PV) cells and wind turbines alone [19]. Mentis et al. developed a tool for planning and maximizing the desalination unit capacities in the Greek Islands using wind and solar renewable energy. Different parameters were taken into account in this work, such as the demand for water and the resources needed for desalination. Expert knowledge was also applied for technological efficiency, usable capital, and economic calculations. The modeling results revealed that prices per cubic meter of water produced ranged from €1.45 for large islands to €2.60 for small islands, far below the actual water price (€7.00–9.00) in these regions. Wind turbines are chosen for large islands, though PV cells may also be added to a coupling device. The estimation is much more difficult on smaller islands, as the solution is not as straightforward since the cost per kilowatt of small wind turbines is much higher than for big wind turbines. The use of PV cells is competitive in these islands and is only limited by the configuration of the cells and their positioning.

The country that is most primed for the use of solar and wind energy is Jordan, where a return on investment of six years is seen for certain projects using wind energy and 2.3 years for PV cells [17]. The findings demonstrate that it is possible to expand the optimum use of renewable energy for power production by up to 76%, contributing to a large decrease in fossil fuel consumption and CO_2 emissions. Up to 36% economic and environmental savings and benefits can be obtained by using a hybrid energy system for freshwater development on an island.

Smaoui et al. investigated the optimum scale for an autonomous hybrid energy facility that uses PV, wind, and hydrogen to provide the energy needed for desalinating seawater in the south of Tunisia. This power is used in the electrolysis unit to produce hydrogen when the electricity produced by wind and solar sources exceeds the load. In this scenario, when the integrated device faces a lack of input power, the needed power for the desalination method would be generated by fuel cells. Hydrogen storage was chosen in this work due to the low operating costs, low noise, and the potential to work in extreme environments [36]. During the winter testing season, the introduced scheme provided freshwater for 14 400 people. The mean annual intake of water is 193.6 m^3/hr, while the peak water consumption for the region is at 530 m^3/hr. The combination of wind and PV systems effectively improves the system's power stability and eliminates the need for energy storage, based on their findings (thus reducing installation costs).

Moh'd et al. also introduced an advanced wind energy technology to fit a desalination process. In their operation, thermal energy was created via wind power to preheat the water. The unit depends on the thermal energy produced during winter as well as at night [37]. The ultimate goal of the project is to use a hybrid system to create an autonomous desalination plant in remote areas. Frictional heat energy in a wind turbine increases the temperature of the oil and then transfers the thermal energy to seawater in a heat exchanger. The benefit of this approach is that the power and performance of the desalination plant are increased and will continue to run at night and on days with reduced direct sunlight. The prospect of using a combination wind and diesel system that involves a water desalination unit on the Isle of Man has been explored by Henderson et al. The device was designed to supply power during peak hours for production of potable water. It was observed that the most technically appealing application for long-term energy storage and power conservation is seawater desalination. There are various studies on the integration of wind energy and PV panels for the generation of freshwater, and the consensus of the findings suggests that power supply can be assured throughout the year by the use of hybrid systems [38].

10.3.3 Nuclear Energy

The IDA made a desalination capability goal of 120 million m^3/d by 2020 with nuclear energy as part of the offering in reaching this milestone. Development in both nuclear power plants and desalination systems have grown independently from one another, with participation by multiple countries. There are no obvious limitations for the choice of nuclear reactors process or the desalination process to be powered by, as any reactor could potentially be coupled with any desalination systems [39, 40]. At its simplest operation, the heat generated from the nuclear reactor is used directly to power the equipment needed for desalination. The cost of nuclear-powered seawater desalination is approximately equal to fossil energy. In India, the first nuclear-powered MSF-RO unit was installed. This unit is based on natively developed multistage flash (MSF) technology, carrying a capacity of about 6300 m^3/day [41].

The freshwater generation could be further expanded through the development and implementation of hybrid technologies, one such being the inclusion of nuclear power into desalination systems. Over the past two decades, such expansion priorities have been looked into by the International Atomic Energy Agency (IAEA) [42], with various integrations of small-size desalination units with nuclear systems [43]. Such implementations include a hybrid method consisting of a low-temperature MED + RO system nuclear heating reactor (NHR) and a low-temperature MED/VC + MED NHR [44]. Performance reviews of the integration between MED methods and NHR suggest that it could perform favorably for various conditions, such as for fulfilling drinking water requirements in certain arid areas [45].

A number of nuclear reactors coupled to different desalination plants were also reported in the analysis. Recently, the development of a hypothetical model was undertaken in Egypt to measure the productive parameters of a RO method powered by a cogenerating nuclear reactor [46]. The cost of nuclear desalination per cubic meter of freshwater output is estimated to be in the range of US$0.40–1.80, depending on the form of reactor and desalination process. The efficiency of the coupled heat pump system for the purpose of space cooling and the assessment of the ground source heat pump system has been tested in this case [47].

10.3.4 Geothermal Energy

Geothermal energy is commonly used worldwide and can be used both in thermal and membrane desalination processes for direct heating and power generation. Steam and hot water can be provided by geothermal energy. Globally, geothermal resources rank seventh of all renewables used for energy generation. In Iceland, however, 86% of space heating and 16% of power generation is supplied by geothermal resources [48]. By the end of 2004, geothermal energy supplied 57 TWh as electricity and 76 TWh as heat in Iceland. The electricity supplied by this source is more reliable than other green sources. Kalogirou showed that the temperature remains constant throughout the year at a given depth from the earth's surface [26]. There are three types of geothermal resources, based on temperature and corresponding to depth zones: (i) low-temperature surface sources with temperatures below 100°C, (ii) average temperature shallow sources with temperatures between 100 and 150°C, and (iii) high-temperature deep sources with temperatures of over 150°C. For desalination purposes, wells at depths as low as 100 meters are adequate.

The primary benefits of geothermal energy in desalination are the reliability and consistency of the energy source. Unlike wind or solar resources, geothermal resources provide a stable energy supply energy at a continuous rate. Thus, heat storage to guarantee energy during intermittency is no longer an issue. Kalogirou showed that the temperature remains constant throughout the year at a given depth from the earth's surface [26]. This also creates a certain flexibility in how geothermal resources are used.

There are many options for the geothermal system to be paired with desalination units, each of which relies on a variation of the MED process. The normal geothermal unit temperature is in the range of 70–90°C, which is highly suitable for the MED phases [49]. Alternatively, geothermal energy can be used to produce the electricity demand of RO desalination. There are also various ways of removing hot steam from the ground to circulate the energy-producing turbine while also directing the steam emitted from the turbine to heat a MED desalination facility. A RO–MED hybrid process can fully utilize the geothermal source in places with high water salinity, such as the Persian Gulf and the Red Sea. Several theoretical and laboratory experiments have studied such geothermal powered hybrid processes. Goosen et al. performed a thorough analysis of the use of geothermal energy for desalination in Greece, Algeria, and Mexico, illustrating the efficiency of geothermal powered desalination systems in these locations [50].

A case study of a geothermal desalination unit (75–90°C) producing a volume equivalent to 600–800 m^3/d was done by Calise et al. The evaluation was done to investigate the efficiency of a multipurpose hybrid-generation (concentrated solar, PV, and geothermal) system capable of providing electricity, space heating and cooling, domestic hot water, and freshwater from MED desalination [51]. The MED desalination facility consists of a range of accessories such as lithium bromide absorption chillers, a water tank, and a heat exchanger. Solar collectors produced electricity to the grid and thermal energy of 100°C. Low-temperature geothermal energy (80°C), in combination with solar-based thermal energy, powered desalination in the MED unit and produced domestic hot water at 45°C. In the summer, solar power was used to fire an absorption chiller. However, solar energy declines dramatically in the winter, resulting in less heat and power output. Most of the energy needed for desalination, for this reason, comes from geothermal energy in winter. Energetic analysis of the system showed high efficiency, and economic analysis shows the greatest potential when hot water demand is high. The inclusion of a heat storage tank is suggested to satisfy the nighttime energy and desalination demands [52].

10.3.5 Wave Energy

The use of wave power is an emerging high-capital-cost system with known limitations to implementation. Sustainable use of this energy source in desalination (RO especially) requires new strategies to reduce costs and improve conversion efficiency in lower energy waters. The cumulative power generated globally by waves is estimated at 2 TWh [53] with high spatial and temporal variability. The effects of such variabilities on desalination output can be substantial. Temporal fluctuations, in particular, are highly disruptive for desalination unit sustainability [54].

Davies [55] demonstrated the large spatial variability of the resource by relating wave energy potential to freshwater shortfalls (calculated as the volume of water needed to bring supplies up to a minimum per capita of 1700 m^3/y) for a number of arid countries. Fully developed and exploited wave energy resources could power desalination to supplement an estimated 2–9.7 km^3/y of a country's water debit, depending on local annual wave resource potentials. Wave resource potential is dependent on a number of variables, the most critical of which is wave height. The energy generated from waves 1 m high could irrigate a strip of land 0.8 m wide, whereas doubling the wave height increases the irrigation area by a factor of five.

A pilot desalination plant developed in Garden Island, Australia, is powered by wave energy harnessed through a technology named CETO. The method utilizes an array of submerged buoys that convert wave motion to electricity to power the high-pressure pumps that drive seawater into the RO unit with a processing capacity of 150 m^3/day [56]. The same technology powers the world's first commercial-scale wave energy array connected to the power grid, which is also located in Australia [24].

The generation of electric power on Pantelleria Island in Italy, with a wave power potential equivalent to 7 kW/m, was achieved with the help of converters developed by the Department of Energy and Information Engineering and Mathematical Models (DEIM) of Palermo University. The installed wave farm eliminated a greenhouse gas (GHG) emission footprint equivalent to burning 1391×10^3 kg diesel fuel (4406×10^3 kg CO_2 emissions) annually. Thus, the island's dependency on fossil fuels was effectively eliminated [57].

The energy criteria and structural characteristics affecting the efficiency of wave energy converters still occupy most of the research space within this technology [58]. A preliminary investigation of a deep-sea RO desalination project to deliver freshwater while reducing CO_2 emissions has shown that wave energy can substantially reduce energy consumption compared to other power systems. A potentially salient feature of such a device is the longer membrane lifespan [59].

10.4 Hybrid Technologies

The incorporation of two or more desalination systems, usually in a bid to increase the overall efficiency relative to that of the individual parts and minimize costs, is termed a hybrid desalination process. Hybrid systems use a synergistic approach to overcome the shortcomings of individual processes, maximize overall system efficiency, and fulfill additional requirements as needed. As hybrid systems need significant investment, an important part of the process is the optimization of hybrid configurations. This section seeks to objectively analyze the available hybrid desalination technologies, both at the lab-scale and pilot-scale.

10.4.1 Forward Osmosis–Nanofiltration Hybrid Systems

Thermal and RO desalination have also been proposed for draw solution recovery in forward osmosis (FO). The value of the thermal desalination application is that isolation is independent of the concentration of the feed. On the other hand, compared to thermal systems, RO uses less energy and needs little capital. Nanofiltration (NF) is used for the regeneration process in the hybrid forward osmosis–nanofiltration (FO–NF) method for seawater desalination [60]. Seven solutes of NaCl, KCl, $MgCl_2$, $CaCl_2$, $MgSO_4$, Na_2SO_4, and $C_6H_{12}O_6$ were tested for filtering. For FO, commercial cellulose triacetate membrane was used, while NF was added to a thin-film composite membrane from GE Osmotics. The rejection of solute in NF relied on the concentration of the solution for dilute drawing. As a draw solution, only Na_2SO_4 sustained > 90% solvent rejection even as the dilute drawing solution concentration increased. Though methods were evaluated individually, optimization of the hybrid system performance, such as finding the right draw solute concentration and an energy usage analysis, was not done.

Following Tan and Ng [61], a related FO–NF method was developed, which requires less strain, is less vulnerable to fouling than standalone RO, and exhibits better flux recovery after washing. All of these improvements decreased the system energy consumption even though the study does not present specific energy consumption (SEC) values.

Major research efforts have also explored fertilizer-drawn forward osmosis (FDFO), in which fertilizers are used as a solution for drawing and direct irrigation of desalinated water. However, it has been noted that the fertilizer content in the diluted solution could still be too high for use specifically as fertilizer [61, 62]. Therefore, FDFO hybrid systems with NF separation processes have been studied in the laboratory and at pilot scales. An optimized FDFO–NF pilot was investigated in the Mildura region of Australia by Kim et al. [63, 64]. The study system consisted of two spiral wound FO modules and one spiral wound NF module. The effects of NF feed flow rate, feed concentration, and applied pressure on water flux and ammonium sulfate rejection were analyzed, with feed concentration found to be the most influential of the three parameters. Diluting the fertilizer solution to 70% of the original concentration during FDFO is recommended to eliminate the use of a more costly two-pass NF while still allowing for a significantly lower nutrient concentration in the product water and making it ideal for direct irrigation.

Generally, the pretreatment of microfiltration and ultrafiltration (UF) systems can reduce the SEC of RO systems. Therefore, it can be surmised from the specifications for membrane cleaning and replacement that the FDFO–NF hybrid system performs at a lower SEC than the standalone RO. FO–NF hybrid systems have also been used for the disposal and reuse of wastewater [65, 66].

10.4.2 Electrodialysis–Reverse Osmosis Hybrid Systems

Due to problems related to scaling and retentate salinity, RO systems, especially those handling difficult feeds, operate at restricted recovery rates [67, 68]. Increasing the recovery of RO systems involves several steps, thus dramatically increasing the capital and operating costs. ED devices, on the other hand, can be run at greater recovery speeds [69], but this is ultimately constrained by the formation of scales. Even electrical desalination systems such as ED cannot match the high salt rejection of RO membranes. Given how the quantity of salt rejected scales up with ED, the magnitude of the added current, and thus the energy absorbed, increases. However, the high baseline recovery rates possible in ED, combined with the high salt removal rates in RO, offer the potential to generate high-quality permeate at high recovery and low SEC [70]. The notion of an electrodialysis–reverse osmosis hybrid process (ED–RO) was first studied by Schmoldt et al. in 1981. As a second step to monitor permeate efficiency, they suggested the use of ED. However, the absence of high-flux, high-selectivity membranes at the time resulted in high energy consumption of 7.94 kWh/m^3 [71]. McGovern et al. modeled ED–RO systems for desalination systems of two types of salted solutions: a brackish saline mixture and an extremely saline brine. The cost of ED can be decreased by hybridization for brackish water desalination. At 50% recovery, the single-stage RO unit handles feed streams containing 3000 ppm TDS, and at 6000 ppm the concentrate source from RO becomes the feed to the ED. Two benefits of the hybrid method are: (i) a decrease in the overall membrane region as higher current densities (and thus salt removal rates) are feasible at higher salinities; and (ii) because the final product is a mixture of RO and ED permeates, the salinity of the ED permeate needed could be higher than the salinity of the final product [72]. When both feed and product salinity are poor, the percentage reduction in cost is highest when salt removal happens at high salinity. If the specifications of product water are more relaxed, the savings of implementing a hybrid system are much smaller. Since ED is a method of salt removal as opposed to water removal, the treatment of RO concentrate with ED results in a highly concentrated brine that can be used further for the processing of salt [73]. A similar configuration of counterflow ED systems with RO did not manage to achieve high recovery rates when applied to treatment brine with 120 000 ppm TDS due to concentration variations between sources, higher osmotic pressure, and lower performance.

Following their modification study of a UF–RO method for handling secondary effluent from a wastewater treatment facility, Zhang et al. implemented the integration of ED in an RO-focused plant [74]. The content of the ED effluent was similar to the method without RO, but with greater total organic carbon (TOC), at a higher average recovery rate of 95%. This study did not, however, include energy usage results. Thampy et al. showed a hybrid method of ED–RO for desalination of brackish water (feed of 2000–4000 ppm TDS) with fast recovery to allow for the ED's permeate to become the RO feed [70]. In the high salinity area, where high solution conductivity converts into greater efficiency, the ED method is used as a pretreatment. In the low salinity area, where the osmotic pressure and thus the energy needed for RO is low, the RO unit could perform well. For all feed with TDS between 2000 and 4000 ppm, water was produced with TDS between 50 and 120 ppm at recoveries of greater than 50%. At 10–20% regeneration, RO alone takes 7.8 kWh/m^3 to generate freshwater. The hybrid ED–RO system operates at 8–10 kWh/m^3 but at a much higher recovery rate of 50-60%. The hybrid ED–RO system's energy consumption is equal or marginally better than RO alone, but with three to five times the standalone RO recovery. These high rates of regeneration are not possible in standalone RO at the SEC equivalent. A comparative framework that extends the use of ED as pretreatment has been proposed by Galama et al. for desalination of seawater [60].

The ED–RO systems explored so far have seen the ED portion used to either directly purify the RO permeate or to desalt the permeate obtained in the second stage of RO, and the generated brine is combined with the RO permeate. ED allows extraction of any salt from the stream of RO concentrate and also decreases polarization of concentration by ion electrophoretic movement [75]. Based on the transport properties of both RO membranes and ED ion exchange membranes, the resulting TDS and product recovery values for feed water generation could yield more energy savings compared to using RO alone. A commercial prototype of such a hybrid device has yet to be developed; however, much of the research regarding this technology focuses on its implementation as an additional treatment method following ED or RO, with only a few experiments done in the form of a single-stage hybrid model. Furthermore, the heavy need for optimization of the ED–RO systems for industrial use makes it an unattractive choice for conventional use.

10.4.3 Reverse Osmosis–Membrane Distillation Hybrid Systems

The low capital costs, high effectiveness, and high recovery capacities make membrane distillation (MD) an obvious choice for hybrid separation technology [76]. The hybridization of MD and RO, where MD is used to handle the concentrate stream from the RO process, has been investigated in a few studies. Choi et al. examined the economic viability of a hybrid RO–MD process for seawater desalination and found that when the recovery and flux are greater than that of RO, and the thermal energy supplied for MD is at a comparatively low cost, the RO–MD hybrid system or an MD standalone system should be able to match with conventional RO [77]. Naidu et al. [78] added direct contact MD to post-treatment of RO wastewater, with a water recovery rate of 85%. Granular activated carbon was used in pretreatment of the brine to mitigate organic fouling. The energy use for RO–MD systems remains largely unknown and requires additional study to see further progress within this particular technology.

10.4.4 Forward Osmosis–Membrane Distillation Hybrid Systems

Although FO is known to be a low energy technique that relies on the spontaneous osmosis method to transfer water from the feed solution to the condensed draw solution, the expense of retrieving the draw solute is often underestimated. Practical FO desalination implementation requires the application of a quick, low-cost separation method to obtain clean water from the diluted draw solute. By eliminating the need for regular substitution of the draw solute, constant reconcentration of the drawing solution will also keep operating costs down. MD integration was proposed in 2007 by Cath et al. [79]. Since then, research into hybrid FO–MD has complemented the parallel development of FO and MD with novel membranes, new extraction solutes for FO, and an effort to minimize energy usage.

10.4.5 Forward Osmosis–Electrodialysis

Electrodialysis is not widely used in recovering drawn solutes from FO desalination. A hybrid FO–ED method for wastewater treatment was demonstrated by Zou et al., with an energy consumption of 0.72 kWh/m^3 [80]. However, a more concentrated drawing solution is required to resolve the high osmotic pressure of seawater by extending it to seawater desalination. The retrieval of such a solution contributes to the elevated cost of electricity.

Bitaw et al. [81] compared SECs for selected hybrid FO systems and offered a novel hybrid FO–ED–RO method using FO to increase the electrical conductivity of the ED draw solution, thereby reducing the energy demand of the ED–RO system while maintaining the advantages of ED. The ED retrieves the draw solute, and its outlet is fed into the RO as feed. The relative energy consumed by ED and RO was dependent on a higher concentration of the intermediate concentration at the ED outlet/RO inlet, increasing RO's contribution to energy consumption.

10.4.6 Freeze Desalination–Membrane Distillation Hybrid System

The benefit of MD is that the efficiency of separation does not depend on the salinity of the feed, and even with hypersaline feed, close to 100% separation efficiency is feasible. In the treatment of concentrated feed by hybridization with MD, the limitations of other processes, such as freeze desalination (FD) and RO, can be overcome [82]. Such hybridized schemes help to realize zero liquid discharge [83, 84]. In addition, excess cold power can be used both for FD freezing and for MD distillate cooling. These features have led to hybrid FD–MD systems being built, wherein MD produces freshwater from the FD residual concentrate.

Wang and Chung [85] demonstrated the FD–MD hybrid concept for saltwater desalination. Seawater was first processed via FD, producing filtered water from the melted ice. The FD brine by-product was then processed by condensing water vapor on the distillate side of an MD membrane, producing high-quality water and a decreased volume of more concentrated brine. A liquefied natural gas (LNG) vaporizer was used to provide cold energy to both the FD crystallizer and MD distillate. This system reports a combined freshwater recovery of 71.5% and reduced energy consumption.

Chang et al. tested the viability of the FD–MD hybrid system, obtaining a total water recovery of 74% using vacuum membrane distillation (VMD) for seawater desalination [86]. LNG regasification provided cold energy to FD while heating the VMD feed used both regasification process waste energy and solar panels. The incorporation of heat between feed, clean ice, and vaporized steam reduced the cooling energy requirement by 27%.

Lu et al. [87] present a mathematical model for a hybrid zero liquid discharge desalination system using membrane distillation crystallization (MDC) and FD. The MDC scheme itself is an integrated process in which MD and crystallization are combined to recover both filtered water and essential resources in crystal form [88]. The effects of different parameters on energy consumption, such as the temperature of the feed MD, feed TDS, temperature of the distillate, and FD recovery ratio for the hybrid FD–MDC method, were measured. MD feed temperature of 80°C, distillate temperature of 5°C, feed TDS of 26 500 ppm were the optimum parameters for reducing SEC and achieving a performance of 30% regeneration. However, the hybrid FD–MDC device remains energy intensive at 810 kWh/m^3 hot energy and 830 kWh/m^3 cold energy with heat integration.

10.4.7 Gas Hydrates–Reverse Osmosis Hybrid System

The gas hydrate (GH) desalination method involves a liquid (saline water) to solid (GH) phase transition to exclude salts, followed by physical separation of the GH from the residual brine. However, some salt may remain occluded on the surface of the GH. Studies report salt rejection by the GH process ranging from 60.5 to 93% depending on GH process details, such as hydraulic pressures and hydrate formers [89, 90]. Therefore, another desalination process should be implemented post-treatment to produce fresh water. Purity can be improved by washing the GH slurries with water, compressing the GH slurries [91]. According to the literature [92, 93], the GH process energy consumption ranges from 1.58 to 47.9 kWh/m^3.

The energy consumption of the RO process is predicted to be reduced by lowering the salinity of the input water. Hence, from the perspective of the RO process, the GH process can be viewed as a pretreatment alternative to reduce the salinity of saltwater [95], thereby increasing energy efficiency and ensuring that RO design parameters and drinking water standards are met [94]. The concentrate of the RO process should be returned to the feed stream of the GH process to enhance the overall recovery of the GH and RO hybrid system.

Lee et al. [96] simulated the hybrid GH–RO process to discover an optimum RO recovery method that reduced RO energy usage. Optimum RO recovery values of 0.6, 0.8, and 0.8 were determined for GH salt rejections of 78%, 84%, and 90%, respectively. The simulation results showed that the maximum permissible energy consumption of the GH process to overcome the saltwater RO process with an energy recovery device is 1.4 kWh/m^3 with GH salt rejection of 78% or 1.9 kWh/m^3 with 90% salt rejection. The findings revealed that combining these processes could reduce energy usage.

10.4.8 Gas Hydrate–Capacitive Deionization Processes (CDI)

Capacitive deionization (CDI) is another form of desalination. This technology, which was first investigated in the 1960s, is based on electro-adsorption on porous carbon electrodes and has several advantages, including low energy consumption, high water recovery, low chemical usage, and high environmental friendliness [97–99]. Cohen et al. [100] found that the energy consumption of the CDI process is dependent on the salinity and that it is much higher for saltwater desalination than for RO. As a result, this approach is more suited and cost-effective for low to medium salinity levels. Because each conventional desalination methods have its own set of benefits and drawbacks, several researchers have proposed hybrid procedures to improve desalination efficiency while lowering costs.

In the hybrid process of GH + CDI [101], the activated carbon electrodes are made up of activated carbon, a binder (polyvinyl alcohol (PVA)), and a cross-linking agent (glutaric acid) to avoid dissolution of PVA in an aqueous solution. Temperature and pressure are recorded by the data acquisition system during hydrate formation. The formed hydrates are then compressed to pellet form with minimal brine trapped inside the pores, and CDI is used post-treatment to improve salt removal efficiency. The dissociated hydrate is passed through CDI to enhance the water quality by applying a voltage through the electrodes. The hybrid desalination process is reported to remove around 82% of Na^+ and 100% of K^+, Ca^{2+}, and Mg^{2+}.

10.5 Future Prospects

From the cases seen, has certainly been progressive growth in desalination systems, even within those that emphasize the use of renewable energies. The rate of development for such technologies is surely intensified by the depletion of fossil fuel supplies and increasing questions regarding environmental and economic aspects.

By considering the effect of bio-organic molecules on the adsorption of membrane materials, the potential cost of desalination energy, material lifespan, and complexity of cleaning techniques can be lowered [88]. Further examinations of membrane materials should rely on the practical use of high-performance FO membranes in functional desalination processes to show their performance. The findings demonstrate that both the minimization of pressure-retarded osmosis (PRO) membrane fouling and the realization of long-term output reliability should be considered in future efforts. In addition, hybrid device stability analysis consisting of RO and PRO technologies driven by renewables will yield invaluable results [102].

In the oil, economic, and environmental fields, achieving sustainability will be the cornerstone of future growth. Part of this effort in reaching sustainability includes the use of state-of-the-art materials for membranes. Thermally powered desalination systems, another prospective implementation, can show efficiency increases by up to 25%. The membrane industry requires around 15 years to reach the requisite expertise in implementing more sophisticated blends, but the thermal sector will only require one or two years as material research is not a limiting factor [103]. Its primary goal is also much simpler: increase water recovery and reduce GHG emissions [104]. The solution to sustainability in the energy field can also be realized by eradicating heavy energy use in desalination systems [105]. In addition to solar powered FOs and dew evaporation, emerging trends focus on the use of hybrid processes powered by dual or multi-resource renewable energy systems. Upcoming solar desalination opportunities require moving to other types of desalination that are less explored.

Dew evaporation in solar powered desalination is a revolutionary strategy in which saturated steam is used as a carrier gas for the dew evaporation process to vaporize water as a distillate from the saline feed. Focusing on using this method in hybrid renewables will enhance the performance of desalination systems [106]. Another incentivized method for increasing the productivity of desalination processes is available in FD systems. The key benefit of indirect FD is that refrigerant-free water is manufactured, which outlines the potential use of product water [107]. A groundbreaking method for generating freshwater is the use of ocean thermocline resources for the desalination system. In this process, freshwater will be created by exploiting the low-temperature difference between surface hot water and deep-sea cold water [108, 109].

10.6 Conclusion

The demand for access to freshwater is now increasing as supplies of clean freshwater dwindle due to overconsumption, pollution, and climate change. The problem is especially acute in arid areas. Desalinating saltwater and brackish water to satisfy the needs of cities is one means of overcoming this problem, albeit an energy intensive solution. Currently, nearly all of the energy demand in desalination is provided by fossil fuels. Most of the desalination plants currently in operation are based in the Persian Gulf, where fossil fuels are sufficiently abundant and low-cost. However, the environmental implications of fossil fuel consumption amid increasing climate change impacts on freshwater supplies make the continued use of these energy systems unsustainable. Fossil fuel resources are not equally distributed, so areas outside of such fuel-rich regions must contend with cost volatilities on the global market and the difficulties of fuel transport and power transmission, in addition to the environmental concerns. Therefore, there is an urgent need in desalination to identify and invest in alternative energy systems.

Renewable and onsite energy sources that take the best advantage of the local resource availability and efficiency, whether solar, wind, geothermal, or other types, provide a solution. Currently, most desalination plants that involve renewable energy use primarily solar or wind power; however, the temporal variability of these resources can be extremely disruptive to desalination processes. Multiple, synchronistic generation sources, such as a solar and wind integrated system, are a proven alternative to single-source systems for reliable continuous power generation.

RO and MED processes are the leading technologies in desalination and are primed for energy and process efficiency innovations. Research has made strides in identifying innovative hybrid processes powered by renewable resources that increase overall process efficiency. Continued research on hybrid processes, however, is needed to improve productivity and minimize the investment costs associated with these systems.

References

1 Bennett, A. (2015). Developments in desalination and water reuse. *Filtration and Separation* 52: 28–33. doi: 10.1016/S0015-1882(15)30181-6.

2 Sauvet-Goichon, B. (2007). Ashkelon desalination plant–A successful challenge. *Desalination* 203: 75–81. doi: 10.1016/j.desal.2006.03.525.

3 Amin, I., Ali, M.E.A., Bayoumi, S., Oterkus, S., Shawky, H., and Oterkus, E. (2020). Conceptual design and numerical analysis of a novel floating desalination plant powered by marine renewable energy for Egypt. *Journal of Marine Science and Engineering* 8: 1–23. doi: 10.3390/jmse8020095.

4 Jiménez Cisneros, B.E., Oki, T., Arnell, N.W., Benito, G., Cogley, J.G., Döll, P., Jiang, T., and Mwakalila, S.S. (2014). Freshwater resources. In: Climate Change 2014: Impacts, Adaptation, and Vulnerability. Part A: Global and Sectoral Aspects. Contribution of Working Group II to the Fifth Assessment Report of the Intergovernmental Panel on Climate Change (eds. C.B. Field, V.R. Barros, D.J. Dokken, K.J. Mach, M.D. Mastrandrea, T.E. Bilir, M. Chatterjee, K.L. Ebi, Y.O. Estrada, R.C. Genova, B. Girma, E.S. Kissel, A.N. Levy, S. MacCracken, P.R. Mastrandrea, and L.L. White), 229–269. Cambridge, United Kingdom and New York, NY, USA: Cambridge University Press.

5 Caldera, U., Bogdanov, D., and Breyer, C. (2016). Local cost of seawater RO desalination based on solar PV and wind energy: A global estimate. *Desalination* 385: 207–216. doi: 10.1016/j.desal.2016.02.004.

6 Feria-Díaz, J.J., López-Méndez, M.C., Rodríguez-Miranda, J.P., Sandoval-Herazo, L.C., and Correa-Mahecha, F. (2021). Commercial thermal technologies for desalination of water from renewable energies: A state of the art review. *Processes* 9: 1–22. doi: 10.3390/pr9020262.

7 Chandrashekara, M. and Yadav, A. (2017). Water desalination system using solar heat: A review. *Renewable and Sustainable Energy Reviews* 67: 1308–1330. doi: 10.1016/j.rser.2016.08.058.

8 Esmaeilion, F. (2020). Hybrid Renewable Energy Systems for Desalination. 10: Springer International Publishing. doi: 10.1007/s13201-020-1168-5.

9 Lund, H., Østergaard, P.A., and Towards, S.I. (2011). 100% renewable energy systems. *Applied Energy* 88: 419–421. doi: 10.1016/j.apenergy.2010.10.013.

10 Koutroulis, A.G., Tsanis, I.K., Daliakopoulos, I.N., Jacob, D. (2013). Impact of climate change on water resources status: a case study for Crete Island, Greece. *Journal of Hydrology* 479: 146–158. doi: 10.1016/j.jhydrol.2012.11.055.

11 Manios, T. and Tsanis, I.K. (2006). Evaluating water resources availability and wastewater reuse importance in the water resources management of small Mediterranean municipal districts. *Resources, Conservation and Recycling* 47 (3): 245–259. doi: 10.1016/j.resconrec.2005.11.001.

12 Latorre, F.J.G., Báez, S.O.P., and Gotor, A.G. (2015). Energy performance of a reverse osmosis desalination plant operating with variable pressure and flow. *Desalination* 366: 146–153. doi: 10.1016/j.desal.2015.02.039.

13 Wheeler, D. and Ummel, K. (2008). Desert Power: The Economics of Solar Thermal Electricity for Europe, North Africa, and the Middle East. Working Paper Number 156. Center for Global Development.

14 Al-Karaghouli, A., Renne, D., and Kazmerski, L.L. (2009). Solar and wind opportunities for water desalination in the Arab regions. *Renewable and Sustainable Energy Reviews* 13: 2397–2407. doi: 10.1016/j.rser.2008.05.007.

15 Mokheimer, E.M.A., Sahin, A.Z., Al-Sharafi, A., and Ali, A.I. (2013). Modeling and optimization of hybrid wind-solar-powered reverse osmosis water desalination system in Saudi Arabia. *Energy Conversion and Management* 75: 86–97. doi: 10.1016/j.enconman.2013.06.002.

16 Zhou, Y. and Tol, R.S.J. (2005). Evaluating the costs of desalination and water transport. *Water Resources Research* 41: 1–10. doi: 10.1029/2004WR003749.

17 Mentis, D., Karalis, G., Zervos, A., Howells, M., Taliotis, C., Bazilian, M. et al. (2016). Desalination using renewable energy sources on the arid islands of South Aegean Sea. *Energy* 94: 262–272. doi: 10.1016/j.energy.2015.11.003.

18 Soshinskaya, M., Crijns-Graus, W.H.J., van der Meer, J., and Guerrero, J.M. (2014). Application of a microgrid with renewables for a water treatment plant. *Applied Energy* 134: 20–34. doi: 10.1016/j.apenergy.2014.07.097.

19 Liu, L.Q. and Wang, Z.X. (2009). The development and application practice of wind-solar energy hybrid generation systems in China. *The Mission of Renewable and Sustainable Energy Reviews* 13: 1504–1512. doi: 10.1016/j.rser.2008.09.021.

20 Ahmadi, A., Esmaeilion, F., Esmaeilion, A., Ehyaei, M.A., and Silveira, J.L. (2020). Benefits and limitations of waste-to-energy conversion in Iran. 1: 27–45. doi: 10.22044/RERA.2019.8666.1007.

21 Ayoub, G.M. and Malaeb, L. (2012). Developments in solar still desalination systems: A critical review. *Critical Reviews in Environmental Science and Technology* 42: 2078–2112. doi: 10.1080/10643389.2011.574104.

22 Hosseini, R., Soltani, M., and Valizadeh, G. (2005). Technical and economic assessment of the integrated solar combined cycle power plants in Iran. *Renew Energy* 30: 1541–1555. doi: 10.1016/j.renene.2004.11.005.

23 Kalogirou, S.A. (2004). Solar thermal collectors and applications. 30: doi: 10.1016/j.pecs.2004.02.001.

24 Alkaisi, A., Mossad, R., and Sharifian-Barforoush, A. (2017). A review of the water desalination systems integrated with renewable energy. *Energy Procedia* 110: 268–274. doi: 10.1016/j.egypro.2017.03.138.

25 Yilmaz, I.H. and Söylemez, M.S. (2012). Design and computer simulation on multi-effect evaporation seawater desalination system using hybrid renewable energy sources in Turkey. *Desalination* 291: 23–40. doi: 10.1016/j.desal.2012.01.022.

26 Kalogirou, S.A. (2005). Seawater desalination using renewable energy sources. *Progress in Energy and Combustion Science* 31: 242–281. doi: 10.1016/j.pecs.2005.03.001.

27 Goosen, M., Mahmoudi, H., Ghaffour, N., and Sablani, S. (2011). Application of renewable energies for water desalination. Desalination, Trends Technol. doi: 10.5772/14011.

28 Paulsen, K. and Hensel, F. (2005). Introduction of a new energy recovery system – Optimized for the combination with renewable energy. *Desalination* 184: 211–215. doi: 10.1016/j.desal.2005.03.060.

29 Ghaffour, N., Lattemann, S., Missimer, T., Ng, K.C., Sinha, S., and Amy, G. (2014). Renewable energy-driven innovative energy-efficient desalination technologies. *Applied Energy* 136: 1155–1165. doi: 10.1016/j.apenergy.2014.03.033.

30 Park, G.L., Schäfer, A.I., and Richards, B.S. (2009). Potential of wind-powered renewable energy membrane systems for Ghana. *Desalination* 248: 169–176. doi: 10.1016/j.desal.2008.05.053.

31 García-Rodríguez, L., Romero-Ternero, V., and Gómez-Camacho, C. (2001). Economic analysis of wind-powered desalination. *Desalination* 137: 259–265. doi: 10.1016/S0011-9164(01)00235-1.

32 Peñate, B., Castellano, F., Bello, A., and García-Rodríguez, L. (2011). Assessment of a stand-alone gradual capacity reverse osmosis desalination plant to adapt to wind power availability: a case study. *Energy* 36: 4372–4384. doi: 10.1016/j.energy.2011.04.005.

33 Segurado, R., Costa, M., Duić, N., and Carvalho, M.G. (2014). Integrated analysis of energy and water supply in islands. Case study of S. Vicente, Cape Verde. *Energy* 92: 639–648. doi: 10.1016/j.energy.2015.02.013.

34 Gökçek, M. and Gökçek, Ö.B. (2016). Technical and economic evaluation of freshwater production from a wind-powered small-scale seawater reverse osmosis system (WP-SWRO). *Desalination* 38: 47–57. doi: 10.1016/j.desal.2015.12.004.

35 Ismail, T.M., Azab, A.K., Elkady, M.A., and Abo Elnasr, M.M. (2016). Theoretical investigation of the performance of integrated seawater desalination plant utilizing renewable energy. *Energy Conversion and Management* 126: 811–825. doi: 10.1016/j.enconman.2016.08.002.

36 Smaoui, M., Abdelkafi, A., and Krichen, L. (2015). Optimal sizing of stand-alone photovoltaic/wind/hydrogen hybrid system supplying a desalination unit. *Solenergy Systems* 120: 263–276. doi: 10.1016/j.solener.2015.07.032.

37 Al-Nimr, M.A., Kiwan, S.M., and Talafha, S. (2016). Hybrid solar-wind water distillation system. *Desalination* 395: 33–40. doi: 10.1016/j.desal.2016.05.018.

38 Nagaraj, R., Thirugnanamurthy, D., Rajput, M.M., and Panigrahi, B.K. (2016). Techno-economic analysis of hybrid power system sizing applied to small desalination plants for sustainable operation. *International Journal of Sustainable Built Environment* 5: 269–276. doi: 10.1016/j.ijsbe.2016.05.011.

39 Khan, S.U.D., Khan, S.U.D., Haider, S., El-Leathy, A., Rana, U.A., Danish, S.N. et al. (2017). Development and techno-economic analysis of small modular nuclear reactor and desalination system across Middle East and North Africa region. *Desalination* 406: 51–59. doi: 10.1016/j.desal.2016.05.008.

40 Belkaid, A., Amzert, S.A., Bouaichaoui, Y., and Chibane, H. (2012). Economic study of nuclear seawater desalination for Mostaganem site. *Procedia Engineering* 33: 134–145. doi: 10.1016/j.proeng.2012.01.1186.

41 Dittmar, M. (2012). Nuclear energy: Status and future limitations. *Energy* 37: 35–40. doi: 10.1016/j.energy.2011.05.040.

42 Mansouri, N.Y. and Ghoniem, A.F. (2017). Does nuclear desalination make sense for Saudi Arabia? *Desalination* 406: 37–43. doi: 10.1016/j.desal.2016.07.009.

43 Belessiotis, V., Papanicolaou, E., and Delyannis, E. (2010). Nuclear desalination: A review on past and present. *Desalin Water Treat* 20: 45–50. doi: 10.5004/dwt.2010.1942.

44 Kim, H.S. and NO, H.C. (2012). Thermal coupling of HTGRs and MED desalination plants, and its performance and cost analysis for nuclear desalination. *Desalination* 303: 17–22. doi: 10.1016/j.desal.2012.07.004.

45 Misra, B.M. and Kupitz, J. (2004). The role of nuclear desalination in meeting the potable water needs in water scarce areas in the next decades. *Desalination* 166: 1–9. doi: 10.1016/j.desal.2004.06.053.

46 Kavvadias, K.C. and Khamis, I. (2010). The IAEA DEEP desalination economic model: A critical review. *Desalination* 257: 150–157. doi: 10.1016/j.desal.2010.02.032.

47 Al-Othman, A., Darwish, N.N., Qasim, M., Tawalbeh, M., Darwish, N.A., and Hilal, N. (2019). Nuclear desalination: A state-of-the-art review. *Desalination* 457: 39–61. doi: 10.1016/j.desal.2019.01.002.

48 Schorr, M. (ed.) (2011). Desalination Trends and Techniques. Rijeka: InTech. doi: 10.1080/19443994.2012.683289.

49 Gude, V.G., Nirmalakhandan, N., and Deng, S. (2010). Renewable and sustainable approaches for desalination. *Renewable and Sustainable Energy Reviews* 14: 2641–2654. doi: 10.1016/j.rser.2010.06.008.

50 Goosen, M., Mahmoudi, H., and Ghaffour, N. (2010). Water desalination using geothermal energy. *Energies* 3: 1423–1442. doi: 10.3390/en3081423.

51 Calise, F., Cipollina, A., d'Accadia, M.D., and Piacentino, A. (2014). A novel renewable polygeneration system for a small Mediterranean volcanic island for the combined production of energy and water: Dynamic simulation and economic assessment. *Applied Energy* 135: 675–693. doi: 10.1016/j.apenergy.2014.03.064.

52 Missimer, T.M., Kim, Y.D., Rachman, R., and Ng, K.C. (2013). Sustainable renewable energy seawater desalination using combined-cycle solar and geothermal heat sources. *Desalin Water Treat* 51: 1161–1170. doi: 10.1080/19443994.2012.704685.

53 Sharmila, N., Jalihal, P., Swamy, A.K., and Ravindran, M. (2004). Wave powered desalination system. *Energy* 29: 1659–1672. doi: 10.1016/j.energy.2004.03.099.

54 Davies, P. A. (2005). Wave-powered desalination: resource assessment and review of technology. *Desalination* 186 (1–3): 97–109.

55 Foteinis, S. and Tsoutsos, T. (2017). Strategies to improve sustainability and offset the initial high capital expenditure of wave energy converters (WECs). *Renewable and Sustainable Energy Reviews* 70: 775–785. doi: 10.1016/j.rser.2016.11.258.

56 Viola, A., Franzitta, V., Trapanese, M., and Curto, D. (2016). Nexus water & energy: A case study of wave energy converters (WECs) to desalination applications in Sicily. *International Journal of Heat and Technology* 34: 379–386. doi: 10.18280/ijht.34S227.

57 Franzitta, V., Curto, D., Milone, D., and Viola, A. (2016). The desalination process driven by wave energy: A challenge for the future. *Energies* 9: 1–16. doi: 10.3390/en9121032.

58 Goosen, M.F.A., Mahmoudi, H., and Ghaffour, N. (2014). Today's and future challenges in applications of renewable energy technologies for desalination. *Critical Reviews in Environmental Science and Technology* 44: 929–999. doi: 10.1080/10643389.2012.741313.
59 Dashtpour, R. and Al-zubaidy, S.N. (2012). Fig. 2. Comparisons between the submerged reverse osmosis system current schemes. 3 (4): 339–345.
60 Tan, C.H. and Ng, H.Y. (2010). A novel hybrid forward osmosis – Nanofiltration (FO-NF) process for seawater desalination: Draw solution selection and system configuration. *Desalin Water Treat* 13: 356–361. doi: 10.5004/dwt.2010.1733.
61 Phuntsho, S., Shon, H.K., Hong, S., Lee, S., and Vigneswaran, S. (2011). A novel low energy fertilizer driven forward osmosis desalination for direct fertigation: Evaluating the performance of fertilizer draw solutions. *Journal of Membrane Science* 375: 172–181. doi: 10.1016/j.memsci.2011.03.038.
62 Ahmed, F.E., Hashaikeh, R., and Hilal, N. (2020). Hybrid technologies: The future of energy efficient desalination – A review. *Desalination* 495: 114659. doi: 10.1016/j.desal.2020.114659.
63 Kim, J.E., Phuntsho, S., Chekli, L., Hong, S., Ghaffour, N., Leiknes, T.O. et al. (2017). Environmental and economic impacts of fertilizer drawn forward osmosis and nanofiltration hybrid system. *Desalination* 416: 76–85. doi: 10.1016/j.desal.2017.05.001.
64 Kim, J.E., Phuntsho, S., and Shon, H.K. (2013). Pilot-scale nanofiltration system as post-treatment for fertilizer-drawn forward osmosis desalination for direct fertigation. *Desalin Water Treat.* 51: 6265–6273. doi: 10.1080/19443994.2013.780804.
65 Corzo, B., de la Torre, T., Sans, C., Escorihuela, R., Navea, S., and Malfeito, J.J. (2018). Long-term evaluation of a forward osmosis-nanofiltration demonstration plant for wastewater reuse in agriculture. *The Chemical Engineering Journal* 338: 383–391. doi: 10.1016/j.cej.2018.01.042.
66 Giagnorio, M., Ricceri, F., Tagliabue, M., Zaninetta, L., and Tiraferri, A. (2019). Hybrid forward osmosis-nanofiltration for wastewater reuse: System design. *Membranes (Basel)* 9: 8–12. doi: 10.3390/membranes9050061.
67 Turek, M., Mitko, K., Piotrowski, K., Dydo, P., Laskowska, E., and Jakóbik-Kolon, A. (2017). Prospects for high water recovery membrane desalination. *Desalination* 401: 180–189. doi: 10.1016/j.desal.2016.07.047.
68 Hoek, E.M.V., Allred, J., Knoell, T., and Jeong, B.H. (2008). Modeling the effects of fouling on full-scale reverse osmosis processes. *Journal of Membrane Science* 314: 33–49. doi: 10.1016/j.memsci.2008.01.025.

69 Stover, R.L. (2013). Industrial and brackish water treatment with closed circuit reverse osmosis. *Desalin Water Treat* 51: 1124–1130. doi: 10.1080/19443994.2012.699341.

70 Thampy, S., Desale, G.R., Shahi, V.K., Makwana, B.S., and Ghosh, P.K. (2011). Development of hybrid electrodialysis-reverse osmosis domestic desalination unit for high recovery of product water. *Desalination* 282: 104–108. doi: 10.1016/j.desal.2011.08.060.

71 Doornbusch, G.J., Tedesco, M., Post, J.W., Borneman, Z., and Nijmeijer, K. (2019). Experimental investigation of multistage electrodialysis for seawater desalination. *Desalination.* 464: 105–114. doi: 10.1016/j.desal. 2019.04.025.

72 McGovern, R.K., Zubair, S.M., and Lienhard, V.J.H. (2014). The benefits of hybridising electrodialysis with reverse osmosis. *Journal of Membrane Science* 469: 326–335. doi: 10.1016/j.memsci.2014.06.040.

73 Jiang, C., Wang, Y., Zhang, Z., and Xu, T. (2014). Electrodialysis of concentrated brine from RO plant to produce coarse salt and freshwater. *Journal of Membrane Science* 450: 323–330. doi: 10.1016/j. memsci.2013.09.020.

74 Zhang, Y., Ghyselbrecht, K., Meesschaert, B., and Pinoy, L., Van der Bruggen, B. (2011). Electrodialysis on RO concentrate to improve water recovery in wastewater reclamation. *Journal of Membrane Science* 378: 101–110. doi: 10.1016/j.memsci.2010.10.036.

75 Pellegrino, J., Gorman, C., and Richards, L. (2007). A speculative hybrid reverse osmosis/electrodialysis unit operation. *Desalination* 214: 11–30. doi: 10.1016/j.desal.2006.09.024.

76 Wang, P. and Chung, T.S. (2015). Recent advances in membrane distillation processes: Membrane development, configuration design and application exploring. *Journal of Membrane Science* 474: 39–56. doi: 10.1016/j.memsci.2014.09.016.

77 Choi, Y.J., Lee, S., Koo, J., and Kim, S.H. (2016). Evaluation of economic feasibility of reverse osmosis and membrane distillation hybrid system for desalination. *Desalin Water Treat.* 57: 24662–24673. doi: 10.1080/19443994. 2016.1152648.

78 Naidu, G., Jeong, S., Choi, Y., and Vigneswaran, S. (2017). Membrane distillation for wastewater reverse osmosis concentrate treatment with water reuse potential. *Journal of Membrane Science* 524: 565–575. doi: 10.1016/j.memsci.2016.11.068.

79 WO2007147013A1.pdf (n.d.).

80 Zou, S. and He, Z. (2017). Electrodialysis recovery of reverse-fluxed fertilizer draw solute during forward osmosis water treatment. *The Chemical Engineering Journal* 330: 550–558. doi: 10.1016/j.cej.2017.07.181.

81 Bitaw, T.N., Park, K., and Yang, D.R. (2016). Optimization on a new hybrid Forward osmosis-Electrodialysis-Reverse osmosis seawater desalination process. *Desalination* 398: 265–281.

82 Martínez-Díez, L. and Florido-Díaz, F.J. (2001). Desalination of brines by membrane distillation. *Desalination* 137: 267–273. doi: 10.1016/S0011-9164(01)00228-4.

83 González, D., Amigo, J., and Suárez, F. (2017). Membrane distillation: Perspectives for sustainable and improved desalination. *Renewable and Sustainable Energy Reviews* 80: 238–259. doi: 10.1016/j.rser.2017.05.078.

84 Nakoa, K., Rahaoui, K., Date, A., and Akbarzadeh, A. (2016). Sustainable Zero Liquid Discharge Desalination (SZLDD). *Solenergy Systems* 135: 337–347. doi: 10.1016/j.solener.2016.05.047.

85 Wang, P. and Chung, T.S. (2012). A conceptual demonstration of freeze desalination–membrane distillation (FD–MD) hybrid desalination process utilizing liquefied natural gas (LNG) cold energy. *Water research* 46 (13): 4037–4052.

86 Chang, J., Zuo, J., Lu, K.J., and Chung, T.S. (2019). Membrane development and energy analysis of freeze desalination-vacuum membrane distillation hybrid systems powered by LNG regasification and solar energy. *Desalination* 449: 16–25. doi: 10.1016/j.desal.2018.10.008.

87 Choi, Y., Naidu, G., Nghiem, L.D., Lee, S., and Vigneswaran, S. (2019). Membrane distillation crystallization for brine mining and zero liquid discharge: Opportunities, challenges, and recent progress. *Environmental Science: Water Research & Technology* 5: 1202–1221. doi: 10.1039/c9ew00157c.

88 Lu, K.J., Cheng, Z.L., Chang, J., Luo, L., and Chung, T.S. (2019). Design of zero liquid discharge desalination (ZLDD) systems consisting of freeze desalination, membrane distillation, and crystallization powered by green energies. *Desalination* 458: 66–75. doi: 10.1016/j.desal.2019.02.001.

89 Han, S., Shin, J.Y., Rhee, Y.W., and Kang, S.P. (2014). Enhanced efficiency of salt removal from brine for cyclopentane hydrates by washing, centrifuging, and sweating. *Desalination* 354: 17–22.

90 Park, K.N., Hong, S.Y., Lee, J.W., Kang, K.C., Lee, Y.C., Ha, M.G., and Lee, J.D. (2011). A new apparatus for seawater desalination by gas hydrate process and removal characteristics of dissolved minerals (Na+, Mg2+, Ca2+, K+, B3+). *Desalination* 274: 91–96.

91 Kang, K.C., Linga, P., Park, K., Choi, S.J., and Lee, J.D. (2014). Seawater desalination by gas hydrate process and removal characteristics of dissolved ions (Na+, K+, Mg2+, Ca2+, B3+, Cl−, SO4 2−). *Desalination* 353: 84–90.

92 Youssef, P.G., AL-Dadah, R.K., and Mahmoud, S.M. (2014). Comparative analysis of desalination technologies. *Energy Procedia* 61: 2604–2607.

93 Javanmardi, J. and Moshfeghian, M. (2003). Energy consumption and economic evaluation of water desalination by hydrate phenomenon. *Applied Thermal Engineering* 23: 845–857.

94 Vu, H.H. and Cho, B.Y. (2011). A study on boron removal by mineral cluster coagulant for seawater desalination application. *Environmental Engineering Research* 16 (4): 227–230.

95 Valavala, R., Sohn, J., Han, J., Her, N., and Yoon, Y. (2011). Pretreatment in reverse osmosis seawater desalination: A short review. *Environmental Engineering Research* 16 (4): 205–212.

96 Lee, H., Ryu, H., Lim, J.H., Kim, J.O., Dong Lee, J., and Kim, S. (2016). An optimal design approach of gas hydrate and reverse osmosis hybrid system for seawater desalination. *Desalination and Water Treatment* 57 (19): 9009–9017.

97 Anderson, M.A., Cudero, A.L., and Palma, J. (2010). Capacitive deionization as an electrochemical means of saving energy and delivering clean water. Comparison to present desalination practices: Will it compete? *Electrochimica Acta* 55 (12): 3845–3856.

98 Porada, S., Zhao, R., Van Der Wal, A., Presser, V., and Biesheuvel, P.M. (2013). Review on the science and technology of water desalination by capacitive deionization. *Progress in Materials Science* 58 (8): 1388–1442.

99 Yu, T.H., Shiu, H.Y., Lee, M., Chiueh, P.T., and Hou, C.H. (2016). Life cycle assessment of environmental impacts and energy demand for capacitive deionization technology. *Desalination* 399: 53–60.

100 Cohen, I., Avraham, E., Soffer, A., and Aurbach, D. (2013). Water desalination by capacitive deionization-advantages limitations and modification. *ECS Transactions* 45 (17): 43.

101 Falahieh, M.M., Bonyadi, M., and Lashanizadegan, A. (2021). A new hybrid desalination method based on the CO2 gas hydrate and capacitive deionization processes. *Desalination* 502: 114932.

102 Ridgway, H.F., Orbell, J., and Gray, S. (2017). Molecular simulations of polyamide membrane materials used in desalination and water reuse applications: Recent developments and future prospects. *Journal of Membrane Science* 524: 436–448. doi: 10.1016/j.memsci.2016.11.061.

103 Wang, Z., Wang, Z., Lin, S., Jin, H., Gao, S., Zhu, Y. et al. (2018). Nanoparticle-templated nanofiltration membranes for ultrahigh performance desalination. *Nature Communications* 9 (1): 1–9. doi: 10.1038/s41467-018-04467-3.

104 Shahzad, M.W., Burhan, M., Ang, L., and Ng, K.C. (2017). Energy-water-environment nexus underpinning future desalination sustainability. *Desalination* 413: 52–64. doi: 10.1016/j.desal.2017.03.009.

105 Fane, A.G.T. (2018). A grand challenge for membrane desalination: More water, less carbon. *Desalination* 426: 155–163. doi: 10.1016/j.desal.2017.11.002.

106 Lee, S., Choi, J., Park, Y.G., Shon, H., Ahn, C.H., and Kim, S.H. (2019). Hybrid desalination processes for beneficial use of reverse osmosis brine: Current status and future prospects. *Desalination* 454: 104–111. doi: 10.1016/j.desal.2018.02.002.

107 Ahmed, F.E., Hashaikeh, R., and Hilal, N. (2019). Solar powered desalination – Technology, energy and future outlook. *Desalination* 453: 54–76. doi: 10.1016/j.desal.2018.12.002.

108 Najim A. Discussion on"Kalista, B., Shin, H., Cho, J., and Jang, A. (2018). Current development and future prospect review of freeze desalination. *Desalination* 447: 167–181.

109 Ng, K.C. and Shahzad, M.W. (2018). Sustainable desalination using ocean thermocline energy. *Renewable & Sustainable Energy Reviews* 82: 240–246. doi: 10.1016/j.rser.2017.08.087.

11

Waste Brine Management

Sirisha Nallakukkala and Bhajan Lal

11.1 Introduction

Though desalination offers potable water, the residual saline byproduct from desalination is a grave environmental concern [1] owing to the pretreatment chemicals existing in the residual solution, leading to higher salinity levels along with an increase in temperature. The heavy metal ions prevailing in the rejected brine solution, such as Cu, Ni, Fe, Cr, Zn, etc., pose an additional issue due to their corrosiveness and retention within the environment. The characteristics of the rejected brine solution usually rely on the quality of the inlet water, recovered water, systems implemented for pretreatment, and the treatment technique itself [2]. The damages to the environment caused by such waste include eutrophication, pH instabilities, and heavy metal ions, which can severely degrade the marine environment. Adopting an ecologically amicable brine management framework requires consideration of numerous technical aspects, including the nature of the waste brine rejected, composition of the concentrate, topography, accessibility to and feasibility of disposal, capital and operational expenses, waste storage, and waste transport [3, 4]. These considerations are reflected in improvements in brine waste treatment technology.

Modern plant designs account for discharge protocols and strategic maintenance of the economic balance related to waste mitigation tactics. The brine waste typically requires some sort of treatment before disposal; however, the treatment required will depend upon stream contaminants, mode of discarding, and associated environmental release protocols. The waste brine is either recycled or treated for disposal. For example, since mixtures containing high salt content decrease thermal conductivity, waste brine can be recycled and used as a cooling agent for steel heat exchangers in power plants. These brines are treated to eliminate dissolved oxygen and

Gas Hydrate in Water Treatment: Technological, Economic, and Industrial Aspects,
First Edition. Bhajan Lal and Sirisha Nallakukkala.

other corrosive impurities that would otherwise damage the power plant's machinery and piping. Depending on salt concentrations in the brine stream, the temperature, pressure, and other threshold limits need to be modified during production to confirm the appropriate purification steps. This step is often the most laborious due to the high level of maintenance required.

11.2 Waste Brine Watercourse Constituents

Waste brine characteristics vary notably based on the constituents present, such as suspended solids, antiscaling agents, heavy metal ions, oil and grease, microorganisms, and organic materials. All of these carry some form of risk to human health and the ecosystem. When discharged to the environment as untreated, the most obvious danger to marine life is the higher sodium content. In such situations, diluting the brine before discharging it is the best solution. Most waste brine discharge protocols entail reductions in sodium, total suspended solids (TSS), and pollutants. The techniques employed to achieve such reductions depend on the characteristics of the receiving watercourse and disposal method.

11.3 Waste Brine Discharge Methods

11.3.1 Sewer Discharge

Releasing waste brine to local sewers is usually a low-cost discard option, depending on accessibility and site location. However, the waste brine requires treatment to satisfy the discharge protocols of the sewer. Diluting the waste brine may be sufficient, depending on the volume and the size of the sewer.

11.3.2 Surface Water Discharge

Release of waste brine to surface waters is a reasonably economical solution. However, transport of the brine must obey regulatory protocols and, depending on the characteristics of the brine, it may need to undergo either treatment to eliminate impurities or dilution to alter its salinity. Surface water discharges may be allowed with slight or no treatment if accepted by the governing agencies.

11.3.3 Evaporation

Waste brine treatment is necessary if it is present in large volumes and contains toxic contaminants. In this method, the waste brine accumulates in evaporation ponds and is heated by renewable solar energy, leaving behind salts. Evaporation incurs minimal operational, maintenance, and energy costs; however, the land area needed to house the evaporative ponds can mean high capital investments. Novel technologies, like wind-aided evaporation, can reduce the footprint and speed evaporation.

The water collected during evaporation is reusable after neutralization with acid and aeration. The acid treatment reduces scaling in downstream equipment, and aeration releases non-condensable gases (e.g. oxygen, and carbon dioxide) to protect equipment from corrosion. Later, the evaporator waste is channeled to a crystallizer until all the impurities in the water crystallize. If the quantity of waste brine is low, then the evaporation process can be neglected, and the waste brine can be sent directly for crystallization. This method is more commonly used in services committed to zero liquid discharge (ZLD) or subject to strict discharge regulations.

11.3.4 Recycling and Reuse

Waste brine is often recycled for supplementary uses like industrial cooling and acid and caustic production. Some companies have collaborated with local municipalities and external agencies to recycle waste brine for use in road de-icing and irrigation of salt-tolerant land and vegetation. Recycled brine waste always needs to be treated prior to reuse to remove or reduce organic impurities.

11.3.5 Deep Well Injection

Deep well injection pumps waste brine into porous rock deep underground and is generally only used when a large volume of waste brine is to be discarded, like those produced during fracturing operations. This technique requires brine treatment to remove various pollutants that may damage the well, including TSS, oils, and greases. Deep well injection is generally reserved for high-volume, high-salinity wastes due to exorbitant costs associated with the permitting, drilling, and monitoring of the well.

When injection is cost-prohibitive, it might be better to consider ZLD systems, which eliminate the liquid to minimize the waste volume. ZLD systems consist of a brine concentrator and crystallizer, which also incurs relatively large capital costs, though the trade-offs are in minimizing liquid waste disposal costs and recuperating value-added materials.

11.3.6 Environmental Impact and Regulatory Compliance

The spent brine poses a risk to the environment due to the high concentration of salts, which can impair soils, flora, and marine life when inappropriately discharged to surface waterways. Because of this, waste brine releases are stringently regulated. As with most environmental protection systems, it is always best to inspect and monitor the discharged wastes to ensure that the available waste brine treatment solution effectively controls the discharge to meet environmental quality protocols.

11.4 Waste Brine Management Methods

The rejected waste brine might be valuable for the production of inorganic salts and metals. Production of significant value-added products like acids, bases, and reagents from the byproduct of desalination is a promising source for supplementing the brine management framework costs. Retrieval of salt from rejected waste brine solution is being extensively studied for re-utilization [5–8]. At present, the most common means for segregating salts from the brine are crystallization, followed by evaporative cooling. Membrane-based techniques, due to their close association with reverse osmosis, are also prominent. Furthermore, electrodialysis, chemical treatment, eutectic freezing, profit-oriented accessible procedures like SAL-PROC, and selective salt recovery are established methods of salt retrieval from the concentrated waste saline solution.

SAL-PROC is a patented, integrated process for the sequential extraction of dissolved elements from saline effluents in the form of valuable salts and chemical compounds. This method consists of several controlled chemical reactions aided by non-hazardous chemicals and evaporation/cooling steps for causing precipitation and crystallization.

Industries that demand brine, such as hydrometallurgy and sodium hypochlorite (NaOCl), lithium carbonate, and chlor-alkali manufacturing industries, alongside various applications for agriculture/irrigation or de-icing, show a clear motivation for creating value from the leftover brine. Generally, the waste brine produced from desalination techniques contains contaminants like heavy metal ions, organic compounds, sulfates (SO_4^{2-}), nitrates (NO_3^-), phosphates (PO_4^{3-}), and suspended solids. The waste brine must undergo pH adjustment and treatment with coagulants, chelating resins, polymers, and other additives to separate out the metal ions and other suspended solids that could foul the membranes. Some of the various waste brine treatments and disposal possibilities accessible are:

- Chemical precipitation
- Adsorption
- Bipolar membrane electrodialysis (BMED)
- Crystallization and evaporation
- Acid and caustic production
- Regeneration of wetlands and agricultural applications
- Softener brine recycling
- De-icing and dust control
- Recovery and trade of solid salts.

11.4.1 Chemical Precipitation

Chemical precipitation is feasible for salt retrieval from the rejected waste saline brine. This method uses precipitants, e.g. sodium phosphates and sodium carbonate (Na_2CO_3), to retrieve salts from saline water and reverse osmosis (RO) reject brine [9]. Magnesium and calcium salts are retrieved by chemical precipitation, and although small concentrations of Ca^{2+} Mg^{2+} ions are also present in the byproduct's stream, they are considered not reusable as they still contain sodium chloride (NaCl). However, these byproducts of Ca and Mg can be used to control pH and retrieve phosphate for use in fertilizers.

11.4.2 Adsorption

Waste brine is a value-added resource for metals, specifically those that are seizable by the adsorption method. The factors influencing the metal retrieval process are the physical and chemical features of the metal, the use of adsorbents, and the economic and ecological benefits of heavy metal retrieval [10, 11]. Nevertheless, the technologies for metal recovery through this method could still use further enhancements in the extraction process and exploration into scaling up the retrieval process [12]. The extraction of metals like cesium, lithium, rubidium, and uranium using commercial selective sorbents (e.g. hexacyanoferrate-based extracting medium CsTreat, zirconium phosphate, and S910) from reject brine produced by desalination technique are some examples that have been appraised in the literature [13]. CsTreat can obtain a high sorption capacity for rubidium, alongside reasonable sorption of cesium. S910 has exhibited good sorption for uranium, though the sorption of lithium by zirconium phosphate has not been proven effective. Another study reported the extraction of > 80% rubidium using 4-tert-butyl-2-(α-methylbenzyl) phenol (t-BAMBP) and 4-sec-butyl-2-(α-methylbenzyl) phenol (BAMBP) from residual waste saline brine. The extraction of phosphorus for limiting the effects of eutrophication was also tested using the same extractant, but it was not economically fruitful.

11.4.3 Bipolar Membrane Electrodialysis

BMED has been used for recycling RO reject brine into usable hydrochloric acid (HCl) and sodium hydroxide (NaOH) base products [14]. The production of NaOCl from the waste brine stream by electrochlorination has also been investigated, where the waste brine is subjected to a DC source nominal voltage supply [7]. NaOCl can be reused in the main desalination process as an antifoulant and disinfectant, reducing material costs. As an added advantage, NaOCl production also releases chlorine gas (Cl_2), thereby reducing or excluding harmful Cl_2 in the logistic cycle. These benefits make the process eco-friendly. Hence, BMED is an attractive solution for managing waste brine in RO frameworks and enhancing the desalination process economically and environmentally.

11.4.4 Crystallization and Evaporation

Evaporation is one of the best options for treating waste brine streams, especially when evaporation ponds are convenient for the location. Through vacuum evaporation, the boiling point of the effluent stream is minimized (thus saving energy) with a crystallized mass of salt and purified watercourse. This technique is useful for drying brine filtrates and can supplement other treatments to enhance overall efficiency.

One of the methods to retrieve the salt from waste saline brine is SSR (solid-state recrystallization), also known as fractional crystallization, as it entails the precipitation of dissolved salts. Fractional crystallization involves a series of methods to recover salts based on their differences in solubility, ability to be coprecipitated, or adsorbed, finally resulting in commercial grade products [15–18]. The main factor in SSR efficiency is the composition of the targeted salt, though pressure and pH are also important factors. Studying the metastability of the salts can be a significant aspect of the kinetics to investigate. Salt precipitation is governed by pH, temperature, residence time, antiscalants, mixing efficiency, and crystal shape, size, and purity. Salts like gypsum ($CaSO_4.2H_2O$), calcium carbonate ($CaCO_3$), magnesium hydroxide ($Mg(OH)_2$), sodium sulfate (Na_2SO_4), and chlorides of Na and Ca can be recycled from waste brine through SSR using the SAL-PROC process.

Solar ponds have also been used to evaporate the waste brine to produce potassium chloride (KCl), potassium sulfate (K_2SO_4), Na_2SO_4, lithium sulfate

(Li_2SO_4), and boric acid (H_3BO_3) [19]. Recovery of Ca, Mg, and Si from waste brine by pretreatment using soda lime was also investigated using isothermal evaporation [20].

11.4.5 Acid and Caustic Production

In some cases, NaOH and Cl_2 can be captured through the electrolysis of large quantities of NaCl within the brine. Multiple electrolysis cycles are carried out to enhance the overall efficiency. Few industries, however, can actually make use of the recovered waste brine stream, much less manufacture chlorine and alkali compounds for retail, through this means.

11.4.6 Regeneration of Wetlands and Agricultural Applications

Rejected waste brine from desalination processes could be used to renew wetlands and brackish environments that suffer deficiencies owing to drought. This option's economic and ecological effects rely on the distinctive features of the waste brine and the receiving wetland. As a long-term benefit, the value and accessibility of water resources can be improved, leading to more opportunities for agricultural activities, stockbreeding, and industrial growth [6, 18, 21, 22].

Rejected waste brine can also be utilized for watering vegetations, wildlife, or microbes that are tolerant of high salt concentrations. Two of the most often considered options are the breeding of *Artemia sp.* (brine water shrimp) that appears in high saline water systems (70–340 g/L) and the microalgae *Dunaliella salina* [21, 23]. *Artemia* is used as food for larvae and fish, and *D. salina* can store large amounts of carotenoids for use in medical and food products [24]. Plants that are permissive of high salinities (8000 ppm), such as beetroot, sugar cane, cotton, dates, barley, spinach, asparagus, wheat, and certain grasses such as paspalum, might also be watered with waste brine [21]. Further investigations are required to examine the capability of the biological cells to grow in highly saline brine environments with and without additional pretreatment, in tandem with its economic viability.

In some situations, the waste brine stream can be utilized for irrigation or agricultural activities. When implemented for agriculture activities, it should be free from contaminants and have low total dissolved solids (TDS). In certain regions, reuse of waste brine for salt-tolerant plants can be a possibility. When the brine waste stream contains a mixture of sodium and hard minerals, it is used to produce liquid fertilizer by treatment via ion exchange to eliminate sodium, thereby retaining the hardness that is valuable for plant development.

11.4.7 Softener Brine Recycling

Waste brines resulting from the desalination techniques can be treated (if needed) and reused in water softening treatments, consequently lowering the water and salt utilization. The brine recycling process is one way to enhance the efficiency of the softening treatments, but it is limited to softeners that utilize a high salt dose and have higher NaCl concentrations following an active cycle. Additionally, use in water softening may result in more concentrated waste, the disposal of which can be more problematic and expensive.

11.4.8 De-icing and Dust Control

In an effort to lower disposal expenses, waste brine can be sent to municipalities and cities for maintaining roads covered by snow and dirt. However, this is only available in certain regions. Not all roads are suitable for waste brine due to the potential presence of naturally occurring radioactive materials (NORM) and chemicals produced during fracturing, which may harm the local environment or the water supply.

11.4.9 Recovery and Trade of Solid Salts

Salts can be retrieved from waste brine by implementing the ZLD: (i) brine concentration, (ii) crystallization, followed by (iii) dewatering of solids. Though ZLD can be costly, it is highly effective in simultaneously discarding waste and extracting salts. The produced salts can be utilized or sold for industrial operations such as manufacturing chemicals, detergents, curing, dyeing, and de-icing.

11.4.10 Cooling

With appropriate treatment, recycled brine can be used in industries that require large volumes of water for cooling processes, such as power generation, oil and gas production and processing, and chemical production. Recycling brine for these purposes will generally demand water treatment to minimize scaling and corrosion of equipment. Brine use for cooling operations is constrained by the salt content and other pollutants, as these may inhibit heat transfer by causing the formation of lather.

11.4.11 Brine Incineration

Waste brine can be sent to an incinerator facility, where it is typically mixed with other solid wastes for processing. Incineration evaporates the water, while the salts in the brine become part of the residual ash that requires further management. Incineration is popular in countries with limited availability of land for landfills.

11.5 Waste Brine Valorization: Opportunities and Challenges

The increasing utilization of desalination techniques to produce freshwater aligns with the detrimental effects of increasing salinity in the sea brought about by irresponsible brine discharge. Recycling of the waste brine obtained through the desalination process to yield beneficial compounds could become an economically viable solution. One pathway to achieve this objective is to recover residual saline solution into valuable synthetic chemical compounds, such as NaOH, which can be re-utilized by the gas hydrate technology, and HCl. It is noted, however, that the methods are complicated by the presence of many impurities, making segregation and purification difficult [25].

11.5.1 Waste Brine as a Resource and Its Application in Gas Hydrates

NaOH and HCl are utilized together in water treatment and have numerous industrial applications. NaOH is used in gas hydrate technology to improve the performance of anti-agglomerates by changing pH and neutralizing the effect of gases such as carbon dioxide (CO_2) in natural gas [26]. The presence of CO_2 in the natural gas tends to lower the pH, resulting in foaming. A lower pH also changes the interaction of surfactant molecules with hydrates [27]. Hence, the addition of NaOH not only enhances the performance of anti-agglomerates but also aids as a neutralizing inhibitor. Neutralizing inhibitors reduce corrosion by eliminating H^+ ions in the aqueous phase [28].

NaOH also aids in reducing the viscosity of hydrate slurries, improving the adsorption of anti-agglomerate molecules on the surface of hydrates, thus guiding toward an easier hydrate formation condition. The pH has a strong influence on the rate of corrosion [29]. At higher pH values, reduction in the solubility of iron carbonate leads to scale formation on the surface of steel [30], thus causing corrosion. As a whole, the addition of small quantities of alkali can eliminate foaming, maximize anti-agglomerate performance, and reduce corrosion in steel pipes. NaOH is preferred for its lower molecular weight and cost.

It has also been shown that the presence of salt and alkaline in waters can decrease the induction time, improve the growth rate of hydrate, and reduce the performance of anti-agglomerates of crude oil relative to that observed in pure water [31]. At a pH of 9, the borate anion becomes boron which is tough to reject [32]. Similar changes are noted for the case of silica [33] and dissolved inorganic carbon [34]. In membrane-based desalination techniques, NaOH is typically used in cleaning membrane fouling [35].

More than 99% of NaOH generated from waste brine is done so by the chlor-alkali technique [36]. Direct electricity is supplied through brine solution, initiating an electrochemical reaction to produce NaOH, H_2, and chlorine (Cl_2). NaOH is implemented for desalination technologies, while H_2 has the potential to be utilized as fuel for processes requiring hydrogenation reactions. By using an oxidation reaction in Cl_2 generation, HCl can be recovered. Overall, the supplementary recovery technologies within desalination can assist in reducing the environmental impact of the waste and convert it into an economic opportunity by recovering valuable synthetic compounds like NaOH and HCl from brine [37, 38].

11.5.2 Membrane-based Process for Valorization of Waste Brine

By combining membrane-based processes, NaCl can be valorized into valuable compounds like NaOH and HCl [39, 40]. The waste brine has to be pretreated through nanofiltration (NF), followed by electrodialysis (ED) and, subsequently, electrodialysis through bipolar membranes (BMED) to generate products, as shown in Figure 11.1 [41]. Each of the pretreatment approaches delivers a perceptible direction for valorizing the waste brine. NF is used to purify concentrated brine such that it primarily constitutes univalent ions. Selective electrodialysis (SED) segregates the univalent and multivalent ions, and ED generates concentrated ionic salts. The implementation of NF, SED, or ED as a pretreatment boosts the efficiency of a BMED system, leading to the generation of valuable and profitable compounds and decreasing the disposal costs.

Figure 11.1 shows the processing technique to recover NaOH and HCl. The residual brine produced after desalination is first purified by NF, by which SO_4^{2-}, Ca^{2+} and Mg^{2+} ions are removed. A higher concentration of SO_4^{2-} may cause fouling [42] and elimination of Ca^{2+}, Mg^{2+} ions in electrodialysis (ED). The permeate of NF is further concentrated in ED by evaporating to the electrolyzer cell requirements for NaCl saturation. TDS are then eliminated by ion exchange (IX) and softening methods. Finally, the obtained concentrated brine is acidified by HCl and sent to a membrane electrolyzer to yield NaOH, Cl_2, and H_2.

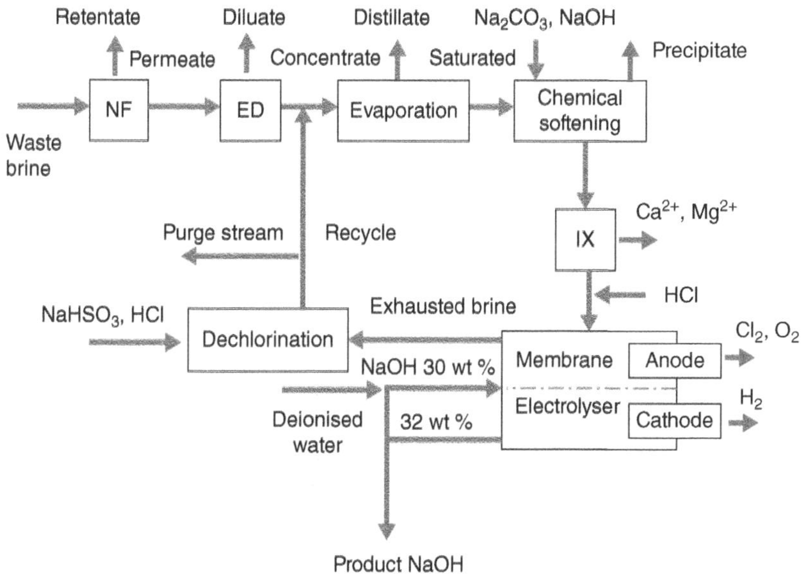

Figure 11.1 Waste brine to NaOH process flow diagram.

In the integrated NF–ED mechanism [43] and ED [44], the raw material is fragmented into dilutant and concentrate streams, where ED conveys NaCl and water from the dilutant to the concentrate channels. The outlet concentration matches that of seawater and permits the ED dilutant to be further recycled back as feed to the desalination unit. The integration of ED and evaporation might be significantly more economical compared to evaporation alone [45]. After saturation, brine is sent to a softening unit where the addition of Na_2CO_3 and NaOH leads to precipitate ions of Ca^{2+} and Mg^{2+} to their respective salts for further precipitation and later removal through filtration. The softening method produces a level of mineral concentration hardness which does not meet the electrolyzer cell requirements [36]. Thus, a cation IX is implemented to reduce the hardness from ppm levels to ppb. The processes of softening, IX, and electrolysis are conventional techniques in a typical chlor-alkali plant [36]. Cl_2, H_2, and NaOH are produced in the electrolyzer, along with an exhaust brine of 20 wt % NaCl. The exhausted brine must be dechlorinated before being discharged to the environment [46] or recycled. The exhausted NaCl is sent to a splitter and further fragmented into purge and recycle channels. The former leaves the system, thereby avoiding impurities from gathering in the system. Alternative means of recovering valuable products is by direct electrosynthesis, where NaOH and HCl can be produced. In this method, the water-splitting reaction mechanism produces H^+ and OH^-, which associate with the brine to yield NaOH and HCl [47].

11.5.3 Recovery of Calcium and Its Conversion to Calcium Sulfate

Recovery of calcium starts treating the waste brine in a continuously stirred batch reactor with oleic acid (65–70% $C_{18}H_{34}O_2$) with the addition of 1 N NaOH to neutralize the solution. The solution is then agitated at 1000 rpm, resulting in the formation of calcium oleate wet scum, which is retrieved, filtered, and purified by washing in absolute alcohol. At food-grade qualities, calcium oleate is a key constituent in face cream and other aesthetic medicine.

Further treatment of the calcium oleate scum can also produce other value products. Reaction with potassium chloride (KCl) forms brown rings containing both organic and inorganic parts. The inorganic parts are eliminated by adding 1 N sulphuric acid (H_2SO_4), leading to the formation of a white precipitate that is observed to be calcium sulfate ($CaSO_4$), commonly known as gypsum. Alternatively, 1 N HCl can be added to yield calcium chloride ($CaCl_2$), which is widely used as fertilizer [48].

11.5.4 Retrieval of Magnesium Oxide from Waste Brine

Magnesium oxide (MgO) has applications in food, face creams, pharmaceuticals, and construction [49, 50]. The vast majority of MgO produced globally is processed from mineral ores, with a smaller share retrieved from saline surface and groundwaters. Waste brines from desalination processes can be used to produce substantial amounts of MgO with lower costs and less environmental damage.

Dong et al. [49] evaluated the economic feasibility of reactive MgO retrieval from reject brine using NaOH and compared the NaOH production route to that of other alkalinity sources, e.g. ammonia (NH_3) and calcium oxide (CaO). The total cost considered for the production of MgO consisted mainly of material costs of alkalinity sources required for synthesis and the cost of energy needed for calcination. Since NaOH is produced from the waste product of the desalination process, the material cost of reject brine utilization is presumed to be zero. Conveyance of feed, size reduction, and packaging of MgO was not considered as it does not contribute significantly to the total cost. A possible process complication is in the formation of complex compounds like sodium oleate and magnesium oleate, owing to the use of NaOH for neutralization.

The retrieval of MgO from waste brine is permitted by the reaction of Mg^{2+} with NaOH, leading to precipitation of $Mg(OH)_2$ and slight quantities of

$CaCO_3$. $Mg(OH)_2$ can be further calcined at 500°C for two hours to yield purer MgO. At higher calcination temperature and time, the reactivity and specific surface area of MgO decreases [49]. The reactions involved during the process are given in Eqs (11.1–11.4).

$$NaOH(aq) \rightarrow Na^{+} + OH^{-} \quad (11.1)$$

$$Mg^{2+} + 2OH^{-} \leftrightarrow Mg(OH)_2 \quad (11.2)$$

$$NaOH(aq) + HCO_3^{-} \leftrightarrow CO_3^{2-} + Na^{+} + H_2O \quad (11.3)$$

$$Ca^{2+} + CO_3^{2-} \leftrightarrow CaCO_3 \quad (11.4)$$

11.5.5 Opportunities and Challenges

- Reduction of effluent; The effluent has a higher concentration compared to waste brine and, by using ZLD, the concentration can further be enhanced with the help of solar ponds.
- Reutilizing ED dilutant; As the ED produces concentrated brine hence, the dilutant can be recycled by hydrate-based desalination.
- Evading of concentration and dilution cycle; This method evades the concentration and dilution of NaCl that are seen in chlor-alkali and desalination processes.
- Potential returns from byproducts; The byproducts of the chlor-alkali process, H_2, and Cl_2, can provide revenue. However, the obtained chlorine contains O_2, water, nitrogen, and carbon dioxide that can be wholesaled for a lower cost. Additionally, NaOCl, a bleaching chemical, can be produced from NaOH and Cl_2.
- Decrease in conveyance costs and supply of waste heat for concentration process through power plants.

Certain innate challenges like the technical knowledge of individual components in a combined system, its operation, control parameters, and issues during start-up and shut-down, of course, must naturally be considered. Some of the features of MgO production that can be improved are the scaling up of the process and environmental issues. The production of alkalis is an energy-intensive process due to the step in precipitating $Mg(OH)_2$. Hence, identification of energy-efficient routes as well as the incorporation of renewable energy sources will enhance the sustainability of the product with a lower environmental risk.

11.6 Evolving Waste Brine Treatment Strategies and Sights for Future Feasibility

Overall, it is evident that waste brine can be detrimental to the environment. Hence diluting waste brine with expended cooling water obtained from industries prior to discharge is suggested [51, 52]. Waste brine management has been revealed to only slightly decrease the expenditure of desalination plants, though its overall production cost has also been reduced recently. The practice of using adsorbents or resins to extract metals like cesium, rubidium, uranium, and phosphorus is the most likely to be economic across the many brine management frameworks [11, 53]. The feasibility of recovering salts using precipitants has shown a promising trajectory in investigations and could soon reach levels of practical profitability [5, 9, 54]. However, it must not be forgotten that ecological sustainability is the principal obligation in brine management. Implementing integrated desalination methods with renewable energy to treat waste brine, such as the case of solar dryer evaporations, should be a highly valued consideration for its innate sustainability [55].

11.7 Conclusion

Though immediate discharge of waste brines is still technically an industrial option, the conditions surrounding such disposal are heavily regulated. The growing concern toward water scarcity and degradation of ecosystem services has encouraged higher levels of scrutiny in brine disposal methods. Wherever environmental protocols are concerned, waste brine reuse and recycling tactics are beneficial in reducing disposal volumes or enhancing production capacity. In selecting a proper brine management framework, a treatability study must be carried out to gain a complete understanding of waste brine composition, including salinity levels, metals present and their exclusion, TDS, and all other pollutants present. A techno-economic analysis is also necessary to make an informed decision on whether waste brine treatment for recycling and reuse is a cost-effective option or not.

Though the generation of potable water from these desalination methods is critical to solving one of humanity's potential crises to come, the detrimental effects resulting from the handling, managing, and discharging of wastes are a serious threat on their own. The waste brines are generally at higher temperatures and contain substances that are perilous to marine life. Other effects of waste brine disposal comprise changes to the dissolved oxygen (DO) content, pH, nutrients, alkalinity, and TDS of the general vicinity within the disposal area. Hence brine management must be practiced to conserve the environment and safeguard the ecosystem. This can be accomplished by reducing the quantity of the disposed waste brine or by implementing ZLD technology by which the produced salts can be safely disposed to land or water. Value-added salts and metals can also be retrieved from waste brines and can be retailed for profit, among many other chemicals that are extractable. The classification of waste brine as a "resource" rather than "waste" is the growing approach for incorporating the principle of sustainability as part of the brine managing framework.

References

1 Al-Bazedi, G., Ettouney, R. S., Tewfik, S. R., Sorour, M. H., and El-Rifai, M. A. (2014). Salt recovery from brine generated by large-scale seawater desalination plants. *Desalination and Water Treatment* 52 (25-27): 4689–4697.

2 Chelme-Ayala, P., Smith, D.W., and El-Din, M.G. (2009). Membrane concentrate management options: A comprehensive critical review. *A Paper Submitted to the Journal of Environmental Engineering and Science Canadian Journal of Civil Engineering* 36: 1107–1119.

3 Einav, R., Hamssib, K., and Periyb, D. (2002). The footprint of the desalination processes on the environment. *Desalination* 152: 141–154.

4 Ahmed, M. and Anwar, R. (2012). An assessment of the environmental impact of brine disposal in marine environment. *International Journal of Modern Engineering Research* 2: 2756–2761.

5 Pérez-González, A.M., Urtiaga, R., Ibáñez, I., and Ortiz (2012). State of the art and review on the treatment technologies of water reverse osmosis concentrates. *Water Research* 46: 267–283.

6 Kim, D.H. (2011). A review of desalting process techniques and economic analysis of the recovery of salts from retentates. *Desalination* 270: 1–8.

7 Tanaka, Y., Ehara, R., Itoi, S., and Goto, T. (2003). Ion-exchange membrane electrodialytic salt production using brine discharged from a reverse osmosis seawater desalination plant. *Journal of Membrane Science* 222: 71–86.

8 Jiang, C., Wang, Y., Zhang, Z., and Xu, T. (2014). Electrodialysis of concentrated brine from RO plant to produce coarse salt and freshwater. *Journal of Membrane Science* 450: 323–330.

9 Sorour, M.H., Hani, H.A., Shaalan, H.F., and Al-Bazedi, G.A. (2015). Schemes for salt recovery from seawater and RO brines using chemical precipitation. *Desalination and Water Treatment* 55 (9): 2398–2407.

10 Peterskovà, M., Valderrama, C., Gibert, O., and Cortina, J.L. (2012). Extraction of valuable metal ions (Cs, Rb, Li, U) from reverse osmosis concentrate using selective sorbents. *Desalination* 286: 316–323.

11 Jeppesen, T., Shu, L., Keir, G., and Jegatheesan, V. (2009). Metal recovery from reverse osmosis concentrate. *Journal of Cleaner Production* 17: 703–707.

12 Morilo, J., Usero, J., Rosado, D., El Bakouri, H., Riaza, A., and Bernaola, F.J. (2014). Comparative study of brine management technologies for desalination plants. *Desalination* 336: 32–49.

13 Gilbert, E. (2012). ZDD – Achieving maximum water recovery. WaterWorld, (01 Dec). https://www.waterworld.com/technologies/article/16193078/zero-discharge-desalination-holds-promise-of-maximum-water-recovery.

14 Ibáñez, R., Pérez-González, A., Gómez, P., Urtiaga, A.M., and Ortiz, I. (2013). Acid and base recovery from softened reverse osmosis (RO) brines. Experimental assessment using model concentrates. *Desalination* 309: 165–170.

15 Mickley, M. (2008). *Treatment of Concentrate*. Springfield VA: The National Technical Information Service.

16 Mickley, M. (2008). *Survey of High-recovery and Zero Liquid Discharge Technologies for Water Utilities*. WateReuse Foundation.

17 Ahmed, M., Arakel, A., Hoey, D., Thumarukudy, M.R., Goosen, M.F., Al-Haddabi, M., and Al-Belushi (2003). Feasibility of salt production from inland RO desalination plant reject brine: A case study. *Desalination* 158 (1-3): 109–117.

18 Giwa, A., Dufour, V., Al Marzooqi, F., Al Kaabi, M., and Hasan, S.W. (2017). Brine management methods: Recent innovations and current status. *Desalination* 407: 1–23.

19 Ossandón, K., Pinto, P., and Cisternas, L.A. (2010). Planning and scheduling of salt harvest in solar evaporation ponds. *Computers & Chemical Engineering* 34: pp. 620–630.

20 Mohammadesmaeili, F., Badr, M.K., Abbaszadegan, M., and Fox, P. (2010). Mineral recovery from inland reverse osmosis concentrate using isothermal evaporation. *Water Research* 44: 6021–6030.

21 Rodríguez-DeLaNuez, F., Franquiz-Suárez, N., Santiago, D.E., Veza, J.M., and Sadhwani, J.J. (2012). Reuse and minimization of desalination brines: A review of alternatives. *Desalination and Water Treatment* 39 (1-3): 137–148.

22 Everest, W.R. and Murphree, T. (1995). Desalting residuals: A problem or a beneficial resource. *Desalination* 102: 107–117.

23 Zhuang, X., Han, Z., Bai, Z., Zhuang, G., and Shim, H. (2010). Progress in decontamination by halophilic microorganisms in saline wastewater and soil. *Environmental Pollution* 158: 1119–1126.

24 Raja, R., Hemaiswarya, S., and Rengasamy, R. (2007). Exploitation of Dunaliella for carotene production. *Applied Microbiology Biotechnology* 74: 517–523.

25 Du, F., Warsinger, D.M., Urmi, T.I., Thiel, G.P., Kumar, A., and Lienhard, J.H. V. (2018). Sodium hydroxide production from seawater desalination brine: Process design and energy efficiency. *Environmental Science & Technology* 52 (10): 5949–5958.

26 Zhao, H., Sun, M., and Firoozabadi, A. (2016). Anti-agglomeration of natural gas hydrates in liquid condensate and crude oil at constant pressure conditions. *Fuel* 180: 187–193.

27 Sun, M. and Firoozabadi, A. (2014). Natural gas hydrate particles in oil-free systems with kinetic inhibition and slurry viscosity reduction. *Energy & Fuels* 28 (3): 1890–1895.

28 Abd El-Lateef, H.M., Abbasov, V.M., Aliyeva, L.I., and Ismayilov, T.A. (2012). Corrosion protection of steel pipelines against CO2 corrosion-a review. *Chemistry Journal* 2 (2): 52–63.

29 Nešić, S., Nordsveen, M., Nyborg, R., and Stangeland, A. (2003). A mechanistic model for carbon dioxide corrosion of mild steel in the presence of protective iron carbonate films – part 2: A numerical experiment. *Corrosion* 59 (6): 489–497.

30 Nešić, S. (2007). Key issues related to modelling of internal corrosion of oil and gas pipelines – A review. *Corrosion Science* 49 (12): 4308–4338.

31 Sinquin, A., Arla, D., Prioux, C., Peytavy, J.L., Glenat, P., and Dicharry, C. (2008). Gas hydrate formation and transport in an acidic crude oil: Influence of salt and pH. *Energy & Fuels* 22 (2): 721–728.

32 Pastor, M.R., Ruiz, A.F., Chillon, M.F., and Rico, D.P. (2001). Influence of pH in the elimination of boron by means of reverse osmosis. *Desalination* 140 (2): 145–152.

33 Ning, R.Y. (2003). Discussion of silica speciation, fouling, control and maximum reduction. *Desalination* 151 (1): 67–73.

34 Milstead, C.E., Riedinger, A.B., and Lonsdale, H.K. (1971). Rejection of carbon dioxide and pH effects in reverse osmosis desalination. *Desalination* 9 (3): 217–223.

35 Redondo, J.A. and Lomax, I. (1997). Experiences with the pretreatment of raw water with high fouling potential for reverse osmosis plant using FILMTEC membranes. *Desalination* 110 (1–2): 167–182.

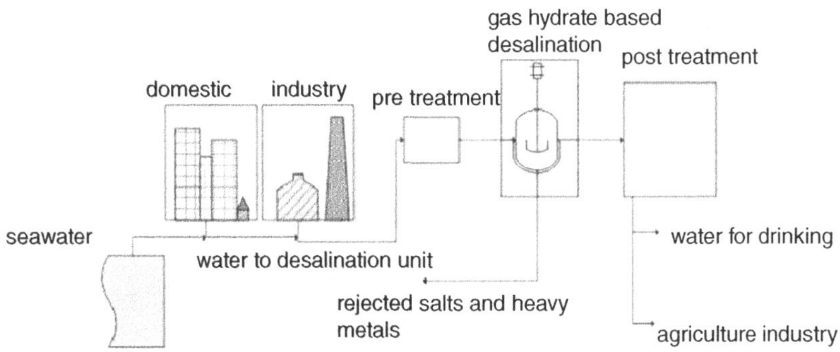

Figure 12.1 General scheme for the production of clean water from seawater/wastewater.

sterilized or dechlorinated at some point before entering the receiving water distribution system [5].

Birnhack et al. studied industrial remineralization post-treatment procedures such calcite and direct chemical dose addition [6,7]. This chapter discusses the three targets of post-treatment: safeguarding human health, protecting economic growth, and corrosion protection.

12.2 Safeguarding Human Health

The World Health Organization (WHO) continually updates and releases the latest standards for drinking water. Desalinated water gained favor early on as an alternative freshwater source owing to growing concern over water scarcity. As early as 1985, WHO initiated consultations to set up a road map for bringing desalinated water to widespread use. [8] WHO also indicated that certain essential minerals, such as calcium and magnesium, need to be replenished in product waters. Many studies have claimed that drinking desalinated product water directly may be harmful to health [6,9–15].

12.2.1 Desalinated Water Quality

Magnesium is a fundamental element that participates in a wide range of reactions, including muscle and nerve activity and stabilization of genes and RNA. In the long term, magnesium deficiency can lead to many health problems. It is estimated that adults need a daily intake of at least 260 mg of magnesium for females and 340 mg for males. However, most studies do not include lower and upper limits for magnesium in drinking water.

Magnesium is derived from both food and drinking water. However, if magnesium content is low in the water supply used for food preparation, it could restrict many people's magnesium consumption. The mineral content of food has fallen 5–40% over the past 50 to 70 years. The WHO reports that 40% of the US population does not meet their minimum required daily intake for calcium and magnesium [16]. An increase from 3.2 mg/L magnesium to 17 mg/L can decrease the odds of death from myocardial infarction by 37%. While often overlooked, magnesium deficiency is a widespread concern in developed countries. Since magnesium has many physiological roles in humans, deficiencies may affect multiple organs and systems, including the heart, vascular system, pancreas, brain, and bones.

Neurological developments in human beings are also affected by the deficiencies of iodine and corresponding compounds. Though iodide and bromide elements are present in seawater in much higher quantities than needed, the desalination process reduces their concentrations substantially. The problems caused by iodine deficiency are reported and posted in a number of articles and research studies among the people of Singapore due to the consumption of desalinated water for drinking purposes. However, adding iodine as a post-treatment solution is not advisable because it may lead to harmful disinfection by-products (DBPs) [7,17].

Although fluoride is present in all kinds of soils and water, it is usually too low to detect in desalinated water [18]. The deficiency of fluoride has a serious impact on dental health and affects the growth of bones. If the fluoride concentrations are too low in drinking water, the introduction of fluoride into the water supply source is highly suggested. In cases where the deficiency is already present and symptomatic, the use of fluoride tablets may be necessary. In 2011, the WHO recommended 0.2 mg/L as the minimum concentration of fluorides in their report, *Safe Drinking water from Desalination* [19].

12.2.2 Remineralization of Desalinated Water for Consumption

Minerals necessary for human health are deficient in the water-produced desalination process, prompting the artificial introduction of calcium and magnesium in the post-treatment processes. The calcium to magnesium ratio should be in the two to three range [20], while iodine should not be added to the desalinated water, as it could result in health and environmental problems. The addition of fluoride to desalinated water is recommended, but it is not considered mandatory. Some districts do not adopt fluoridation due to the risk of developing dental fluorosis.

12.3 Impacts on Agriculture

Agriculture nourishes the population and forms the basis of nearly all economies and, thus, straddles the domains of safeguarding human health and economic growth. The Food and Agriculture Organization [21] has suggested using seawater [22], brackish water [23], wastewater, and oilfield-produced brine [24] as sources of water for agricultural applications. These alternative sources generally require desalination, which extracts the minerals from the saline solution to produce pure water. Monterrey-Viña et al. [25] elaborated on the use of desalinated water sources for agriculture in the Canary Islands through systematically defined aspects, many of which apply to agricultural use of desalinated water more generally [26].

12.3.1 Effects of Desalinated Water on the Irrigation System

Depending on the sources of desalinated water, the composition of the water received for irrigation can differ greatly. Water quality influences the soil nutrient availability, soil structure, crop yield, and crop nutritional values. This section discusses some of the parameters influencing water quality from an irrigation water composition point of view [26]. A few considerable parameters are:

- Sodium adsorption ratio
- Necessary nutrients like Ca, Mg, and SO_4
- Ca to Mg Ratio
- Phytotoxicity of Cl, Na, and boron

Crops take in water through osmosis, with the intensity of the process dependent on the salinity levels of the soil. At high salinities, clay dispersal may be substantially impacted, depending on the compositions of the water and receiving soil. Monovalent ions (Na^+, K) aggregate soil particles, whereas *divalent* ions (Ca^+ and Mg^+) cluster particles. A well aggregated soil allows for better water infiltration and drainage, whereas clustering of soil particles impedes drainage, resulting in surface crusting. Thus, imbalances in the relative abundance of monovalent ions in irrigation waters can degrade soil structure [27] and, in turn, cut off water to plant root systems. However, extracting salts by desalination is also problematic as it generally results in ionic imbalances, which impair plant growth [28]. Post-treatment of desalination seeks to restore the mineral balances.

- *Sodium adsorption ratio*

Soil samples can be analyzed using a simple sodium adsorption ratio (SAR) test and an exchangeable sodium percent test (ESP). The SAR test is an indicator of ionic balance. The combined effect of the two parameters indicates how likely soil and crops are to be affected. Desalinated waters generally have high SAR values due to depletion of Ca^{+} and Mg^{+}, which can be remedied by adding these minerals back during post-treatment. High salinity wastewaters must be desalinated or diluted before being used in irrigation to avoid damage to crops.

- *Necessary nutrients like Ca, Mg, and SO4*

Nitrogen, phosphorus, and potassium are the primary nutrients in plant growth and are generally supplemented through fertilization. Calcium (Ca^{+}), magnesium (Mg^{+}) and sulfate (SO_4) are the secondary minerals required for plant growth and are generally taken up from the irrigated soil. Their deficiency can influence yields and affect the nutritional value of food crops. Since desalinated water has no nutritional value on its own to supplement soil mineral taken up by harvested crops, its use in irrigation may lead to depletion of these minerals in the soil. Plants not receiving any natural rainwater and separated from soil nutrients, such as those within a greenhouse setting or in hydroponic setups, will be most affected because the product water lacks both the necessary minerals and iron. According to Martínez-Alvarez et al. [29], horticultural crops must contain around 80–120 mg/L of calcium and 24–36 mg/L of magnesium.

- *The ratio of calcium to magnesium*

Based on the findings of Qadir et al. [30], irrigation waters high in Mg relative to Ca disperse clay particles (reduce aggregation) when applied to soils with >25% exchangeable magnesium, leading to porosity impairment and reduced water infiltration to crop root systems. Therefore, any volume of irrigation water should contain more calcium than magnesium.

- *Cl^{-}, Na^{+} and boron phytotoxicity*

After desalination, sodium and chloride are the most concentrated ions used in agriculture. There were high incidences of boron and phytotoxicity in a study when the sodium concentration was above 210 mg/L and the chloride concentration was higher than 350 mg/L. Plants are susceptible to boron concentrations above 150 ppm. Pandey et al. [31] managed to explain the functions of boron in plants effectively: As the plant ages, it seems to suffer impairments in the membrane function, and boron curbs this by aiding in carbohydrate synthesis, nodule growth, and plant reproduction. Therefore, it needs to be added in minute amounts but not in larger concentrations. Depending on the crop type being irrigated, if the boron level is too high, an additional desalination phase is to be taken as post-treatment.

12.3.2 Remineralization Needed for Irrigation

Alongside the need for sulfate for growth, plants also require calcium and magnesium. These three ions are essential for a healthy crop and must be added to desalinated product water. Bar-Tal et al. [33] found that the optimal concentration required to make vegetable fertilizers is in the range of 1.5-2.5 mg/dL calcium and 0.25 mg/dL magnesium. In cases where the concentrations of sodium, chloride, and boron are still too high after the desalination process, an additional desalination pass must be added to lower their concentration. Other such components are generally added separately to fertilizers.

12.4 Corrosion of Distribution Network and Scaling

Whenever fluid transport is involved, the reality of corrosion is unavoidable. Barton et.al [33], in their assessment of the UK water distribution systems, found that the documented piping materials were mostly cast iron, steel, ductile iron, polyvinyl chloride, polyethylene, and, for more recent installations, polypropylene. In 2006, WHO reported that many household plumbing materials could cause unnecessary health problems [34]. The two main classes of material available for piping are metals, found in early installations, and nonmetals. Metallic materials comprise galvanized steel, copper, lead, and lead compounds. The non-metallic category, which is employed in more recently installed systems, is made up of polyvinylchloride and polyethylene. Metals such as copper, lead, iron, zinc, cement, or asbestos reinforced concrete with even the slightest corrosion are vulnerable to mineral leaching when in contact with very soft waters, such as desalinated water [35–45].

12.4.1 Corrosion Problems Involving Desalinated Water

A detailed understanding of this problem requires consideration of the factors used to gauge whether water is corrosive or not. The most widely used parameters are the LSI and the CSMR.

There are very few published case studies regarding the corrosion of industrial water networks. Though there are common incidents, like those which occurred in Flint, Michigan, they are not highlighted in the international media. Marangou and Savvides [46] identified the corrosion effects of desalinated product water on the distribution systems of the first desalination plant in Cyprus. The water at the furthest end of the drinking pipe turned brown due to the high iron content. The ratio of carbon dioxide and lime was tuned in an attempt to solve the condition, thus increasing the alkalinity of the product water. Thus, The higher pH

significantly decreased the level of iron in the water. However, LSI values increased, increasing the alkalinity further increased turbidity. The solution was to add magnesium sulfate ($MgSO_4$) to increase hardness. Their conditioning consisted of 45 mg/L of $CaCO_3$ for alkalinity and 40 mg/L $MgSO_4$ to restore hardness, with a pH of 8.5.*Corrosivity to steel*
The combination of variables that impact the corrosivity of water is pH, alkalinity, hardness, dissolved inorganic carbon, sulfate, chloride, and total dissolved solids. Correcting the water's pH to >7, and usually 8–8.5pH, is standard practice, as alkalinity reduces the tendency of water to accept electrons [47].*Corrosion of lead and copper-containing networks*

Rusting of pipes containing lead and copper occurs in a water distribution systems piping made up of different metals that allow galvanic corrosion. Since it releases heavy metals, galvanic corrosion can cause serious health concerns [45].

12.4.2 Remineralization Needed for Corrosion Control

To avoid corrosion problems, steps need to be taken to change the water chemistry during post-treatment processes [48,49].

- Carbon dioxide for buffering.
- Calcium enrichment for steel, lead, copper, and zinc corrosion control.
- Lower sulfate and chloride concentrations for steel corrosion control.
- Balance sulfate and chloride concentrations to at least a 2:1 ratio for lead and copper corrosion control.

12.5 Remineralization Techniques

As early as 1981, Gabrielli [49] introduced the possibility of using carbon dioxide and natural carbonate rocks to reach basic desalinated water stability parameters. Today, this is the most popular technique for remineralization. Figure 12.2 presents a schematic of the desalination process, followed by remineralization and pH adjustment. Desalinated water can also be mixed with groundwater for a simpler process.

12.5.1 Emerging Techniques

A recent desalination plant post-treatment system is shown in Figure 12.3. The process starts with using desalinated water from gas hydrate treatment. It also involves the addition of lime and carbon dioxide, alongside chlorine (stored as chlorine gas) and fluorosilicic acid (FSA) for water fluoride care (if required by

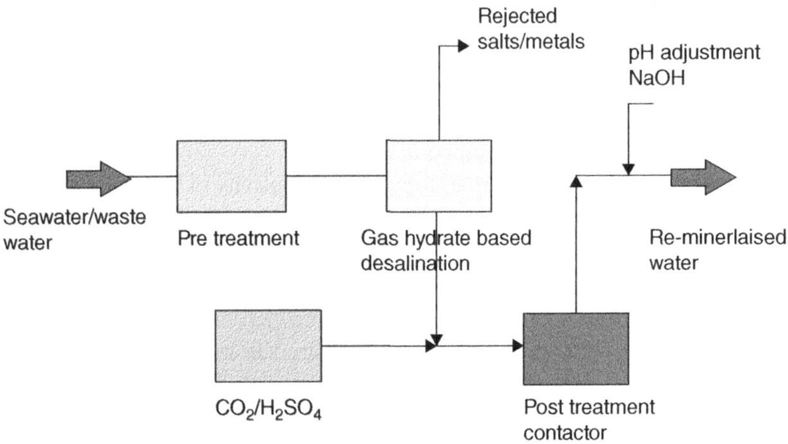

Figure 12.2 Conventional scheme of water treatment from saline to fresh water.

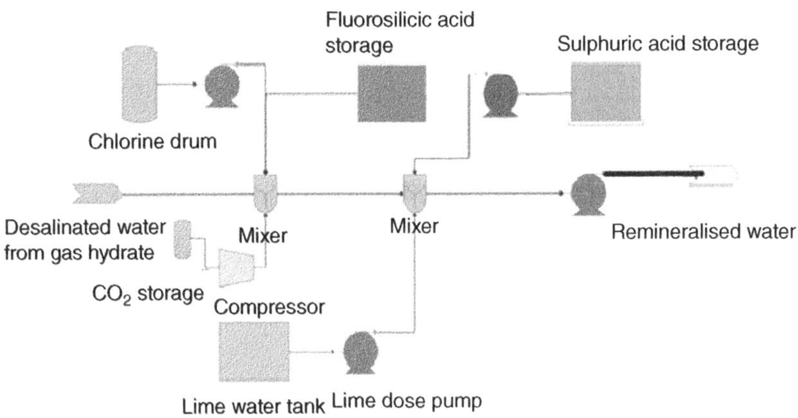

Figure 12.3 Emerging technique of remineralization to be integrated with the gas hydrate desalination plant.

local regulations). The material pH is modified by adding sulfuric acid. The chemicals were added before the lime to dramatically decrease the pH.

Adding lime improves the alkalinity and hardness of the returned product water. First, if the calcium hardness of the product water is relatively high (>100 mg/L), the mix of lime and reverse osmosis permeate may have a pH that is above the usual target range of 7.6 to 8.3. Adding sulfuric acid to the concentrated raw water would lower the pH of the finished water. Reducing pH will prevent residual turbidity from being produced.

The remineralization by-product is normally supplied and processed by the desalination plant as either powdered hydrated lime or as pebble-lime (CaO), which is then slaked to produce hydrated lime $Ca(OH)_2$. Pulverized lime is processed as 25-kg bags in smaller desalination plants. In this way, lime is fed into the lime slurry (milk of lime). This lime slurry is mixed with fresh water provided by the desalination process and blended in lime saturator tanks to create saturated limewater [50].

12.6 Conclusion

Since the beginning of the Industrial Revolution in the nineteenth century, rapid population growth and rising urbanization have led to the growing demand for drinking fresh water, irrigation, and industrial use. Overconsumption of natural freshwater sources has spurred the development of water reuse and water desalination strategies. However, desalinated water is not suitable for direct use, as it is corrosive and negatively affects human and environmental health. Desalinated water is slightly acidic, lacks minerals, and cannot be used unbuffered, thus making remineralization an important component downstream of desalination.

While post-treatment literature does exist, fundamental gaps remain in the design and optimal water quality objectives of proper post-treatment. Moreover, the available literature frequently fails to classify post-treatment processes by source water type (seawater, brackish sea water, brackish soil water, cool soil water, industrial wastewater) or desalination process.

To conclude, the remineralization step can be integrated into gas hydrate desalination plants through a variety of modes and approaches. The choice of method results in different water quality results and costs, thus influencing the end-use of the treated water.

References

1 Parker, A. (1942). Potable water from sea-water. *Nature* 149 (3772): 184–186.

2 Cha, J.-H. and Seol, Y. (2013). Increasing gas hydrate formation temperature for desalination of high salinity produced water with secondary guests. *ACS Sustainable Chemistry & Engineering* 1 (10): 1218–1224. doi: 10.1021/sc400160u.

3 Han, S., Rhee, Y.-W., and Kang, S.-P. (2017). Investigation of salt removal using cyclopentane hydrate formation and washing treatment for seawater desalination. *Desalination* 404: 132–137. doi: 10.1016/j.desal.2016.11.016.

4 Ghalavand, Y., Hatamipour, M.S., and Rahimi, A. (2015). A review on energy consumption of desalination processes. *Desalin. Water Treat* 54 (6): 1526–1541. doi: 10.1080/19443994.2014.892837.

5 Sürmelihindi, G., Passchier, C.W., Spötl, C., Kessener, P., Bestmann, M., Jacob, D.E., and Baykan, O.N. (2013). Laminated carbonate deposits in Roman aqueducts: Origin, processes and implications. *Sedimentology* 60 (4): 961–982. doi: 10.1111/sed.12000.

6 Birnhack, L. and Lahav, O. (2018). *Post-Treatment of Desalinated Water-Chemistry, Design, Engineering, and Implementation*. Elsevier Inc.

7 Lesimple, A., Ahmed, F.E., and Hilal, N. (2020). Remineralization of desalinated water: Methods and environmental impact. *Desalination* 496: no. June, 114692. doi: 10.1016/j.desal.2020.114692.

8 Kozisek, F. (2005). Health risks from drinking demineralised water. *Nutrients in Drinking Water* 1 (1): 148–163.

9 Nriagu, J., Darroudi, F., and Shomar, B. (2016). Health effects of desalinated water: Role of electrolyte disturbance in cancer development. *Environmental Research* 150: 191–204.

10 Ovadia, Y.S., Gefel, D., Aharoni, D., Turkot, S., Fytlovich, S., and Troen, A.M. (2016). Can desalinated seawater contribute to iodine-deficiency disorders? An observation and hypothesis. *Public Health Nutrition* 19 (15): 2808–2817.

11 Rosborg, I. and Kozisek, F. (2016). *Drinking Water Minerals and Mineral Balance*. Springer.

12 Rygaard, M., Arvin, E., and Binning, P.J. (2009). The valuation of water quality: Effects of mixing different drinking water qualities. *Water Research* 43 (5): 1207–1218.

13 Saini, R.D. (2017). Health risks from long term consumption of reverse osmosis water. *International Journal of Applied Chemistry* 13: 293–301.

14 Shlezinger, M., Amitai, Y., Goldenberg, I., and Shechter, M. (2016). Desalinated seawater supply and all-cause mortality in hospitalized acute myocardial infarction patients from the Acute Coronary Syndrome Israeli Survey 2002–2013. *International Journal of Cardiology* 220: 544–550.

15 Naser, A.M., Shamsudduha, M., Clasen, T.F., and Narayan, K.M.V. (2019). Letter to the editor regarding, 'the unintended consequences of the reverse osmosis revolution'. *Environmental Science & Technology* 53 (13): 7173–7174.

16 Cotruvo, J. (2006). Health aspects of calcium and magnesium in drinking water. *Water Conditioning & Purification* 48: 6–40.

17 Kim, D., Amy, G.L., and Karanfil, T. (2015). Disinfection by-product formation during seawater desalination: A review. *Water Research* 81: 343–355.

18 Abouleish, M.Y.Z. (2016). Evaluation of fluoride levels in bottled water and their contribution to health and teeth problems in the United Arab Emirates. *The Saudi Dental Journal* 28 (4): 194–202.

19 Organization, W.H. (2011). *Safe Drinking-water from Desalination.* World Health Organization.

20 Rizk, Z.S. (2009). Inorganic chemicals in domestic water of the United Arab Emirates. *Environmental Geochemistry and Health* 31 (1): 27–45.

21 Rosanoff, A. (2013). Changing crop magnesium concentrations: Impact on human health. *Plant and Soil* 368 (1): 139–153.

22 Rosanoff, A., Weaver, C.M., and Rude, R.K. (2012). Suboptimal magnesium status in the United States: Are the health consequences underestimated? *Nutrition Reviews* 70 (3): 153–164.

23 Martínez-Alvarez, V., González-Ortega, M.J., Martin-Gorriz, B., Soto-García, M., and Maestre-Valero, J.F. (2018). Seawater desalination for crop irrigation—Current status and perspectives. In: *Emerging Technologies for Sustainable Desalination Handbook*, (ed. V.G. Gude), 461–492. Elsevier. doi: 10.1016/B978-0-12-815818-0.00014-X.

24 Moustafa, K. (2018). Seawater for sustainable agriculture. In: *Emerging Technologies for Sustainable Desalination Handbook*, (ed. V.G. Gude), 493–514. Elsevier. doi: 10.1016/B978-0-12-815818-0.00015-1.

25 Monterrey-Viña, A., Musicki-Savic, A., Díaz-Peña, F.J., and Peñate-Suárez, B. (2020). Technical and agronomical assessment of the use of desalinated seawater for coastal irrigation in an insular context. *Water* 12 (1): 272.

26 Martínez-Alvarez, V., Maestre-Valero, J.F., González-Ortega, M.J., Gallego-Elvira, B., and Martin-Gorriz, B. (2019). Characterization of the agricultural supply of desalinated seawater in Southeastern Spain. *Water* 11 (6): 1233.

27 Kumar, R., Ahmed, M., Bhadrachari, G., and Thomas, J.P. (2018). Desalination for agriculture: Water quality and plant chemistry, technologies and challenges. *Water Science and Technology: Water Supply* 18 (5): 1505–1517.

28 Burn, S. et al. (2015). Desalination techniques—A review of the opportunities for desalination in agriculture. *Desalination* 364: 2–16.

29 Martínez-Alvarez, V., González-Ortega, M.J., Martin-Gorriz, B., Soto-García, M., and Maestre-Valero, J.F. (2017). The use of desalinated seawater for crop irrigation in the Segura River Basin (south-eastern Spain). *Desalination* 422: 153–164.

30 Qadir, M. et al. (2018). High-magnesium waters and soils: Emerging environmental and food security constraints. *The Science of the Total Environment* 642: 1108–1117.

31 Pandey, N. (2018). Role of plant nutrients in plant growth and physiology. In: *Plant Nutrients and Abiotic Stress Tolerance*, (eds. M. Hasanuzzaman, M. Fujita, H. Oku, K. Nahar, B. Hawrylak-Nowak), 51–93. Singapore: Springer. doi: 10.1007/978-981-10-9044-8_2.

32 Bar-Tal, A., Yermiyahu, U., Ben-Gal, A., Schwartz, A., Faingold, I., and Seligmann, R. (2017). Optimization of calcium and magnesium concentrations for fertigation of tomato with desalinated water. *The Israel Journal of Plant Sciences* 64 (3–4): 80–91.

33 Barton, N.A., Farewell, T.S., Hallett, S.H., and Acland, T.F. (2019). Improving pipe failure predictions: Factors affecting pipe failure in drinking water networks. *Water Research* 164: 114926.

34 Organization, W.H. and Council, W.P. (2006). *Health Aspects of Plumbing*. World Health Organization.

35 Guo, X. et al. (2016). Simple combination of oxidants with zero-valent-iron (ZVI) achieved very rapid and highly efficient removal of heavy metals from water. *Water Research* 88: 671–680.

36 Mi, Z., Zhang, X., and Chen, C. (2016). Iron release in drinking water distribution systems by feeding desalinated seawater: Characteristics and control. *Desalin. Water Treat.* 57 (21): 9728–9735.

37 Mahmoud, M.T., Hamouda, M.A., Al Kendi, R.R., and Mohamed, M.M. (2018). Health risk assessment of household drinking water in a district in the UAE. *Water* 10 (12): 1726.

38 Deng, A., Xie, R., Gomez, M., Adin, A., Ong, C.N., and Hu, J. (2014). Impact of pH level and magnesium addition on corrosion of re-mineralized seawater reverse osmosis membrane (SWRO) product water on pipeline materials. *Desalination* 351: 171–183.

39 Liang, J. et al. (2014). Impact of elevated Ca2+/Mg2+ concentrations of reverse osmosis membrane desalinated seawater on the stability of water pipe materials. *Journal of Water and Health* 12 (1): 24–33.

40 Liang, J. et al. (2013). Impact of flow rate on corrosion of cast iron and quality of re-mineralized seawater reverse osmosis (SWRO) membrane product water. *Desalination* 322: 76–83.

41 Kabir, F., Chowdhury, S., Mazumder, M.A.J., Zahir, H., and Alhooshani, K. (2018). Effects of plumbing premise on the occurrences and variability of heavy metals in desalinated and blended tap water. *Desalin. Water Treat.* 107: 257–271.

42 Al-Mudhaf, H.F., Al-Khulaifi, N.M., Al-Hayan, M.N., and Abu-Shady, A.-S.I. (2012). Effects of household storage and plumbing systems on the levels of trace elements in desalinated drinking water in Kuwait.

43 Lee, H., Rasheed, U., and Kong, M. (2018). A study on the comparison of corrosion in water supply pipes due to tap water (TW) and reclaimed water (RW). *Water* 10 (4): 496.

44 Gonzalez, S., Lopez-Roldan, R., and Cortina, J.-L. (2013). Presence of metals in drinking water distribution networks due to pipe material leaching: A review. *Environmental Toxicology and Chemistry* 95 (6): 870–889.

45 Li, M., Liu, Z., Chen, Y., and Korshin, G.V. (2020). Effects of varying temperatures and alkalinities on the corrosion and heavy metal release from low-lead galvanized steel. *Environmental Science and Pollution Research* 27 (2): 2412–2422.

46 Marangou, V.S. and Savvides, K. (2001). First desalination plant in Cyprus—product water aggresivity and corrosion control. *Desalination* 138 (1–3): 251–258.

47 Birnhack, L. and Lahav, O. (2018). Post-Treatment of Desalinated Water—Chemistry, Design, Engineering, and Implementation. In: *Emerging Technologies for Sustainable Desalination Handbook*, (eds. V.G. Gude, 305–350. Elsevier. doi: 10.1016/B978-0-12-809240-8.00008-3.

48 Lesimple, A., Ahmed, F.E., and Hilal, N. (2020). Remineralization of desalinated water: Methods and environmental impact. *Desalination* 496: 114692.

49 Gabbrielli, E. (1981). A tailored process for remineralization and potabilization of desalinated water. *Desalination* 39: 503–520.

50 Voutchkov, N. (2012). *Desalination Engineering: Planning and Design*. McGraw Hill Professional.

Index

Note: Page numbers with italic *f* and *t* denote figures and tables respectively.

Gas Hydrate in Water Treatment: Technological, Economic, and Industrial Aspects,
First Edition. Bhajan Lal and Sirisha Nallakukkala.

d

z

Printed and bound by CPI Group (UK) Ltd, Croydon, CR0 4YY

17/09/2026

14964252-0003